RICE BRAN AND RICE BRAN OIL

RICE BRAN AND RICE BRAN OIL

Chemistry, Processing and Utilization

Edited by

LING-ZHI CHEONG

XUEBING XU

ELSEVIER

ACADEMIC PRESS
An imprint of Elsevier

AOCS PRESS

Academic Press and AOCS Press
Academic Press is an imprint of Elsevier
125 London Wall, London EC2Y 5AS, United Kingdom
525 B Street, Suite 1650, San Diego, CA 92101, United States
50 Hampshire Street, 5th Floor, Cambridge, MA 02139, United States
The Boulevard, Langford Lane, Kidlington, Oxford OX5 1GB, United Kingdom

Published in cooperation with American Oil Chemists' Society www.aocs.org
Director, Content Development: Janet Brown

Library of Congress Cataloging-in-Publication Data
A catalog record for this book is available from the Library of Congress

British Library Cataloguing-in-Publication Data
A catalogue record for this book is available from the British Library

ISBN 978-0-12-812828-2

Publisher: Andre Gerhard Wolff
Acquisition Editor: Nancy Maragioglio
Editorial Project Manager: Susan Ikeda
Project Manager: Nilesh Kumar Shah
Cover Designer: Alan Studholme

Typeset by SPi Global, India

CONTENTS

CONTRIBUTORS

Md. Alauddin
Department of Nutrition and Food Technology, Jessore University of Science and Technology, Jessore, Bangladesh

Pradosh Prasad Chakrabarti
Centre for Lipid Science and Technology, CSIR-Indian Institute of Chemical Technology, Hyderabad, India

Nurhan Turgut Dunford
Oklahoma State University, Department of Biosystems and Agricultural Engineering, Robert M. Kerr Food & Agricultural Products Center, Stillwater, OK, United States

Nisi Gao
Wilmar Global Research and Development Center, Shanghai, China

Umar Garba
Department of Agro-Industry, Faculty of Agriculture, Natural Resources and Environment, Naresuan University, Phitsanulok, Thailand; Department of Food Science and Technology, Faculty of Agriculture, Bayero University, Kano, Nigeria

Wei Han
Academy of State Administration of Grain, Beijing, China

Dongping He
Wuhan Polytechnic University, Wuhan, China

Md Zakir Hossen Howlader
Department of Biochemistry and Molecular Biology, University of Dhaka, Dhaka, Bangladesh

Jahidul Islam
Laboratory of Nutrition, Department of Science of Food Function and Health, Graduate School of Agricultural Science, Tohoku University, Sendai, Japan

Jörg J. Jacoby
Wilmar Biotechnology Research and Development Center, Shanghai, China

Ram Chandra Reddy Jala
Centre for Lipid Science and Technology, CSIR-Indian Institute of Chemical Technology, Hyderabad, India

Yuanrong Jiang
Wilmar Biotechnology Research & Development Center, Shanghai, China

Michio Komai
Laboratory of Nutrition, Department of Science of Food Function and Health, Graduate School of Agricultural Science, Tohoku University, Sendai, Japan

Oi-Ming Lai
Department of Bioprocess Technology, Faculty of Biotechnology and Biomolecular Sciences, Universiti Putra Malaysia, Selangor, Malaysia

Wee-Ting Lai
Institute of Bioscience, Universiti Putra Malaysia, Selangor, Malaysia

Wai-Fun Leong
Department of Food Science, The Pennsylvania State University, College of Agricultural Sciences, University Park, PA, United States

Lingyi Liu
Wuhan Polytechnic University, Wuhan, China

Sadia Rahman
Department of Biochemistry and Molecular Biology, University of Dhaka, Dhaka, Bangladesh

Hitoshi Shirakawa
Laboratory of Nutrition, Department of Science of Food Function and Health, Graduate School of Agricultural Science, Tohoku University, Sendai, Japan

Riantong Singanusong
Centre of Excellence in Fats and Oils; Department of Agro-Industry, Faculty of Agriculture, Natural Resources and Environment, Naresuan University, Phitsanulok, Thailand

Yong Wang
Wilmar (Shanghai) Biotechnology Research & Development Center Co., Ltd, Shanghai, China

Juan Wu
Wilmar Global Research and Development Center, Shanghai, China

Tiankui Yang
Wilmar Global Research and Development Center, Shanghai, China

Tie Yin
Academy of State Administration of Grain, Beijing, China

Baoru Yin
Wilmar Global Research and Development Center, Shanghai, China

Xiaolin Zhang
Academy of State Administration of Grain, Beijing, China

Yan Zheng
Wilmar Global Research and Development Center, Shanghai, China

Xuhui Zhuang
Academy of State Administration of Grain, Beijing, China

Yanping Zou
Wilmar Global Research and Development Center, Shanghai, China

PREFACE

Rice bran is the byproduct of paddy rice milling. After harvesting from the rice fields, paddy rice is collected for further handling and transportation. Paddy rice is the individual rice kernels in their natural and unprocessed state. Paddy rice contains the protective hull (around 20% of the dry base). The hull should be removed during the initial milling processing. The hull is not usually edible, but it can be used for boiler burning, where the ash can also be used for silica or activated carbon extraction. The dehulled rice kernel can be used for consumption as brown rice. Although brown rice is currently recommended for healthy eating, many rice producers are moving onto further processing to produce white rice. The brown skin and rice germ is removed as rice bran (around 10% of the dry paddy). The rice bran mainly contains fiber, protein, oil, and residual starch. Rice bran is mainly used as feedstuff or as raw material for oil extraction. The whole rice bran can be considered for edible uses if the quality is controlled. This possibility is currently under exploration in the industry due to high health values in terms of fiber, protein, and oil-soluble bioactive compounds such as oryzanols, phytosterol, tocopherols, tocotrienols, etc. However, rice bran contains active lipase that can hydrolyze the oil in the bran. This leads to poor quality for use in food and brings challenges in rice bran oil processing for edible uses. Because the rice milling industry varies from country to country, and due to cultural variation for food preferences, there is a huge technical challenge in general for the better use of rice bran and rice bran oil for food or higher value-added applications. This is an area facing high technology and operation difficulty, and there is a need for intensive technology and operation innovations.

Rice is one of the oldest cultivated plants in the world, cultivated mainly in tropical and subtropical regions. It is impossible to state precisely where it originated, but wild varieties suggest that it must have come from Asia, Africa, or the Americas. Rice was cultivated in China and India as far back as 700 BC From these countries, cultivation extended to Japan, Indonesia, and as far as Persia. It was introduced to Virginia (United States) in 1647 and to Brazil in 1750. Today, 90% of global production originates in Eastern Asia. In Europe, rice is cultivated in Italy, Spain, and Portugal. Rice can be cultivated between 45° North and 40° South, but it requires a temperature range of 25–30°C. High precipitation (wet rice) or irrigation (watered rice) and soil rich in humus are also common requirements.

From the Food and Agriculture Organization report in 2018, global paddy production is around 740–750 million metric tons (MMT), whereas the number was around 600 MMT in 2000. On the country base, the production in each country (production amount, percentage) is listed as follows: China (211 MMT, 28.4%), India (159 MMT, 21.4%), Indonesia (77 MMT, 10.4%), Bangladesh (53 MMT, 7.1%), Vietnam (43 MMT, 5.8%), Myanmar (26 MMT, 3.5%), Thailand (25 MMT, 3.4%), Philippines (18 MMT, 2.4%), Brazil (11 MMT, 1.5%), Pakistan (10 MMT, 1.3%), United States (10 MMT, 1.3%), Cambodia (10 MMT, 1.3%), Japan (8 MMT, 1.1%), and Others (82 MMT, 11.0%). With rice bran production of 10% and oil based on rice bran at 10%–16%, a conservative estimate of rice bran oil produced globally per year should be more than 7.5 MMT. However, a nonofficial estimation of current production of rice bran oil is not more than 2 MMT. For example, only 10%–15% of natural resources of rice bran oil has been explored in China. A large volume of resources has not yet been used well for food applications.

On the other hand, rice bran oil is one of the most nutritious and functional oils in nature. More details have been elaborated in chapters of this book. Furthermore, rice bran after oil extraction has a lot of potential for food and other applications in terms of fibers and proteins in modern food recommendation criteria. This leaves vast potentiality to utilize rice bran in better ways.

Rice bran and rice bran oil is not well recognized by consumers and industry. Because rice is primarily grown in Asia (more than 90%), mainstream research efforts in the western world are not widely pursued compared to many other oil-bearing materials. Even in Asian countries, the efforts are relatively weak and scattered. Technology and applications vary from country to country. Regulation and standardized practice are also varied. More promotion and communication are therefore highly needed and at a critical stage.

With this information in mind, the International Association of Rice Bran Oil (IARBO) was initiated. In 2011, a 1-day symposium on rice bran oil was held in Beijing, where Prof. Miyazawa from Japan and Dr. Prasad from India were invited to give lectures, as well as two nutritionists from Thailand. During the symposium, an idea was proposed to form an organization for better promotion of the awareness of this oil. Although forming an international organization was not high in terms of motivation and inspiration, and it was not concluded as an action point, many participants thought it was a good idea. The conclusion to act was due to the uncertainty of forming an international organization.

In 2012, we made a trip to Japan and participated in a local oils and fats conference in India for the purpose to meet more people and to materialize plans for forming a rice bran oil organization. During the Indian conference, we met Dr. Mehta who introduced the symposium in Thailand on rice bran oil, which was organized by Dr. Singanusong from Naresuan University, Thailand, in 2013. We agreed to join the conference at Naresuan University and engaged all interested parties so far at that time, where the organization was proposed and agreed upon by the participating parties. The general structure was proposed, and the initial board structure was settled. Following the first board meeting in late 2013 in Bangkok, the organization was created and shaped, including approval of the bylaws. Xuebing Xu was selected as the first president of the IARBO. The first annual meeting was proposed in Wuhan, China, in 2014. In the following years, the annual meetings were successfully held in Mumbai (2015), Tokyo (2016), Bangkok (2017), and Hanoi (2018). IARBO has developed into a stable organization and is making a strong impact on the community, in particular in the associate countries. More and more participants are joining the annual conferences. There is better awareness of rice bran oil through the efforts of the organization.

This book is part of the IARBO effort. The editors appreciate the IARBO board for its encouragement and inspiration. Thanks also to AOCS Press and the Elsevier publication team for making this book a reality. Thanks particularly go to Ms. Janet Brown from AOCS for her everlasting effort to push forward the book's publication. Without their efforts, this book would not have come true. Last but not the least, we appreciate the efforts from all the authors. They provided big support for the finishing of this book.

Xuebing Xu
Lingzhi Cheong

CHAPTER 1

Chemistry of Rice Bran Oil

Nurhan Turgut Dunford
Oklahoma State University, Department of Biosystems and Agricultural Engineering, Robert M. Kerr Food & Agricultural Products Center, Stillwater, OK, United States

1. INTRODUCTION

Rice (*Oryza sativa* L.) is a member of the *Poaceae* or *Graminaceae* family native to southeast Asia. It has been cultivated as a food crop for centuries. Rice still is a very important staple food for a large segment of the world's population. It is commonly consumed as milled or white rice, which is produced by removing the hull and bran layers of the rough rice kernel during the dehulling and milling processes, respectively. The bran, which comprises 3%–8% of the kernel and contains pericarp, aleurone, and subaleurone fractions, is a valuable byproduct of rice processing because it contains a high concentration of health beneficial bioactive compounds, including edible lipids.

Although it is not widely used as a cooking oil worldwide, demand for rice bran oil (RBO) as a "healthy oil" in specialty applications and functional food has steadily increased (Ali and Devarajan, 2017).

Processing aspects, nutritional properties, and various applications of RBO are discussed in the other chapters of this book. This chapter specifically focuses on the chemical composition and other properties of RBO.

2. OIL CONTENT OF RICE BRAN

Chemical composition of bran depends on rice variety, treatment of the grain prior to milling, milling technology used, degree of milling, and the downstream processing of bran, that is, fractionation. Typical oil content in rice bran varies between 10% and 23%. Genotype significantly affects the oil content in bran (Goffman et al., 2003). Oil contents of a collection of 204 rice accessions grown in Beaumont, Texas, USA were examined. A genetically diverse germplasm collection including historical and present-day U.S. cultivars, as well as Asian, European, South American, and African rice cultivars, were included in the investigation (Goffman et al., 2003). Oil contents of the genotypes examined varied from 17% to 27%. Over 75% of the lines had oil contents higher than 22% (weight/weight [*w*/w]). Another study examining

Rice Bran and Rice Bran Oil
https://doi.org/10.1016/B978-0-12-812828-2.00001-9

15 rice varieties grown in Ghana (Amissah et al., 2003) revealed that oil content in the samples (13%–20%) was similar to the oil content reported in other varieties (Goffman et al., 2003). Glutinous rice is shown to contain more oil than nonglutinous brown rice (Taira, 1984).

The degree of milling has a significant effect on the oil content of bran (Saunders, 1985). For example, 0%–8% milling produced bran with about 17%–18% oil content, whereas increased milling from 6%–9% to 9%–10% decreased the oil content from 16.5% to 14.2%, respectively. Increased milling contaminates bran with endosperm, which is low in oil content. In general, bran from parboiled rice contains a considerably higher amount of oil than bran from raw rice (Islam et al., 2002; Rao et al., 1965). According to Rao et al. (1965), oil content of parboiled rice bran was higher (28%–34%) than that in raw bran at 5% degree of milling (24%–26%). The researchers speculated that oil in the aleurone layer migrated to the bran during parboiling and increased the oil content in the bran. Also, bran from parboiled rice contains less starch, increasing the oil fraction in the bran.

3. FATTY ACID COMPOSITION OF RICE BRAN OIL

Similar to the other grains and oilseeds, chemical and fatty acid compositions of rice vary substantially with variety, agronomic practices, and environmental conditions. The studies on 24 lowland nonglutinous rice varieties grown on the Hiroshima Agricultural Experiment Station, Japan, in 1976 and 1977 found that variety had a significant effect on stearic, oleic, and linoleic acid contents in bran (Taira et al., 1979). Crop year had the most significant effect on palmitoleic and linolenic acid contents. A significant positive correlation between the daily mean temperature during ripening and palmitoleic, stearic, oleic, and arachidic acid contents was observed. The correlations between myristic, palmitic, linoleic, and linolenic acid contents and daily mean temperatures were negative and significant in year 1976 but not in 1977. The latter results were explained by the lack of significant temperature variation during the 1977 crop year. A significant negative correlation between oleic and linoleic and linolenic acid contents and a positive correlation between linoleic and linolenic acid contents were observed in both years. Although these results indicate the effect of environmental conditions and variety on fatty acid composition, it is important to note that a 2-year study at one location might not be enough to establish reliable correlations.

A study carried out on 204 rice genotypes identified two groups: one with low palmitic acid (<17.5% with a mean of 16%) and one with palmitic acid in the range of 17.5% and 22% (Lugay and Juliano, 1964). The cultivar Indica had a higher saturated/unsaturated fatty acid ratio (S/U) than Japonica. Another study confirmed the latter findings demonstrating that Indica cultivars, Peta and Malagkit Sungsong Puti, had lower iodine value (I.V.) than the Japonica cultivar (Taira, 1984). It has been also reported that glutinous-type rice had higher myristic, palmitic, and stearic acid and lower oleic acid content that those of nonglutinous rice. Lugay and Juliano (1964) examined fatty acid composition of RBO processed in different countries and found significant differences. The reported differences in the I.V. of RBO may be due to the crop variety processed, environmental conditions, and the type and degree of processing. For example, a dewaxing process removes more saturated fatty acids than unsaturated ones.

In general, saturated fatty acid composition of RBO is quite high, between 19% and 35% (Firestone, 1999). Palmitic acid is the major saturated fatty acid. Unsaturated fatty acids comprise most of the fatty acids in RBO (55%–87%). Monounsaturated fatty acid, oleic acid, is the most abundant unsaturated fatty acid (38%–48%) followed by polyunsaturated fatty acid linoleic acid (16%–36%). Many other studies reported similar fatty acid composition for RBO (Rukmini and Raghuram, 1991; Latha and Nasirullah, 2014).

About half of the triacylglycerides (TAG) in RBO was triunsaturated, meaning that all three fatty acids on the glycerol backbone were unsaturated (see Table 1) (Jin et al., 2016). Only 7% of the TAG was monounsaturated. PLO (palmitic-linoleic-oleic), OLL (oleic-linoleic-linoleic), and OOL (oleic-oleic-linoleic) were the most abundant species, at 19.3%, 16.4%, and 18.0%, respectively.

About 50% of the fatty acids on the sn-2 position on the glycerol backbone in RBO TAG was linoleic acid (Table 2) (Berger et al., 2005). Oleic acid on the sn-2 position comprised about 45% of the fatty acids. The most abundant fatty acid on the sn-1,3 position was oleic acid (42.5%), followed by linoleic (29.6%) and palmitic acid (21.5%).

4. FREE FATTY ACID CONTENT OF RICE BRAN OIL AND VARIOUS NEUTRALIZATION APPROACHES

4.1 Free Fatty Acid Content

The shelf life of brown rice is quite short, about 3–6 months. This is partly due to the rapid hydrolysis of lipids in rice grain. Lipases naturally present in the grain hydrolyze TAG generating free fatty acids (FFA), which are not

Table 1 Triacylglyceride composition of RBO

Triacylglyceride[a]	Amount (%)
PPL	3.6
PPO	3.4
Mono-UTAG	7.0
PLL	8.6
PLO	19.3
POO	12.0
Di-UTAG	41.5
LLL	3.8
OLL	16.4
OOL	18.0
OOO	10.3
Tri-UTAG	49.9

[a]Tri-UTAG, triunsaturated triacylglycerols; Di-UTAG, diunsaturated triacylglycerols; Mono-UTAG, monounsaturated triacylglycerols; M, myristic; P, palmitic; S, stearic; O, oleic; L, linoleic; Ln, linolenic. Although TAGs with low levels such as OLLn, LLM, OOLn, PLnO, SLL, and SOO are not listed in the table, they are included in Mono-UTAG, Di-UTAG, and Tri-UTAG.
Adapted from Jin, J., Xie, D., Chen, H., Wang, X., Jin, Q., Wang, X., 2016. Production of Rice bran oil with light color and high oryzanol content by multi-stage molecular distillation. J. Am. Oil Chem. Soc. 93(1), 145–153.

Table 2 Regiospecific distribution of fatty acids on the glycerol backbone of TAG

Fatty acid	sn-1	sn-1,3
16:0	2.8	21.5
18:0	0.8	2.2
18:1	45.1	42.5
18:2	49.9	29.6
18:3	1.1	1.1
20:0	–	0.9
20:1	0.1	0.8

Adapted from Berger, A., Rein, D., Schäfer, A., Monnard, I., Gremaud, G., Lambelet, P., et al., 2005. Similar cholesterol—lowering propertiesof rice bran oil, with varied γ-oryzanol, in mildly hypercholesterolemic men. Eur. J. Nutr. 44(3), 163–173.

desirable in edible oils. FFA accelerates oil quality degradation by producing off-flavors, off-odors, and other oxidation products. Rice needs to be stabilized to minimize lipolytic hydrolysis by inactivating endogenous lipases prior to milling and storage.

FFA content of RBO varies significantly depending on the quality of bran used for oil extraction. In general, FFA content of RBO is between 2% and 5%. However, extremely high FFA contents ranging from 15% to 40% have also been reported (Bhattacharyya and Bhattacharyya, 1989).

Various research groups examined the effects of different rice stabilization techniques on FFA formation during storage (Kim et al., 2014; Ramezanzadeh et al., 1999). Total FFA content of the oil in the bran increased from 2.5% to 54.9% during storage at 25°C under vacuum for 16 weeks (Ramezanzadeh et al., 1999). When bran was stored at a lower temperature (4–5°C), FFA formation slowed down to 25.4%. Microwave treatment of the bran prior to storage at 25°C under vacuum further reduced the FFA formation to 6.9%. Storage of the microwave-treated bran at 4–5°C for 16 weeks retained the initial FFA content at 2.5%. A study carried out by Champagne and Hron (1992) demonstrated that FFA content in brown rice treated with ethanol vapor did not increase significantly during storage at 36°C for 6 months. Ethanol denatures and deactivates endogenous enzymes including lipases and reduces lipolytic hydrolyses reactions that produce FFA. However, ethanol-treated rice kernels were more susceptible to oxidative deterioration due to the disrupted caryopsis coat and kernel fissuring, which increased the porosity of the kernel making it prone to oxidation. Furthermore, heat treatment degrades antioxidants, that is, tocopherols, naturally present in the kernel.

Several research groups demonstrated the efficacy of infrared heating for stabilizing rice bran and reducing FFA formation during storage (Wang et al., 2017a; Yılmaz, 2016; Ding et al., 2015). Other stabilization techniques such as chemical treatment by spraying hydrochloric acid over rice bran to reduce pH (Prabhakar and Venkatesh, 1986), ohmic heating (Lakkakula et al., 2004), and extrusion (Sayre et al., 1985) have been shown to be quite effective in extending the bran quality during storage. However, most of the latter advanced stabilization techniques are still in the research phase. Further research and development work is needed to demonstrate efficacy at the commercial scale and economic viability for the industry to adopt these techniques.

4.2 Rice Bran Oil Neutralization Techniques

FFA content of oils is reported as either a weight percentage of oil or acid value, which is defined as the weight of KOH (mg) needed to neutralize the organic acids present in 1 g of oil. In general, good quality oilseeds produce oil with <1% FFA in crude oil. The voluntary industry standard for refined edible quality oil is 0.05% FFA or less. Usually, refined edible oils sold in the U.S. contain about 0.01% FFA. The Indian Standard specification for refined RBO allows a maximum acid value of 0.5 (0.25% FFA) (Krishna et al., 2006).

Crude RBO goes through a series of refining processes to remove undesirable compounds, meet edible oil quality standards, and extend its shelf life. Usually, neutralization, also referred to as refining or deacidification, is the second step after degumming in conventional crude edible oil refining operations. However, crude RBO contains a relatively high amount of wax, 1%–2% of oil, which interferes with the refining process. Hence, a dewaxing step is included in the RBO refining process preceding degumming, which removes phospholipids (also referred to as gums) (Orthoefer, 1996a). Neutralization process either converts FFA to neutral acylglycerides, which remain in the oil, or removes them from the oil.

The conventional neutralization technique, alkali refining, utilizes caustic soda to convert FFA to sodium or potassium salts of fatty acids (soap stock). Then, soap stock is removed from the oil by centrifugation (Bhattacharyya and Bhattacharyya, 1987). Although the latter method is quite efficient in neutralizing oil, high neutral oil (TAG) loss and large amount of water usage are the major disadvantages of alkali refining.

RBO with very high FFA content is usually deacidified by miscella refining using alcohols, that is, ethanol, isopropanol, or a mixture of two alcohols, after hexane extraction and prior to removal of the solvent from oil (desolventizing) (Bhattacharyya et al., 1986; Rodrigues et al., 2014; Oliveira et al., 2012). Neutralization by solvent extraction, also referred to as liquid-liquid extraction, is based on the difference in the solubility of FFA and TAG in a suitable solvent. The advantages of the liquid-liquid extraction method include mild process temperature and pressure, and minimal TAG loss during neutralization (Rodrigues et al., 2006).

Neutralization of RBO using porous and nonporous membranes with or without solvent addition has been examined (De et al., 1998; Kale et al., 1999; Manjula and Subramanian, 2006). The molecular weights of FFA and TAG are <300 Da and higher than 800 Da, respectively. In theory, a hydrophobic membrane with molecular weight cut of about 300–500 Da could effectively separate FFA from TAG. However, irregularities in pore size of the commercial membranes and very small differences in molecular weights of FFA and TAG lead to inefficient separation using nanofiltration membranes alone. Combination of conventional refining with membrane processing appears to be more effective in reducing FFA in RBO. For example, a combination of solvent extraction of FFA with ethanol followed by membrane separation is shown to be technically feasible (De et al., 1998). Yet, introduction of another solvent in the process and need for solvent-resistant membranes make this method less attractive than direct membrane processing of oil, which has its own disadvantages such as inefficient removal of FFA from TAG and low flux. The neutralization tests run with nonporous membranes were not very successful due to either low selectivity, poor flux, or both (Manjula and Subramanian, 2006). At its current status, membrane refining systems do not appear to be viable commercial options for replacing the existing conventional technology.

Physical refining or neutralization refers to removal of FFA under vacuum and high temperature. During the process, unsaponifiable and odor compounds are also removed (Cvengros, 1995). This process, which will be discussed in detail in another chapter in this book, is suitable for high FFA content oils such as RBO. Physical neutralization eliminates soap stock production, reduces neutral oil losses, and produces a high purity FFA byproduct that can be used as feedstock by oleochemical industry. Physical neutralization utilizes less water, steam, and power, and requires lower capital investment than the conventional refining. The impact of oil refining on the environment is also reduced. However, efficiency of physical FFA removal depends on the quality of incoming oil. Any compound that might go through adverse changes and reactions at high temperature, that is, metals or chlorophyll, needs to be removed prior to physical neutralization to produce a high quality final product. Kim et al. (1985) reported that steam refining was less effective than caustic refining in removing FFA from RBO. Molecular distillation in a wiped film short-path evaporator (Martins et al., 2006) produces a higher quality product than steam-stripped oil, but this process might be cost inhibitive for commodity oils.

Refining of RBO results in oil losses of 20%–50% (*w*/w of total oil) during conventional oil processing (Orthoefer, 1996a, b; Gingras, 2000). Furthermore, conventional refining processes significantly reduce (about 50%) health beneficial bioactive components in refined oil (Orthoefer, 1996a). High-pressure extraction and fractionation technology employing supercritical carbon dioxide (SC-CO_2) is an alternative technique for vegetable oil refining. SC-CO_2 extraction and fractionation of RBO at high pressures and low temperature result in a product with high TAG and low FFA, waxes, and unsponifiable contents due to the lower selectivity of SC-CO_2 for the latter compounds (Zhao et al., 1987).

Dunford and King developed a patented RBO fractionation process that significantly reduces FFA content and increases oryzanol and other phytosterol ester contents in the TAG-rich phase (Dunford and King, 2004). Low-pressure and high-temperature conditions were found to be favorable for minimizing TAG and phytosterol losses during FFA removal from crude

RBO (Dunford and King, 2000). Oil fractions with 1% FFA, about 95% TAG, and 0.35% free sterol with 1.8% oryzanol content could be obtained with the described SC-CO_2 fractionation technique, which utilized a pilot scale packed fractionation column. A later study by the same researchers (Dunford et al., 2002) used a similar approach to the one described earlier (Dunford and King, 2000) but improved the FFA removal and phytosterol enrichment in the TAG phase using a two-step processing scheme. Initially, FFA were removed in the first column, and then a low acidity phytosterol-enriched oil fraction was obtained with a second step fractionation process (Dunford et al., 2002). Low pressure (138 bar) and high temperature (80°C) effectively removed FFA from crude RBO without significant oryzanol loss in the TAG-rich phase. Oryzanol content of the raffinate fraction, low in FFA and high in TAG, was three times higher than that of the original RBO. Phytosterol fatty acid ester content of the raffinate fraction was also increased during the deacidification process; however, enrichment of these moieties was not as high as that found for oryzanol.

Biological neutralization of high FFA content oils can be achieved using either intact microorganisms or isolated enzymes. More detailed discussion on biological neutralization can be found in another chapter in this book. Cho, Kwon, and Yoon (Cho et al., 1990) have shown that *Pseudomonas* strain (BG1) assimilates long-chain fatty acids without secreting extracellular lipases. Unfortunately, this microorganism does not digest short-chain fatty acids, having <12 carbon atoms and linoleic acid. The latter fatty acids sometimes inhibit cell growth. Although butyric, valeric, caproic, caprylic, and capric acids have higher solubility in water than oleic acid, they were not utilized. This could be due to the toxicity of short-chain fatty acids to microorganisms. Using intact microorganisms rather than purified enzymes may improve the economic viability of the process. However, BG1 has not yet been tested with vegetable oils. Furthermore, using intact cells limits mass transfer, adversely affecting reaction kinetics. Hence, this process is still in the proof-of-concept phase.

Neutral oil yield is significantly reduced when conventional RBO neutralization techniques are used for refining, adversely affecting the economic viability of the refined RBO production. There have been attempts to improve the RBO refining process to achieve higher refined oil yield by reesterifiying FFA. Esterification of FFA with glycerol or di- or monoacylglycerides to produce neutral acylglycerides eliminates the need to remove FFA from oil and minimizes oil loss (Bhattacharyya and Bhattacharyya, 1989; Bhosle and Subramanian, 2005). Esterification reactions can be carried out with or without a chemical catalyst or using enzymes. The studies on chemical esterification of FFA at high temperatures, 200–270°C, go back to the 1850s (Bhosle and Subramanian, 2005). There have been a few recent applications of high temperature esterification to RBO. For example, chemical esterification of degummed RBO FFA at 200°C with 70% excess glycerol for 4 h in the presence of 0.2% catalyst ($SnCl_2$) reduced the acid value from 24.3% to 3.0% (Singh and Singh, 2009). Another study (Bhattacharyya and Bhattacharyya, 1987) examined catalytic (stannous chloride and p-toluene sulphonic acid) esterification of FFA in crude RBO with glycerol in a nitrogen atmosphere. The effect of catalyst on the esterification reaction rate was significant only during the initial 2 h. Esterification of RBO containing 15%–30% FFA with glycerol under vacuum in the presence of p-toluene sulphonic acid followed by degumming and dewaxing resulted in a product with 1.6%–4.0% FFA. De and Bhattacharyya (1999) demonstrated that high-temperature (210°C) and low-pressure (1.3 kPa) esterification of degummed, dewaxed, and bleached RBO containing 9.5%–35.0% FFA with monoacylglycerides at 210°C and 1.3 kPa reduced the FFA content to 0.5%–3.5%. It appears that chemical esterification methods have not been adopted by the vegetable oil refining industry, probably due to the very high reaction temperatures required.

Biorefining of edible oils could be more acceptable to consumers wary of residual chemicals in their food and chemical processing in general. Enzymes isolated from microorganisms have been used for esterification of FFA to glycerol, phytosterols, and mono- and diglycerides (Bhattacharyya and Bhattacharyya, 1989; Li et al., 2018, 2017, 2016). Enzymatic neutralization of degummed and dewaxed high acid, 30% FFA, RBO was carried out using *Mucor miehei* lipase (Lipozyme™) in the presence of glycerol in the reaction mixture (Saunders, 1985). FFA content of the RBO could be reduced from 30% to 3.6% in 10 h under the following conditions: theoretical amount of glycerol, 10% enzyme, and 10% water based on the oil weight, at 10 mmHg and 70°C. As expected, lower (50°C) and higher (80°C) reaction temperatures than the optimum temperature for *M. miehei* lipase (70°C) lowered esterification rate, consequently, reducing FFA neutralization efficiency. Addition of 30% excess glycerol over the theoretical amount in the reaction medium did not improve esterification efficiency.

A multistep enzymatic esterification method reduced the FFA content of RBO oil from 45% to 4% (Lakshmanan et al., 1992). Lipozyme IM 20 (*M. miehei* immobilized on an ion exchange resin) was used as a catalyst at 70°C and 10 mmHg. The reaction time was 4 h. First phase of the reaction reduced the FFA content to 15.7%. Fresh enzyme and glycerol additions to the reaction mixture in multiple steps following the first phase further reduced the FFA content to 4% after the third addition.

Several animal and human clinical studies indicated that diacylglycerol (DAG) consumption decreases postprandial TAG levels in serum and suppresses accumulation of TAG in body fat and liver (Hibi et al., 2011; Kawashima et al., 2008). Hence, in an effort to produce DAG-enriched RBO while neutralizing it, degummed, dewaxed, and bleached RBO with 20.2% FFA and 0.3% DAG content was reacted with MAG using lipase RM IM as a biocatalyst (Song et al., 2012). The optimum reaction conditions were determined as 56°C, 4.77% enzyme loading, 5.75 h reaction time, and MAG/RBO ratio of 0.25. A final product with 0.28% FFA and 27.98% DAG content was obtained under the latter reaction conditions. The enzyme could be used nine times with 90% of its original catalytic activity still remaining.

Presence of active lipase in rice bran not only leads to FFA formation but also generates a significant amount of partial glycerides, monoacylglycerol (MAG), and DAG. The latter compounds adversely affect downstream refining, potentially leading to the formation of glycidyl esters during deodorization (Van Hoed et al., 2010; Craft et al., 2012). Glycidyl esters are reported to be carcinogenic and genotoxic (Craft et al., 2012). A relatively new study examined production of MAG- and DAG-free RBO (Li et al., 2017). First, MAG and DAG in RBO were hydrolyzed to FFA using immobilized lipase *Malassezia globose*, SMG1-F278N, followed by esterification of FFA using the same enzyme. Hydrolysis process was carried out at 30°C and pH 6. Final product after hydrolysis contained 66.3% TAG, 33.3% FFA, and 0.1% DAG. Deacidifaction reaction carried out at substrate (ethanol) to FFA molar ratio of 1.5:1 and enzyme loading of 40 Unit/g (based on oil weight) at 30°C resulted in 99.8% deacidification efficiency.

Free phytosterol and its esters have been reported to possess many health benefits including cholesterol-lowering (Miettinen et al., 1996), antiinflammatory, antiatherogenic, and anticancer effects (Rudkowska, 2010; Brufau et al., 2008; Awad and Fink, 2000; Bouic, 2001). Wang et al. (Wang et al., 2016) reported a process that describes enzymatic neutralization of high acid RBO by esterifying FFA to phytosterols. Lipozyme RM IM, which is a sn-1,3-specific lipase, was used for the esterification reaction. The reaction of RBO (10 g) and phytosterol (2.34 g) in hexane at 70°C for 60 h reduced the FFA content from 15.8% to 1.2%. Phytosterol ester content in the oil increased from 0% to 29.3%, and most of the vitamin E naturally present in RBO was retained during the enzymatic neutralization process.

Fatty acid ethanolamides are lipid-signaling molecules that are ubiquitous in nature, found in animal and plant cells. They have been reported to have antiinflammatory, anticancer, antiproliferative, and neuroprotective functions (Kilaru et al., 2007; Calignano et al., 1998). Some ethanolamides can reduce pain sensation and allergic reaction, inhibit mast cell degranulation, and lessen energy homeostasis (Astarita et al., 2006; Lucanic et al., 2011). A recent study examined the effect of type of acyl acceptor, including ethanolamine, on the neutralization of FFA in RBO (Wang et al., 2017b). Optimum conditions for FFA/ethanolamine reaction were determined as follows: 2% Lipozyme 435, 1:1 mass ratio of oil to solvent (hexane), 1:1 M ratio of FFA to ethanolamine, 5% molecular sieve to remove water at 76°C. Acid value of RBO could be reduced from 21.5 to 1.6 mg/g after 4 h reaction. The final oil was rich in fatty acid ethanolamides (11.9%). FFA neutralization reaction using glycerol or MAG as acyl acceptor took longer and residual acidity was higher than that achieved by amidation even after 16 h of reaction. The differences between amidation and esterification reactions are due to the chemical structure of the acyl acceptors used in the reactions. Ethanolamine has both NH_2 and OH groups. NH_2 preferentially reacts with COOH groups on fatty acids forming fatty acid ethanolamine. Although NH_2 group may also form fatty acid esteramines, spontaneous acyl migration rapidly converts fatty acid esteramines to fatty acid ethanolamides.

5. RICE BRAN OIL OXIDATION

RBO is promoted as a stable product due to high concentrations of antioxidants, that is, oryzanol, tocopherols, and tocotrienols, naturally present in the oil (Latha and Nasirullah, 2014; Mishra and Sharma, 2014). Several studies examined the oxidative stability of pure RBO and oil blends containing RBO under various conditions including frying (Latha and Nasirullah, 2014; Mishra and Sharma, 2014; Debnath et al., 2012). Commercial refined, bleached, and deodorized (RBD) RBO containing 22.6% palmitic, 43.7% oleic, and 29.2% linoleic acids was heated at 180°C up to 8 h to determine its stability (Latha and Nasirullah, 2014). The most significant change was observed in its peroxide value (PV), which increased from 0.2 to 2.9 Meq O_2 in 8 h. Under the same experimental conditions, about 9% of the polyunsaturated fatty acids were lost through degradation to oxidation products. No trans-fatty acid was detected in the heated oil. Although a significant decrease in tocopherol content was measured, the oryzanol content did not change during heating at 180°C for 8 h.

Food industry often utilizes vegetable oil blends to optimize thermal and oxidative stability, flavor, nutritional properties, and the cost of the final product. A study on the blends of RBO (initial I.V. = 95.4, and acid value = 0.163) and sunflower oil (initial I.V. = 113.4 and acid value 0.022) demonstrated that, although all the pure and blended oils exhibited good thermal stability during repeated deep fat frying cycles, the oil blend with 60% RBO and 40% sunflower oil was the best mix for frying applications because its specific gravity, refractive index, or acid value did not change significantly (Kumar Sharma et al., 2006). A small change was observed in I.V. due to a slight increase in palmitic and stearic acid percentages and a gradual decrease in linoleic acid content during repeated deep fat frying cycles. Frying tests carried out using fresh potato chips exhibited better stability of 60% RBO and 40% sunflower oil blend as compared to the pure oils (Mishra and Sharma, 2014). More trans-fatty acids were formed in all the pure and blended oil samples when fresh potatoes were used for the tests as compared to the tests carried out using dried potatoes.

Thermal stability of RBD-RBO purchased at a grocery store in the U.S. and its blend with other vegetable oils was examined in our research laboratories (unpublished data). Two types of oil, regular RBO, RBD-RBO, and oryzanol rich RBO, RBD-HO/RBO, were investigated. As expected, fatty acid compositions of both oils were very similar (Table 3). RBD-RBO and RBD-HO/RBO contained about 0.3% and 1% oryzanol, respectively.

Table 3 Fatty acid composition (%) of RBD-RBO and RBD-HO/RBO

Fatty acid name	RBD-RBO	RBD-HO/RBO
Myristic acid (C14:0)	0.29	0.28
Palmitic acid (C16:0)	14.97	14.97
Palmitoleic acid (C16:1)	0.17	0.17
Stearic acid (C18:0)	2.04	1.99
Oleic acid (C18:1n9c)	39.97	41.66
Linolelaidic acid (C18:2n6t)	0.20	0.39
Linoleic acid (C18:2n6c)	38.14	36.03
Arachidic acid (C20:0)	0.67	0.72
g-Linolenic acid (C18:3n6)	0.13	0.19
cis-11-Eicosenoic acid (C20:1)	0.62	0.83
Linolenic acid (C18:3n3)	1.77	1.24
Behenic acid (C22:0)	0.28	0.42
Lignoceric acid (C24:0)	0.38	0.45

Table 4 Quality parameters for the original oils

Oil	PV (meq/Kg)	OSI (h)	FFA (%, w/w)	CV (micromole/g)	p-AV
HO-RBO	1.2 ± 0.032	12.9 ± 0.11	n.d.	n.d.	n.d.
RBD-RBO	1.4 ± 0.001	11.98 ± 0.42	0.014 ± 0.0002	8.95 ± 0.19	37.63
Cotton seed oil	1.11 ± 0.001	6.79 ± 0.03	n.d.	n.d.	n.d.

HO-RBO, high oryzanol RBO; RBD-RBO, regular refined, bleached and deodorized RBO; PV, peroxide value; OSI, oxidative stability index; FFA, free fatty acids; CV, carbonyl value; p-AV, p-anisidine value; n.d., not determined.

Initial oil quality parameters for the oils were within the industry standards (Table 4). Cotton seed oil is widely used for frying snack foods. Its bland taste does not interfere with the food flavor carefully formulated with selected ingredients. The oil heating experiments carried out in our laboratories exhibited superior stability with RBD-RBO and HO-RBO as compared to cotton seed oil (unpublished data) (Table 5). Increasing oryzanol content in RBO from 0.3% to 1% had an adverse effect on the OSI of the oil during heating. In an effort to improve oxidative stability of cotton seed oil, it was blended with HO-RBO. OSI of the blended oil increased from 8.54 to 11.05 h as the HO-RBO percent in the blended oil increased from 25% to 75%, respectively. PV of the blended oils decreased with increasing HO-RBO ratio in the oil from 1.11 (pure cotton seed oil) to 1.59 (25% HO-RBO), 1.50 (50% HO-RBO), and 1.28 meq/Kg (75% HO-RBO).

Table 5 Effect of heating time on OSI of RBD-RBO, HO-RBO and cottonseed oils[a]

Oil/Heating time	OSI (h)
Original oil	11.98 ± 0.2
12 h-RBD-RBO	10.96 ± 0.25
24 h-RBD-RBO	10.68 ± 0.16
48 h-RBD-RBO	9.9 ± 0.33
Original oil	12.9 ± 0.1
12 h-HO-RBO	8.83 ± 0.08
24 h-HO-RBO	6.66 ± 0.36
48 h-HO-RBO	3.74 ± 0.06
Original oil	6.79 ± 0.03
12 h-Cottonseed	4.45 ± 0.33
24 h-Cottonseed	3.18 ± 0.04
48 h-Cottonseed	1.12 ± 0.04

[a]Samples were heated for 0, 12, 24, and 48 h at 180°C and allowed to cool to room temperature prior to OSI tests.

These results indicate that cotton seed oil oxidative stability can be improved by blending it with HO-RBO. After 48 h of heating at 180°C, the OSI of the pure RBD-RBO, HO-RBO, and cotton seed oils decreased from 11, 9.8, 12.9, and 6.79 h to 9.9, 3.74, and 1.12 h, respectively (Table 4). The oil with the highest thermal stability was RBD-RBO with only a 9% decrease in OSI after 48 h heating at 180°C.

6. OTHER CHEMICAL REACTIONS WITH RICE BRAN OIL

High FFA content of crude RBO leads to large oil losses during its refining to edible grade, adversely affecting its economic viability in common food applications. Some argue that utilization of RBO for biofuel production might be a better alternative. A number of studies reported conversion of RBO to biodiesel, and methyl or ethyl esters of fatty acids (Zullaikah et al., 2005; Lin et al., 2009; Kasim et al., 2009). Usually, acid catalysis is used for the transesterification reactions due to high FFA content in the RBO. However, considering that biodiesel is a commodity product that needs to be produced and marketed in large volumes, utilization of food oils such as health beneficial RBO produced in limited quantities for biodiesel production might not be a sustainable approach.

Rogers et al. (Rogers et al., 2009) define organogels as "self-standing, thermoreversible, anhydrous, viscoelastic materials structured by a three-dimensional supramolecular network of self-assembled small molecules in an organic liquid at concentrations no greater than their percolation threshold, usually ≤2% (w/w)." Organogels are getting attention as solid-like, plastic materials, which can be used as *trans*-fat-free margarine, shortening, and spreads. A number of natural food-grade waxes including rice bran wax, carnauba Brazilian wax, fruit wax carnauba wild wax, berry wax, candelilla wax, beeswax, and sunflower wax have been evaluated for their vegetable oil-gelling properties. It has been reported that rice bran wax at up to 5% w/w concentration in RBO exhibited weak gelling behavior (Doan et al., 2015). Carnauba wild wax, berry wax, candelilla wax, beeswax, and sunflower wax were efficient oleogelators forming strong gels at concentration of <2% w/w.

RBO has also been evaluated as a potential base stock for producing biodegradable lubricants (Rani et al., 2015). The physiochemical and thermal properties of RBO were found to be similar to other vegetable oils and the commercial mineral oil, SAE20W40. Although the frictional properties of RBO were comparable to SAE20W40, the wear scar diameter for RBO was larger, indicating the need for antiwear additives in potential lubricant formulations.

7. CONCLUSIONS

RBO is promoted for its high nutritional value due to the high concentrations of health beneficial bioactive compounds naturally present in the oil. The biggest challenge for the economic feasibility of RBO is its very high FFA content, which results in extremely high oil losses during refining. Although chemical and enzymatic esterification reactions have been examined to minimize oil losses during refining, to the best of our knowledge, these processes are still at the development and proof-of-concept phase and have not been widely adapted by the industry. RBO-based industrial products such as biodiesel and lubricants have been researched. However, considering the relatively low market availability and higher cost of RBO as compared to other commodity oils such as soybean and canola oils, it is highly unlikely that RBO could compete with other vegetable oils as feedstock for industrial bioproduct manufacturing. Niche markets such as functional foods and nutraceutical and high value applications in cosmetics and pharmaceutical appear to be more plausible avenues for expanding RBO markets.

REFERENCES

Ali, A., Devarajan, S., 2017. Nutritional and health benefits of rice bran oil. In: Manickavasagan, A., Santhakumar, C., Venkatachalapathy, N. (Eds.), Brown Rice. Springer International Publishing, Cham, pp. 135–158.

Amissah, J.G.N., Ellis, W.O., Oduro, I., Manful, J.T., 2003. Nutrient composition of bran from new rice varieties under study in Ghana. Food Control 14 (1), 21–24.

Astarita, G., Di Giacomo, B., Gaetani, S., Oveisi, F., Compton, T.R., Rivara, S., et al., 2006. Pharmacological characterization of hydrolysis-resistant analogs of oleoylethanolamide with potent anorexiant properties. J. Pharmacol. Exp. Ther. 318 (2), 563–570.

Awad, A.B., Fink, C.S., 2000. Phytosterols as anticancer dietary components: evidence and mechanism of action. J. Nutr. 130 (9), 2127–2130.

Berger, A., Rein, D., Schäfer, A., Monnard, I., Gremaud, G., Lambelet, P., et al., 2005. Similar cholesterol—lowering propertiesof rice bran oil, with varied γ-oryzanol, in mildly hypercholesterolemic men. Eur. J. Nutr. 44 (3), 163–173.

Bhattacharyya, A.C., Bhattacharyya, D.K., 1987. Deacidification of high FFA rice bran oil by reesterification and alkali neutralization. J. Am. Oil Chem. Soc. 64 (1), 128–131.

Bhattacharyya, A.C., Majumdar, S., Bhattacharyya, D.K., 1986. Edible quality rice bran oil from high FFA rice bran oil by Miscella refining. J. Am. Oil Chem. Soc. 63 (9), 1189–1191.

Bhattacharyya, S., Bhattacharyya, D.K., 1989. Biorefining of high acid rice bran oil. J. Am. Oil Chem. Soc. 66 (12), 1809–1811.

Bhosle, B.M., Subramanian, R., 2005. New approaches in deacidification of edible oils—a review. J. Food Eng. 69 (4), 481–494.
Bouic, P.J.D., 2001. The role of phytosterols and phytosterolins in immune modulation: a review of the past 10 years. Curr. Opin. Clin. Nutr. Metabolic Care 4 (6), 471–475.
Brufau, G., Canela, M.A., Rafecas, M., 2008. Phytosterols: physiologic and metabolic aspects related to cholesterol-lowering properties. Nutr. Res. 28 (4), 217–225.
Calignano, A., GL, R., Giuffrida, A., Piomelli, D., 1998. Control of pain initiation by endogenous cannabinoids. Nature 394, 277.
Champagne, E.T., Hron, R.J., 1992. Stabilizing brown rice to lipolytic hydrolysis by ethanol vapors. Cereal Chem. 69 (2), 152–156.
Cho, S.Y., Kwon, T.W., SH, Y., 1990. Selective removal of free fatty acids in oils using a microorganism. J. Am. Oil Chem. Soc. 67 (9), 558–560.
Craft, B.D., Nagy, K., Seefelder, W., Dubois, M., Destaillats, F., 2012. Glycidyl esters in refined palm (Elaeis Guineensis) oil and related fractions. Part II. Practical recommendations for effective mitigation. Food Chem. 132 (1), 73–79.
Cvengros, J., 1995. Physical refining of edible oils. J. Am. Oil Chem. Soc. 72 (10), 1193–1196.
De, B.K., Bhattacharyya, D.K., 1999. Deacidification of high-acid rice bran oil by reesterification with monoglyceride. J. Am. Oil Chem. Soc. 76 (10), 1243–1246.
De, B.K., Das, R., Dutta, B.K., Bhattacharyya, D.K., 1998. Membrane degumming and dewaxing of rice bran oil and its refining. Lipid/Fett. 100 (9), 416–421.
Debnath, S., Rastogi, N.K., Gopala Krishna, A.G., Lokesh, B.R., 2012. Effect of frying cycles on physical, chemical and heat transfer quality of rice bran oil during deep-fat frying of Poori: an Indian traditional fried food. Food Bioprod. Process. 90 (2), 249–256.
Ding, C., Khir, R., Pan, Z., Zhao, L., Tu, K., El-Mashad, H., et al., 2015. Improvement in shelf life of rough and brown rice using infrared radiation heating. Food Bioprocess Technol. 8 (5), 1149–1159.
Doan, C.D., Van De Walle, D., Dewettinck, K., Patel, A.R., 2015. Evaluating the oil-gelling properties of natural waxes in rice bran oil: rheological, thermal, and microstructural study. J. Am. Oil Chem. Soc. 92 (6), 801–811.
Dunford, N.T., JA, T., JW, K., 2002. A continuous countercurrent supercritical fluid deacidification process for phytosterol ester fortification in rice bran oil. Food Res. Int. 36 (2), 175–181.
Dunford, N.T., King, J.W., 2004. Inventorssupercritical fluid fractionation process for phytosterol ester enrichment in vegetable oils. U.S. Patent 6,677,469 B1.
Dunford, N.T., King, W.J., 2000. Phytosterol enrichment of rice bran oil by a supercritical carbon dioxide fractionation technique. J. Food Sci. 65, 1395–1399.
Firestone, D., 1999. Physical and Chemical Characteristics of Oils, Fats and Waxes. AOCS Press, Champaign, IL, p. 152.
Gingras, L., 2000. Refining of rice bran oil. INFORM 11 (11), 1196–1203.
Goffman, F.D., Pinson, S., Brergman, C., 2003. Genetic diversity for lipid content and fatty acid profile in rice bran. J. Am. Oil Chem. Soc. 80 (5), 485–490.
Hibi, M., Sugiura, Y., Yokoyama, R., Takase, H., Shiiba, D., Meguro, S., et al., 2011. The short-term effect of diacylglycerol oil consumption on total and dietary fat utilization in overweight women. Obesity 19 (3), 536–540.
Islam, M.R., Roy, P., Shimizu, N., Kimura, T., 2002. Effect of processing conditions on physical properties of parboiled Rice. Food Sci. Technol. Res. 8 (2), 106–112.
Jin, J., Xie, D., Chen, H., Wang, X., Jin, Q., Wang, X., 2016. Production of rice bran oil with light color and high oryzanol content by multi-stage molecular distillation. J. Am. Oil Chem. Soc. 93 (1), 145–153.
Kale, V., Katikaneni, S.P.R., Cheryan, M., 1999. Deacidifying rice bran oil by solvent extraction and membrane technology. J. Am. Oil Chem. Soc. 76 (6), 723–727.
Kasim, N.S., Tsai, T.-H., Gunawan, S., Ju, Y.-H., 2009. Biodiesel production from rice bran oil and supercritical methanol. Bioresour. Technol. 100 (8), 2399–2403.

Kawashima, H., Takase, H., Yasunaga, K., Wakaki, Y., Katsuragi, Y., Mori, K., et al., 2008. One-year ad libitum consumption of diacylglycerol oil as part of a regular diet results in modest weight loss in comparison with consumption of a triacylglycerol control oil in overweight Japanese subjects. J. Am. Diet. Assoc. 108 (1), 57–66.

Kilaru, A., Blancaflor, E.B., Venables, B.J., Tripathy, S., Mysore, K.S., Chapman, K.D., 2007. The N-acylethanolamine-mediated regulatory pathway in plants. Chem. Biodivers. 4 (8), 1933–1955.

Kim, S.K., Kim, C.J., Cheigh, H.S., Yoon, S.H., 1985. Effect of caustic refining, solvent refining and steam refining on the deacidification and color of rice bran oil. J. Am. Oil Chem. Soc. 62 (10), 1492–1495.

Kim, S.-M., Chung, H.-J., Lim, S.-T., 2014. Effect of various heat treatments on rancidity and some bioactive compounds of rice bran. J. Cereal Sci. 60 (1), 243–248.

Krishna, A.G.G., Hemakumar, K.H., Khatoon, S., 2006. Study on the composition of rice bran oil and its higher free fatty acids value. J. Am. Oil Chem. Soc. 83 (2), 117–120.

Kumar Sharma, H., Kaur, B., Chandra Sarkar, B., Singh, C., 2006. Thermal behavior of pure rice bran oil, sunflower oil and their model blends during deep fat frying. 2006. 57 (4), 6.

Lakkakula, N.R., Lima, M., Walker, T., 2004. Rice bran stabilization and rice bran oil extraction using Ohmic heating. Bioresour. Technol. 92 (2), 157–161.

Lakshmanan, A., Rao, P.V., Jayaraman, K., Lakshmanan, C.M., 1992. Lipase catalyzed deacidification of high free fatty acid rice bran oil. Biotechnol. Tech. 6 (2), 169–172.

Latha, R.B., Nasirullah, D.R., 2014. Physico-chemical changes in rice bran oil during heating at frying temperature. J. Food Sci. Technol. 51 (2), 335–340.

Li, D., Faiza, M., Ali, S., Wang, W., CP, T., Yang, B., et al., 2018. Highly efficient deacidification of high-acid rice bran oil using methanol as a novel acyl acceptor. Appl. Biochem. Biotechnol. 184 (4), 1061–1072.

Li, D., Liu, P., Wang, W., Wang, X., Yang, B., Wang, Y., 2017. An innovative deacidification approach for producing partial glycerides-free rice bran oil. Food Bioprocess Technol. 10 (6), 1154–1161.

Li, D., Wang, W., Durrani, R., Li, X., Yang, B., Wang, Y., 2016. Simplified enzymatic upgrading of high-acid rice bran oil using ethanol as a novel acyl acceptor. J. Agric. Food Chem. 64 (35), 6730–6737.

Lin, L., Ying, D., Chaitep, S., Vittayapadung, S., 2009. Biodiesel production from crude rice bran oil and properties as fuel. Appl. Energy 86 (5), 681–688.

Lucanic, M., Held, J.M., Vantipalli, M.C., Klang, I.M., Graham, J.B., Gibson, B.W., et al., 2011. N-acylethanolamine signalling mediates the effect of diet on lifespan in *Caenorhabditis elegans*. Nature 473, 226.

Lugay, J.C., Juliano, B.O., 1964. Fatty acid composition of rice lipids by gas-liquid chromatography. J. Am. Oil Chem. Soc. 41 (4), 273–275.

Manjula, S., Subramanian, R., 2006. Membrane technology in degumming, dewaxing, deacidifying, and decolorizing edible oils. Crit. Rev. Food Sci. Nutr. 46 (7), 569–592.

Martins, P.F., Ito, V.M., Batistella, C.B., Maciel, M.R.W., 2006. Free fatty acid separation from vegetable oil deodorizer distillate using molecular distillation process. Sep. Purif. Technol. 48 (1), 78–84.

Miettinen, T., Vanhanen, H., Wester, I., 1996. Inventorsuse of a stanol fatty acid ester for reducing serum cholesterol level. U.S. Patent 5,502,045.

Mishra, R., Sharma, H.K., 2014. Effect of frying conditions on the physico-chemical properties of rice bran oil and its blended oil. J. Food Sci. Technol. 51 (6), 1076–1084.

Oliveira, C.M., Garavazo, B.R., Rodrigues, C.E.C., 2012. Liquid-liquid equilibria for systems composed of rice bran oil and alcohol-rich solvents: application to extraction and deacidification of oil. J. Food Eng. 110 (3), 418–427.

Orthoefer, F.T., 1996a. Rice bran oil. In: Hui, T.H. (Ed.), Bailey's Industrial Industrial Oil and Fat Products, Edible Oil and Fat Products; Oils and Oilseeds. 2, fifth ed. John Wiley and Sons, New York, NY, pp. 393–409.

Orthoefer, F.T., 1996b. Rice bran oil: healthy lipid source. Food Techol. 50 (12), 62–64.
Prabhakar, J.V., Venkatesh, K.V.L., 1986. A simple chemical method for stabilization of rice bran. J. Am. Oil Chem. Soc. 63 (5), 644–646.
Ramezanzadeh, F.M., Rao, R.M., Windhauser, M., Prinyawiwatkul, W., Tulley, R., Marshall, W.E., 1999. Prevention of hydrolytic rancidity in rice bran during storage. J. Agric. Food Chem. 47 (8), 3050–3052.
Rani, S., Joy, M.L., Nair, K.P., 2015. Evaluation of physiochemical and tribological properties of rice bran oil—biodegradable and potential base stoke for industrial lubricants. Ind. Crop. Prod. 65, 328–333.
Rao, S.N.R., Ananthachar, T.K., Desikachar, H.S.R., 1965. Oil content of bran from rice milled to different degrees of polishing. J. Food Sci. Technol. (Mysore) 2, 115–116.
Rodrigues, C.E.C., Gonçalves, C.B., Marcon, E.C., EAC, B., AJA, M., 2014. Deacidification of rice bran oil by liquid-liquid extraction using a renewable solvent. Sep. Purif. Technol. 132, 84–92.
Rodrigues, C.E.C., Onoyama, M.M., Meirelles, A.J.A., 2006. Optimization of the rice bran oil deacidification process by liquid-liquid extraction. J. Food Eng. 73 (4), 370–378.
Rogers, M.A., Wright, A.J., Marangoni, A.G., 2009. Oil organogels: the fat of the future? Soft Matter 5 (8), 1594–1596.
Rudkowska, I., 2010. Plant sterols and stanols for healthy ageing. Maturitas 66 (2), 158–162.
Rukmini, C., Raghuram, T.C., 1991. Nutritional and biochemical aspects of the hypolipidemic action of rice bran oil: a review. J. Am. Coll. Nutr. 10 (6), 593–601.
Saunders, R.M., 1985. Rice bran: composition and potential food uses. Food Rev. Int. 1 (3), 465–495.
Sayre, R.N., Nayyar, D.K., Saunders, R.M., 1985. Extraction and refining of edible oil from extrusion-stabilized rice bran. J. Am. Oil Chem. Soc. 62 (6), 1040–1043.
Singh, S., Singh, R.P., 2009. Deacidification of high free fatty acid-containing rice bran oil by non-conventional reesterification process. J. Oleo Sci. 58 (2), 53–56.
Song, Z., Liu, Y., Jin, Q., Li, L., Wang, X., Huang, J., et al., 2012. Lipase-catalyzed preparation of diacylglycerol-enriched oil from high-acid rice bran oil in solvent-free system. Appl. Biochem. Biotechnol. 168 (2), 364–374.
Taira, H., 1984. Lipid content and fatty acid composition of nonglutinous and glutinous varieties of foxtail millet. J. Agric. Food Chem. 32 (2), 369–371.
Taira, H., Taira, H., Maeshige, M., 1979. Influence of variety and crop year on lipid content and fatty acid composition of lowland non-Glutinus brown rice. Jpn. J. Crop Sci. 48 (2), 220–228.
Van Hoed, V., Vila Ayala, J., Czarnowska, M., De Greyt, W., Verhé, R., 2010. Optimization of physical refining to produce rice bran oil with light color and high oryzanol content. J. Am. Oil Chem. Soc. 87 (10), 1227–1234.
Wang, T., Khir, R., Pan, Z., Yuan, Q., 2017a. Simultaneous rough rice drying and rice bran stabilization using infrared radiation heating. LWT Food Sci. Technol. 78, 281–288.
Wang, X., Lu, J., Liu, H., Jin, Q., Wang, X., 2016. Improved deacidification of high-acid rice bran oil by enzymatic esterification with phytosterol. Process Biochem. 51 (10), 1496–1502.
Wang, X., Wang, X., Wang, T., 2017b. An effective method for reducing free fatty acid content of high-acid rice bran oil by enzymatic amidation. J. Ind. Eng. Chem. 48, 119–124.
Yılmaz, N., 2016. Middle infrared stabilization of individual rice bran milling fractions. Food Chem. 190, 179–185.
Zhao, W., Shishikura, A., Fujimoto, K., Arai, K., Saito, S., 1987. Fractional extraction of rice bran oil with supercritical carbon dioxide. Agric. Biol. Chem. 51, 1773–1777.
Zullaikah, S., Lai, C.-C., Vali, S.R., Ju, Y.-H., 2005. A two-step acid-catalyzed process for the production of biodiesel from rice bran oil. Bioresour. Technol. 96 (17), 1889–1896.

CHAPTER 2

Nutritional Studies of Rice Bran Oil

Oi-Ming Lai*, Jörg J. Jacoby†, Wai-Fun Leong‡, Wee-Ting Lai§
*Department of Bioprocess Technology, Faculty of Biotechnology and Biomolecular Sciences, Universiti Putra Malaysia, Selangor, Malaysia
†Wilmar Biotechnology Research and Development Center, Shanghai, China
‡Department of Food Science, The Pennsylvania State University, College of Agricultural Sciences, University Park, PA, United States
§Institute of Bioscience, Universiti Putra Malaysia, Selangor, Malaysia

1. INTRODUCTION

In the present day, there are two different species of domesticated rice grown worldwide, *Oryza sativa* (Asian rice) and *Oryza glaberrima* (African rice). However, African rice yields considerably less than Asian rice and, for this reason, is not widely grown nowadays; even in West Africa, only 20% of rice grown is the African type. The main subspecies of Asian rice cultivated extensively today are *O. sativa japonica* and *O. sativa indica*.

The origins of domesticated rice are still lively debated, especially for the Asian cultivar (Callaway, 2014). African rice was domesticated from *Oryza barthii* in a single region, probably near the Niger River in West Africa, >3000 years ago (Wang et al., 2014). The domestication of Asian rice began even earlier. Archaeological findings of rice bulliform phytoliths at the Shangshan site in China's Lower Yangtze region have a morphology closer to modern *O. sativa* than to wild species, and some researchers suggest that the domestication began in the Lower Yangtze region >9000 years ago (Harris, 2007; Vaughan et al., 2008; Zuo et al., 2017). This is in agreement with a genetic study by Molina et al. advocating a single evolutionary origin of domesticated *japonica* and *indica* rice ~8200–13,500 years ago in the Yangtze River Valley (Molina et al., 2011).

Disagreeing with this hypothesis is a study from 2012 in which the researchers conducted in-depth genetic analysis suggesting that the first domestication of *O. sativa japonica* from the wild rice *Oryza rufipogon* occurred in the middle area of the Pearl River in southern China (Huang et al., 2012). The authors further propose that the *japonica* cultivar spread from the Pear River Valley into South East and South Asia, and that *O. sativa indica* was subsequently developed from crosses between *japonica* rice and local wild rice (Huang, et al., 2012). However, this hypothesis

Rice Bran and Rice Bran Oil
https://doi.org/10.1016/B978-0-12-812828-2.00002-0

remains controversial because of the lack of archaeological findings in support of it. After the spread to the rest of Asia, rice was introduced to Europe through Western Asia, and later to the Americas through European colonization.

In our day, rice is, together with wheat and maize, the most common staple food used in human consumption. Especially in Asia, a large part of the population depends on rice as their primary source of carbohydrates, and it comes as no surprise that most of the rice is grown in Asia. The global rice paddy production is forecasted to reach 756.7 million metric tons (MT) in 2017 (FAO, 2017). Countries in Asia are seen producing 90% (684.2 MT) of the world's production in 2017, led by China (210.2 MT), India (164.2 MT), Indonesia (74.2 MT), Bangladesh (50.8 MT), Vietnam (43.3 MT), Thailand (33.7 MT), and Myanmar (29.5 MT) (Table 1).

The whole rice grain encompasses several components: starchy endosperm (70%), hull (20%), bran (7%–8.5%), and embryo (2%–3%) (Ju and Vali, 2005; Orthoefer, 2005). During the 19th century, large scale rice milling industrialization was established with the help of automated rice hulling and polishing machines invented by Evaristo Conrado Engelberg and Sampson Moore, respectively (Barker et al., 2014). The hulling of the paddy rice produces brown rice, and subsequent polishing separates the starchy

Table 1 Global paddy production in 2017 (FAO, 2017)

Country or region	Rice paddy production (2017 forecast) (MT)
WORLD	756.7
ASIA	684.2
Bangladesh	50.8
Cambodia	10.1
China	210.2
India	164.2
Indonesia	74.2
Japan	10.4
Myanmar	29.5
Pakistan	11.1
Philippines	19.6
Thailand	33.7
Vietnam	43.3
AFRICA	31.1
SOUTH AMERICA	25.5
NORTHERN AMERICA	8.1
EUROPE	4.1

endosperm (white rice) from the rice bran and embryo. Although polishing removes many beneficial nutrients found in brown rice, the great majority of rice around the world is consumed as white rice. Rice bran, the byproduct of rice milling, contains 15%–23% oil, 12%–16% protein, 21%–27% fiber, and many micronutrients (Champagne, 2004; Cho and Samuel, 2009). However, freshly milled rice bran has a very short shelf life because lipases inside the bran converts the lipids into free fatty acids, making the bran unfit for human consumption. Furthermore, rice bran contains wax esters, unsaponifiable matter, and dark pigment, making the refining process challenging (Ju and Vali, 2005). For these reasons, the vast majority (up to 90%) of rice bran is still used as cheap animal feed for cattle, poultry, and swine.

Nevertheless, several methods have been developed to stabilize rice bran by inactivating the lipases inside of it for possible food applications (Gul et al., 2015). Additionally, optimizations of the refinery process in the 1950s made it possible to economically produce high quality rice bran oil (RBO) for human consumption. The latest FAO statistics covering worldwide RBO production available for 2013 (Table 2) shows a worldwide production of around 1.1 MT (FAOSTAT, 2013). The largest producers were India with 780 KT and China with 105 KT, followed by Japan and Thailand

Table 2 Rice bran oil production, supply, and uses in 2013 (FAOSTAT, 2013) and maximum potential for RBO production

Country	RBO production (KT)	Maximum potential for RBO production (KT)[a]
WORLD	1115	10,594
India	780	2299
China	105	2943
Japan	64	146
Thailand	45	472
Myanmar	29	413
Brazil	15	172
Cambodia	11	141
Vietnam	11	606
South Korea	10	74
Nepal	9	67
Sri Lanka	9	35
Laos	5	56
Pakistan	5	155
Bangladesh	3	711

[a]Calculated as 7% bran obtained from paddy production in 2017 (FAO, 2017) and 20% oil from bran.

with 64 KT and 45 KT, respectively. However, the RBO market is still rapidly growing worldwide. In 2014, the global RBO market size was estimated at >1.2 MT, with India producing >900 KT in 2014. China was the second largest market with an estimated production of around 200 KT followed by Japan and Thailand with >80 KT and 50 KT, respectively. By 2016, the Chinese market more than doubled to 440 KT according to the Chinese Cereals and Oils Association estimates. Yet, there is enormous potential for future increases in RBO production. By using the 2017 estimates for global paddy production (756.7 MT) and an assumed yield of 7% bran from the paddy, we can calculate a maximum global production potential of 10.6 MT RBO, assuming a 20% yield from the bran. Especially in China, the prospects of further growths in RBO production are promising, as this country only utilizes 15%–20% of its maximum capacity.

1.1 Composition and Nutritive Value

Since around a century ago, RBO gained popularity in Japan as it is considered a "healthy oil" and is one of the most used vegetable oils in this country. India has also a long tradition of using RBO and, nowadays, the reputation and sale of RBO is rapidly increasing in China, Korea, Thailand, and Vietnam as well. One reason for the rise of RBO as a cooking oil is that the World Health Organization (WHO), the American Heart Association (AHA), the National Institute of Nutrition (NIN) in India, the Indian Council of Medical Research (ICMR), the Japanese Oil Chemists' Society (JOCS), and the Chinese Cereals and Oils Association (CCOA) have endorsed RBO as a "healthy oil". This is due to the balanced fatty acid composition of RBO with a saturated (SFA), monounsaturated (MUFA), and polyunsaturated fatty acid (PUFA) ratio of approximately 0.6 (SFA): 1.1 (MUFA): 1 (PUFA), which is very close to the ratio advocated by these organizations. Palmitic acid is the most abundant SFA in RBO (17.0%–21.5%), whereas myristic and stearic acid are only present at a small percentage (Table 3). Oleic acid is the main MUFA with around 40%, whereas linoleic acid, a ω-6 PUFA, is the main PUFA with around 35%. The ω-3 PUFA linolenic acid is only present in minor amounts (Bakota et al., 2014; Cho and Samuel, 2009; Orsavova et al., 2015; Orthoefer, 2005; Pali, 2013).

Beside the advantageous fatty acid composition, the unsaponifiable components of RBO are also of great nutritional value. The most important one is γ-oryzanol, which is a ferulate (4-hydroxy-3-methoxy cinnamic acid) ester of triterpene alcohols and plant sterols, mostly cycloartenyl ferulate,

Table 3 Fatty acid composition of RBO (summarized from Orthoefer, 2005; Kahlon, 2009; Pali, 2013, Bakota et al., 2014; Orsavova et al., 2015)

Name of FA	%
Myristic	0.4–1.0
Palmitic	17.0–21.5
Stearic	1.0–3.0
Oleic	38.4–42.3
Linoleic	33.1–37.0
Linolenic	0.5–2.2
SFA	18.4–25.5
MUFA	38.4–42.3
PUFA	33.6–39.2

24-methylenecycloartanyl ferulate, and campesteryl ferulate (Rogers et al., 1993). The health benefits of γ-oryzanol have been proven in many animal and human studies, and are therefore subject of a detailed discussion later in this chapter.

The γ-oryzanol content of RBO depends largely on the refining method. Crude RBO has a γ-oryzanol content of around 1.8%–2.2%, whereas physically refined RBO contains 1.1%–1.7%. However, alkali refining reduces γ-oryzanol to just 0.2% (Patel and Naik, 2004). Other important micronutrients of RBO are the vitamin E derivatives tocopherol and tocotrienol. Tocopherols (≈0.04%) are predominantly present as α-tocopherol and γ-tocopherol, whereas tocotrienols (≈0.07%) are primarily present as γ-tocotrienols and, to a lesser part, as α-tocotrienols (De Deckere and Korver, 1996; Orthoefer, 2005; Rogers, et al., 1993). RBO also contains notable amounts of squalene and phytosterols, primarily as β-sitosterol, campesterol, stigmasterol, and isofucosterol (Sugihara et al., 2010).

Even though RBO is mainly used for human consumption, the oil is also used in animal consumption. Examples are usage in weanling and growing pigs (Kang and Kim, 2016; Li et al., 2017), as well as broiler chickens (Kang and Kim, 2016) to improve growth. Feeding dairy cows with 4% RBO helps increase conjugated linoleic acid (CLA) concentrations in milk (Lunsin et al., 2012). These could presumably benefit consumers because CLA is claimed to have anticarcinogenic, antiobese, antidiabetic, and antihypertensive properties (Koba and Yanagita, 2014). It is even used in horses that have a difficult time keeping weight on or that have Equine Metabolic Syndrome (Zimmel, 2011). Furthermore, the oil is claimed to give horses and dogs a rich and shiny look to their coats.

1.2 Clinical Trials in Humans using Rice Bran Oil

The health effect of RBO and its ingredients has been studied extensively over the last several decades in animals and humans. However, this subchapter will focus on the clinical trials done in humans (and nonhuman primates) that used whole RBO (Table 4). Other studies conducted with bioactive compounds of RBO will be discussed later in this chapter.

Although earlier studies have attributed the health effects of rice bran on the fiber found in it (Hegsted et al., 1993), recent studies suggest that the oil extracted from rice bran is responsible for the cholesterol-lowering effect. Most et al. (2005) conducted two studies to examine the cholesterol-lowering abilities of defatted rice bran and RBO. In the first study, the authors recruited 26 healthy subjects that, after a 3-week run-in diet, were randomly assigned for an additional 5 weeks to either a low-fiber diet (16.6 ± 1.8 g dietary fiber/day) as a control or a high-fiber diet containing 56-94 g/day of defatted rice bran, which doubled the amount of dietary fiber. Surprisingly, subjects on the low-fiber control diet had lower LDL cholesterol (LDL-C) and apolipoprotein B (ApoB) levels than subjects had on the defatted rice bran diet (Most et al., 2005). The authors went on to conduct a second study to elucidate the effect of RBO on cholesterol level. In this cross-over study, 14 volunteers were randomized to consume either a control oil or RBO for 5 weeks and then crossed-over to receive the other diet for 5 weeks. The control oil was a blend of peanut oil, olive oil, corn oil, canola oil, palm oil, and butter that had a fatty acid composition similar to RBO. Blood samples collected after each 5-week diet period showed statistically significant lower total cholesterol (TC), LDL-C, and ApoB levels in the subjects consuming one-third of their daily dietary fat as RBO than subjects on the control oil diet. The authors concluded that compounds in RBO played a major role in lowering cholesterol as both oils had a similar fatty acid composition. This observation is in agreement with studies done in cynomologus monkeys fed with different vegetable oils (Nicolosi et al., 1991; Wilson et al., 2000). In these studies, RBO lowered blood lipids more than a blend of oils with a similar fatty acid composition and was as good as canola and corn oil in reducing serum total cholesterol (TC) and LDL-C, despite RBO having a higher SFA and lower MUFA content in comparison to canola oil and a lower PUFA content than corn oil. Similar results were obtained in a cross-over trial with 15 middle-aged and elderly subjects. In the study, either RBO, canola, corn, or olive oil contributed two-thirds of the dietary fat during a 32-day diet period for each oil.

Table 4 Human clinical trials with RBO

Study reference	Number of subjects	Inclusion criteria	Age	Duration (days)	Comparison oil(s)	Amount (daily)	Design
Most et al. (2005)	14	Healthy	18–50	35	*Blend of*: PNO,OO, MZO, CNO,PO, and butter	1/3 of daily fat	Cross-over
Lichtenstein et al. (1993)	15	Hypercholesterolemic	44–78	32	(1) CNO (2)MZO (3) OO	2/3 of fats (20% of calories)	Cross-over
Salar et al. (2015)	75	Diabetic	50–52 (mean)	56	(1) SFO (2) CNO	30 g	Parallel
Kuriyan et al. (2005)	14	Hyperlipidemic	40–60	90	SFO	Not fixed	Cross-over
Lai et al. (2012)	35	Type-2 diabetic	30–80	35	SBO	18 g	Parallel
Raghuram et al. (1989)	21	Hyperlipidemic	NR	30	PO and PNO	35 g	Parallel
Berger et al. (2005)	30	Mildly hypercholesterolemic	40–65	45	PNO	50 g	Run-in PNO Day15–45 RBO
Zavoshy et al. (2012)	50	Hyperlipidemic	25–65	28	NR	30 g (1400 kcal)	Parallel

Continued

Table 4 Human clinical trials with RBO—cont'd

Study reference	Number of subjects	Inclusion criteria	Age	Duration (days)	Blends	Amount (daily)	Design
Suzuki and Oshima (1970)	50	Healthy	19–21	7	RBO:SFFO (1) 100:0 (2) 85:15 (3) 70:30 (4) 50:50 (5) 0:100	60 g	Parallel
Malve et al. (2010)	73	Hyperlipidemic	18–65	90	RBO:SFFO (8:2)	<0.5 L/ month	Parallel
Utarwuthipong et al. (2009)	16	Hypercholesterolemic	44–67	56	(1) SBO (2) PO (3) RBO (4) RBO: PO 3:1	20% of energy intake	Cross-over
Devarajan et al. (2016a)	400	Normoglycemic and newly diagnosed diabetics	50 (mean)	56 days	RBO:SSO 80:20, compared to baseline	4–5 L/2 weeks	Parallel
Devarajan et al. (2016b)	400	Normo- and hypertensive	40–60	60 days	RBO:SSO 80:20, compared to baseline	4–5 L/2 weeks	Parallel

CNO, canola oil; MZO, maize oil; OO, olive oil; PNO, peanut oil; PO, palm oil; RBO, rice bran oil; SBO, soy bean oil; SFFO, safflower oil; SFO, sunflower oil; SSO, sesame oil.

Again, there was no statistically significant differences in TC or LDL-C levels in subjects consuming either RBO, canola, or corn oil. However, RBO significantly lowered TC and LDL-C levels in comparison to the olive oil diet (Lichtenstein et al., 1994).

In a clinical trial comparing RBO, sunflower oil, and canola oil in 75 postmenopausal women with type-2 diabetes using a parallel study design, Salar et al. found that substituting RBO or canola oil for sunflower oil can significantly lower TG, LDL-C, and triacylglyceride levels (Salar et al., 2016). Interestingly, in this trial with type-2 diabetes patients, RBO was even more efficient in lowering plasma lipids than canola oil. The superiority of RBO over sunflower oil in reducing plasma lipids was shown in an additional study using a cross-over design with 14 subjects over 3 months (Kuriyan et al., 2005). In addition, the positive effect of RBO in type-2 diabetic patients was confirmed in a trial where 35 patients consumed either a modified milk drink with 18 g RBO or 18 g soybean oil daily for 5 weeks. In this study, only the patients consuming RBO had a significant reduction in TC and LDL-C (Lai et al., 2012).

The hypocholesterolemic effect of RBO was also seen in 12 subjects with baseline TC > 190 mg/dL when they replaced their normal cooking oil (palm and peanut oil) with RBO for 30 days. In comparison, nine subjects who continued to use their usual cooking oil did not have improved TC levels after 30 days (Raghuram et al., 1989). Another interesting study compared the effect of low and high γ-oryzanol RBO (50 mg/day vs. 800 mg/day) replacing 50% of total fat in the diet after a 2-week run-in period with peanut oil (PNO) (Berger et al., 2005). Notably, although PNO statistically lowered TC and LDL-C, the LDL-C/HDL-C ratio did not change. Both RBO groups had decreased TC levels similar to PNO but were more effective in lowering LDL-C than peanut oil, leading to a significant decrease (−18.9%) in the LDL-C/HDL-C ratio. However, surprisingly, there was no statistical difference between the low γ-oryzanol and high γ-oryzanol groups in lowering blood lipids, indicating that other bioactive components in RBO, such as phytosterols and tocotrienols, might act alone or synergistically with γ-oryzanol to lower cholesterol. The potential of RBO to lower blood lipids was also shown in a parallel-design clinical trial with 50 hyperlipidemic patients consuming a low-calorie diet (1400 kcal/day) with or without 30 g per day of RBO for 28 days. The RBO-consuming group showed a significant reduction in total cholesterol, low-density lipoprotein, and atherogenic ratio of TC/HDL-C in comparison to the control group (Zavoshy et al., 2012).

As early as 1970, researchers tried to further improve the health benefits of RBO by blending several oils considered healthy with RBO. Suziki and Oshima blended RBO with different concentrations of safflower oil (SFFO) and gave it to 10 healthy females per group for 7 days. Subjects consuming a blend of 70% RBO and 30% SFFO had the highest reduction in TC (−26%), whereas subjects given 100% RBO or 100% SFFO had a 15% and 13% decrease in TC, respectively, after 7 days (Malve et al., 2010; Suzuki and Oshima, 1970). The authors suggested that the high linoleic acid of safflower oil together with bioactive components from RBO act synergistically to lower TC levels. A randomized study with 73 hyperlipidemic patients consuming a blend of RBO and SFFO (8:2) or their normal cooking oil for 3 months showed that 82% of the subjects in the RBO/SFFO group had their LDL-C levels lowered to below <150 mg/dL, whereas in the control group only 57% of the patients had LDL-C levels below that number. These clinical trials suggested that a blend of RBO and SFFO are beneficial for human health by lowering TC and LDL-C, known risk factors for cardiovascular disease.

Another interesting study compared a blend of RBO and palm oil (PO) (3:1), with RBO, PO, or soybean oil (SBO). In a cross-over trial, each of the 16 hypercholesterolemic participants consumed a free-choice weight-maintaining control diet (55% energy (en) carbohydrates, 15% en protein, 30% en fat) with or without the four test oils (20% of the total energy) for 8 weeks in each testing period. Interestingly, all test oils except pure PO reduced TC and LDL-C levels in comparison to the control diet, suggesting that PO in a 3:1 RBO/PO blend is not harmful. However, SBO also significantly decreased HDL-C levels by 9%, whereas RBO had no effect on HDL-C levels and RBO/PO slightly increased HDL-C. Therefore the atherogenic index, an important cardiovascular disease risk factor, was only significantly reduced during the study periods when the diet contained RBO (−18%) and the RBO/PO blend (−15%) compared with the control diet, indicating that RBO and a mixture of RBO and PO are superior to SBO and PO (Utarwuthipong et al., 2009).

Two other studies investigated a blend of 80% RBO and 20% sesame oil (SSO) on human health. The first study with 300 patients suffering from type-2 diabetes and 100 normoglycemic subjects showed that, after 8 weeks, patients consuming either RBO/SSO or RBO/SSO + glibenclamide (a diabetes drug) had a significant reduction in fasting and postprandial plasma glucose, HbA1c, TC, LDL-C, and TG levels. Patients using just the drug glibenclamide showed only a significant decrease in their fasting and postprandial plasma glucose and HbA1c levels but not their lipid levels, indicating a potential use of this new blend in managing diabetes and hyperlipidemia (Devarajan et al., 2016a). In a second study by the same authors, the 80:20 blend of RBO and SSO was tested in 300 hypertensive patients and 100 normotensives subjects. The RBO/SSO blend with or without the hypertension drug nifedipine or nifedipine treatment alone lowered systolic and diastolic blood pressure in hypertensive patients after 60 days, with the combination of RBO/SSO and drug having the largest effect, providing support that RBO may also have a blood pressure-lowering effect. However, only patients receiving RBO/SSO (with or without drug) had also significant lower levels of TC, LDL-C, and TG (Devarajan et al., 2016b).

All these studies in humans and nonhuman primates provide evidence that RBO has a hypocholesterolemic and possibly other health effects, and that not only the fatty acid composition is important for the health benefits of RBO but also the bioactive compounds found in the oil, which will be discussed in detail in the subsequent subchapters.

2. ORYZANOL

In 1954, γ-oryzanol was first isolated from RBO by Kaneko and Tsuchiya, and was recognized as a single compound with absorption maxima at 230, 290, and 315 nm. Later, γ-oryzanol was confirmed as a mixture of ferulic acid ester with phytosterols (Tsuchiya and Kaneko, 1957). With the advance of analytical techniques and a great deal of interest in its potential health benefits, many more γ-oryzanol components and their derivatives were detected. In the early days, γ-oryzanol was isolated and purified by column chromatography on alumina, followed by separation and detection using thin layer chromatography and UV (Seetharamaiah and Prabhakar, 1986). In 1993, Roger et al. performed separation and determination of γ-oryzanol compounds using reversed-phase HPLC and identified cycloartenyl ferulate, 24-methylene cycloartanyl ferulate, and campesteryl ferulate as the three major oryzanol ferulates, although later it was confirmed that only two of three peaks are single compound, whereas cycloartanyl ferulate is a mixture of two other oryzanol ferulates. A more comprehensive separation and identification of γ-oryzanol components were reported by Xu and Godber (1999) involving robust sample preparation using low-pressure silica column, normal-phase preparative scale HPLC, and reverse-phase HPLC, and detection using GC/MS successfully recognized a total of 10 oryzanol ferulates in RBO. Chromatography techniques such as HPLC/DAD/FLD,

HPLC/MS, HPLC/MS/MS, LC-GC, and GCMS are currently the popular methods used on γ-oryzanol separation and detection (Aladedunye et al., 2013; Chia et al., 2015; Chotimarkorn et al., 2008; Miller and Engel, 2006; Mingyai et al., 2018).

γ-Oryzanol is a collective of phytosteryl ferulate molecules, where each phytosteryl ferulate is an ester of ferulic acid with phytosterol. Xu and Godber (1999) had identified 10 components of γ-oryzanol in rice (*O. sativa* L.): Δ^7-stigmastenyl ferulate, stigmasteryl ferulate, cycloartenyl ferulate, 24-methylenecycloartanyl ferulate, Δ^7-campestenyl ferulate, campesteryl ferulate, Δ^7-sitostenyl ferulate, sitosteryl ferulate, campestanyl ferulate, and sitostanyl ferulate. Meanwhile Aladedunye et al. (2013) reported the finding of 23 γ-oryzanol components in wild rice (*Zizania palustris*) grown in North America. Among these, ferulic acid ester of 4,4′-dimethylsterols (cycloartenol and 24-methylenecycloartanol) and ferulic acid ester of 4-desmethylsterols (campesteryl) were the three major γ-oryzanols in rice (*O. sativa* L.), accounting for >80% of the total γ-oryzanol in brown rice (Cho et al., 2012; Miller and Engel, 2006; Mingyai, et al., 2018) and in RBO (Krishna et al., 2001; Xu and Godber, 1999). Fig. 1 shows the structure of three major γ-oryzanols in RBO and ferulic acid.

The amount and composition of γ-oryzanol in rice and RBO vary substantially depending on the varietal, cultivars, method of cultivation, method of RBO extraction, and subsequent processing practices. Miller and Engel, 2006 analyzed 30 brown rice samples of various European cultivars grown at different sites and in different seasons, and showed the total γ-oryzanol ranging from 26 to 63 mg/100 g of rice. Similarly Kim et al. (2015) analyzed 16 Korean rice varieties and reported γ-oryzanol concentrations ranging from 26.7 to 61.6 mg/100 g of rice. γ-Oryzanol is particularly concentrated in rice bran. Chotimarkorn et al. (2008) found that the concentration of γ-oryzanol in five varieties of long-grained rice bran cultivated in Thailand ranged from 56 to 108 mg/100 g of rice bran, whereas Cho et al. (2012) reported that γ-oryzanol content in conventional and organic brown rice bran were 60.2 mg/100 g and 65.8 mg/100 g. Krishna et al. (2001) reported that the γ-oryzanol in crude RBO of 18 different Indian paddy cultivars were ranging from 1.63%–2.72% (1630–2720 mg/100 g crude oil), whereas Rogers et al. (1993) reported the γ-oryzanol of five commercially refined RBO were only 11.5–78.8 mg/100 g refined RBO. Sawadikiat and Hongsprabhas (2014) found that in four different brands of commercially refined RBO, the concentration of γ-oryzanol ranged from 248 to 887 mg/100 g, whereas in crude oil the γ-oryzanol ranged from 1599 to 1666 mg/100 g.

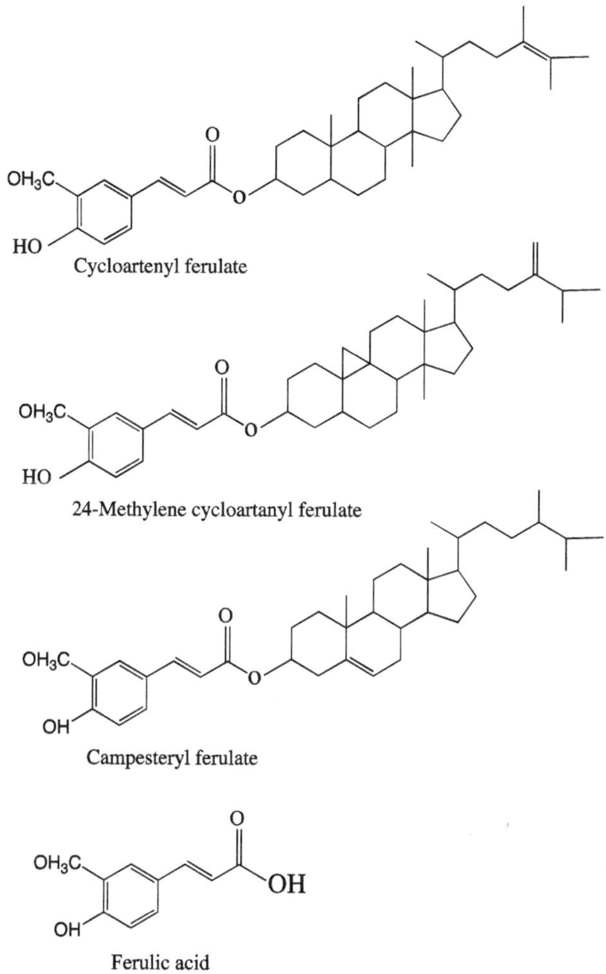

Fig. 1 Structure of the three major phytosteryl ferulates in RBO: cycloartenyl ferulate, 24-methyene cycloatanyl ferulate, and campesteryl ferulate. Ferulic acid is the basic component of phytosteryl ferulates.

Krishna et al. (2001) investigated the effect of various oil refining steps on the retention of γ-oryzanol and revealed that chemical refining caused significant loss of γ-oryzanol but not physical refining. The γ-oryzanol in physically refined RBO ranged from 1.1%–1.74%, whereas for alkaline refined RBO, ranges were 0.19%–0.20%. The author noted that the alkaline neutralization steps in chemical refining was the main culprit of γ-oryzanol degradation, whereas bleaching and deodorization did not have a significant effect on γ-oryzanol, whereas degumming and dewaxing causes minimum loss at 1.1% and 5.5%. Similar observation was reported by Van Hoed et al. (2010), indicating that caustic washing using strong sodium hydroxide (0.65%) causes approximately 90% loss in γ-oryzanol, but they also found that reduction in sodium hydroxide to 0.122% could instead retain 90% of the total γ-oryzanol.

Conventional solvent extraction by hexane is by far the most common method for RBO extraction due to its simplicity, low cost, and better yield. Nevertheless, various methods were explored for their potential to extract major phytonutrients such as γ-oryzanol, phytosterol, and vitamin E. Xu and Godber (1999) reported that optimized supercritical extraction yield four times more γ-oryzanol (5.39 mg/g of rice bran) than the solvent extraction by organic solvent mixture of hexane and isopropanol (1.68 mg/g of rice bran). They also showed that saponification during solvent extraction reduces oryzanol content by half compared to without saponification.

Chia et al. (2015) reported that the crude RBO from subcritical CO_2 extraction contained approximately 10 times more total γ-oryzanol and total tocol compounds, and lower free fatty acids content and peroxide value than hexane extraction. They postulated that the lower extraction temperature and pressure in subcritical CO_2 preserved the antioxidants. However Mingyai et al. (2018) did not observe a similar trend when comparing hexane extraction and supercritical CO_2 extraction on black and white rice bran. They reported lower total γ-oryzanol in supercritical CO_2 extraction (215 mg/100 g oil and 408 mg/100 g oil) than hexane extraction (333 mg/100 g oil and 543 mg/100 g oil).

In 2013, Aladedunye et al. noted the composition of wild rice (*Z. palustris*) γ-oryzanol is different than the commercial brown rice. Wild rice was claimed to have more diverse γ-oryzanol components and higher total γ-oryzanol than brown rice (135.2 mg/100 g vs. 68.8 mg/100 g), although its cycloartenyl ferulte, 24-methylenecycloartanyl ferulate, and campesteryl ferulate account for only 55%–64% of total γ-oryzanol. Recently, Mingyai et al. (2018) investigated three extraction methods (solvent, cold press, supercritical CO_2) with three rice varieties (Hom-nin—black rice, red jasmine—red rice, and Khao Dawk Mali—white rice). In this study, the yield of RBO was not reported, but the extraction methods were instead compared by total γ-oryzanol, total phytosterol, and vitamin E. The cold press extraction was claimed to be best for black rice, whereas solvent extraction by hexane was most efficient in white rice. They reported that black rice RBO contained significantly higher total γ-oryzanol, total phytosterol, and vitamin E than the red rice and white rice. Another study analyzed eight different varieties of rice with various bran colors (one white, two light browns, two browns, one red, and two purples), remarking that the bran color was not associated with the concentration of vitamin E and total γ-oryzanol, but the red and purple rice brans had significantly higher total phenolics and better antioxidant capacity than the brown rice brans (Min et al., 2011). It is unknown if and how the color compound plays a role on γ-oryzanol in rice bran. But it is clear that each variety possesses diverse nutritional properties.

2.1 Functions of Oryzanol

RBO is a complex mixture of its macrocomponents (various lipid compounds) and microcomponents (tocols, oryzanol, phytosterols, policosanol, etc.) where, if combined, exerted a wide spectrum of biological activities. Oftentimes RBO, instead of its individual components, was used in nutritional and functionality research. Studies that focus on the individual component and their fractions may provide in-depth understanding of these components.

2.1.1 Antioxidant

Oxidation causes many problems for the food industry. Lipid oxidation is, by far, one of the most challenging problems faced by the food industry. Although synthetic antioxidants such as butylated hydroxyanisole (BHA) and butylated hydroxytoluene (BHT) have demonstrated efficacy in improving oxidative stability, their usage remains unpopular due to the reported adverse health effects. Hence, there is always a demand for natural antioxidants for use in food, and γ-oryzanol is definitely a great option.

γ-Oryzanol has gained much attention due to its antioxidant property. It was shown to increase oxidative stability of vegetable oils at frying temperature (Gertz et al., 2000) and delayed toxic cholesterol oxide formation in cooked beef at refrigeration temperature (Kim, et al., 2015). Oryzanol fortified biscuit and fat showed good stability during baking and storage (Kumar et al., 2014; Prasanth Kumar et al., 2014). Xu and Godber (2001) evaluated the antioxidant capacity of γ-oryzanol and ferulic acid using the linoleic acid model. They reported that the three major γ-oryzanols (cycloartenyl ferulate, 24-methylene cycloartanyl, and campesteryl ferulate) each shows significant antioxidant activity in a linoleic acid model at concentrations of 0.40 mM and 0.16 mM but not at a lower concentration of 0.08 mM. Antioxidant activity of γ-oryzanol was lower than the free ferulic acid and alpha-tocopherol. The authors hypothesized that the γ-oryzanol, being a larger molecule with less mobility than the free ferulic acid, had lowered its overall antioxidant capacity. The antioxidant property of γ-oryzanol was strongly related to its ferulic acid. Like many phenolic antioxidants, ferulic acid is an effective radical scavenger that traps and stabilizes radical species, thereby minimizing the detrimental effect of reactive oxygen species (ROS) (Srinivasan et al., 2007). Although the ferulic acid has a higher antioxidative capacity, the bound ferulic structure within γ-oryzanol is believed to be more heat stable because the higher molecular number makes it less likely to evaporate at high temperature (Winkler-Moser et al., 2015). The antioxidative activity of the three major RBO γ-oryzanols (cycloartenyl ferulate, 24-methylene cycloartanyl ferulate, and campesteryl ferulate) was also addressed in a cholesterol oxidation study (Xu et al., 2001). In contrast to the linoleic acid model study, this study reported that all γ-oryzanol showed better antioxidant activity than vitamin E components (Xu et al., 2001). In a test tube study, γ-oryzanol was found to have more than four times the effectiveness of vitamin E at stopping tissue oxidation (Hiramitsu and Armstrong, 1991). Antioxidant behavior of each antioxidant compound is a complex dynamic phenomenon, where the affinity of antioxidant (γ-oryzanol or vitamin E) for a specific medium changes its oxidative capacity.

Szczesniak et al. (2016) reviewed some recent animal model experiments on rat, mice, and horses, concluding that γ-oryzanol showed a promising ability to inhibit the formation of new free radicals and to scavenge lipid soluble organic radicals. However, the efficacy of γ-oryzanol against inorganic oxygen-derived radicals has yet to be elucidated. In a recent review, Minatel et al. (2016) echoed the importance of γ-oryzanol and its antioxidant activity. They discussed extensively the complex interaction of antioxidant γ-oryzanol in the biological system, and how redox imbalances are related to various metabolic disorders and chronic diseases. They had carefully mapped out the synergistic interaction of γ-oryzanol with organ-organelles that eventually improved glucose metabolism by downregulating the endoplasmic reticulum stress-responsive gene, improving insulin production, promoting fatty-acid metabolism, and reducing total triglyceride in cells. They also discussed the possible effect of γ-oryzanol in a complex network of inhibiting adipocyte activity and hence preventing obesity. The mechanism of γ-oryzanol in reducing inflammation by radical scavenging activity was also elucidated. Zin the following text are some studies on the health benefits of γ-oryzanol and its metabolites.

2.1.2 Hypolipidemic Effect

Many studies had demonstrated the hypolipidemic effect of RBO, but not many elucidate the exclusive effect of oryzanol. An in vitro study using γ-oryzanol (26 μmol/L) shows decreased cholesterol apical uptake in human Caco-2 cells (HBT-37) (Mäkynen et al., 2012). Another study on human hepatocellular carcinoma (HepG2) cells reported a significant effect of γ-oryzanol (50 μmol/L) on downregulation of hepatic lipogenesis-related genes (Wang et al., 2015). Wilson et al. (2007) evaluated the effect of RBO, ferulic acid, and oryzanol on serum lipids and cholesterols of hamsters. All four experimental groups were fed a hypercholesterolemic diet for 10 weeks, with variation where group 1 (control) was fed with 10% coconut oil, group 2 with 10% RBO, group 3 with 0.5% ferulic acid, and group 4 with 0.5% oryzanol. They found no significant difference between the RBO group and oryzanol group on plasma lipids, plasma cholesterol, and plasma vitamin E. In other words, γ-oryzanol, independent of RBO, shows a significant hypocholesterolemic effect. They also found that the oryzanol group, compared with the ferulic acid group, shows greater effect on lowering serum LDL-C and raising serum HDL-C. They hypothesized that this could be due to the ability of oryzanol, not ferulic acid, in promoting fecal excretion of cholesterol and its metabolics. A more recent study by Bhaskaragoud et al. (2016) found that γ-oryzanol supplementation (0.1% and 1%) improved lipid profile in rats fed a high-fat diet.

Accinni et al. (2006) studied the effect of combined dietary supplementation in 57 dyslipidemic volunteers for 4 months. The experiment groups included: group A on placebo as control; group B on PUFA n-3 (660 mg EPA and 440 mg DHA) and tocopherol (4 mg); and group C the same as group B plus oryzanol (40 mg) and niacin (18 mg). They found that groups B and C showed significant reduction in total serum lipid, increased HDL-C, reduced LDL-C, and improved inflammatory and oxidative status, and group C (with oryzanol) showed better overall improvement than group B. Several clinical studies in the early 1960s and 1980s reported that γ-oryzanol consumption (300 mg/day) showed positive outcomes on serum lipid profile, on thyroid-stimulating hormone, and on menopausal symptoms (Ishihara et al., 1982; Shimomura et al., 1980) A previous review by Cicero and Gaddi (2001) somehow determined that at least 300 mg of oryzanol was required to lower serum LDL, without considering the synergistic effect with other bioactives.

2.1.3 Antidiabetic Effect

The use of oryzanol from crude RBO (50 and 100 mg/kg) has been shown to mitigate the oxidative stress and reduce blood glucose level in streptozotocin-induced diabetic rat models (Ghatak and Panchal, 2012). Kozuka et al. (2012) found that γ-oryzanol from brown rice improves glucose metabolism in mice fed a high-fat diet. A subsequent study reported that γ-oryzanol inhibited the activity of DNA methytransferases (DNMTs) in high-fat diet mice and hence reduced preference on a high-fat diet. They believe that γ-oryzanol could be a promising therapy for obesity-related diabetes syndrome (Kozuka et al., 2017). The hypoglycemic effect of γ-oryzanol was demonstrated by its ability to normalize liver enzymes, regulate the secretion of hormone insulin, and improve hepatic gluconeogenesis and blood glucose (Son et al., 2011). The complex interaction between γ-oryzanol and organ-organelles on glucose metabolism was discussed in a recent review (Minatel et al., 2016).

2.1.4 Anticancer Effect

There are abundant studies on the anticancer effect of crude RBO and RBO, but only a few were on γ-oryzanol or its fractions. Hudson et al. (2000) found that ferulic acid inhibits the growth of human breast and colon cancer cells. Kong et al. (2009) demonstrated that cyloartenylferulate could potentially inhibit growth of human colorectal adenocarcinoma cell line SW480 and trigger apoptosis in early stage colorectal cancer cells. A comprehensive review by Szczesniak et al. (2016) describing the detailed physiology effects of γ-oryzanol in animal models believes that γ-oryzanol, a safe and promising anticancerogenic compound, is more effective in cancer prevention than treatment. They also concluded that the anticancer effect of γ-oryzanol was strongly related to its antioxidant capacity and anti-inflammatory property.

2.1.5 Effect on Menopause

In the 1960s and 1980s, some Japanese research groups had shown interest in the effect of γ-oryzanol in menopausal women. A clinical study incorporating treatment of γ-oryzanol for 38 days was shown to improve menopausal symptoms in 7 out of 13 women with surgical menopause (Murase and Iishima, 1963). A study by Ishihara et al. (1982) on 40 postmenopausal women reported significant improvement in HDL-C, LDL-C, TC, and TG after 4–8 weeks treatment with γ-oryzanol (300 mg/day). A similar effect on lipid profile improvement was reported in a clinical study involving the use of γ-oryzanol (300 mg/day) in patients with elevated serum lipid profiles (Yoshino et al., 1989). Some recent studies reporting the positive effect on postmenopausal-related symptoms focused on rice bran extract (Nam et al., 2017; Nhung et al., 2016).

2.1.6 Effects on Dementia

Ferulic acid and γ-oryzanol are promising components in neurological related diseases. Lee et al. (2013) demonstrated various positive effects of ferulic acid (25 μg/mL) in experimental Schwann cells. They postulated that ferulic acid is a useful treatment of peripheral nerve injury. A clinical study on patients with Alzheimer's disease, vascular dementia, or dementia with Lewy bodies showed that γ-oryzanol supplementation (50 mg, twice daily) improved behavioral and psychological symptoms of dementia, without any side effects (Fujii et al., 2018). The mechanism of action was not investigated in this study, but the authors believe that the γ-oryzanol improved the stress tolerance in the subjects by increasing the catecholamine relatives in the limbic system.

2.1.7 Other Effects

In 1980, Shimomura et al. observed a significant reduction of serum thyroid stimulating hormone (TSH) in patients with elevated serum TSH after a single oral dose (300 mg) of RBO γ-oryzanol, but a similar effect was not observed in healthy normal subjects. They postulated that γ-oryzanol has a direct action at the hypothalamus rather than the pituitary gland in regulating TSH. No other studies were found in the literature on the effect of γ-oryzanol on serum TSH or related to hypothyroidism. Islam et al. (2008) reported the anti-inflammatory effect of phytosteryl ferulate in a colitis mice model. They suggested the possible use of phytosterol ferulate in prevention of gastrointestinal inflammatory diseases.

3. PHYTOSTEROLS AND SQUALENE

Phytosterols, also known as plant sterol, are cholesterol-like compounds found exclusively in plants. Like cholesterol, phytosterol plays an important role in plant functionality such as regulating cell membrane fluidity and permeability, and supporting cell structure. A few hundreds of different phytosterols have been reported in the literature, where each can be classified into three different classes of sterol structure based on methylation at the four carbons in the sterol structure: 4,4-dimethyl-, 4-methy-, and desmethyl- (Piironen et al., 2000). Phytosterols are important nonsaponifiable constituents found in RBO, where it could exist in the form of bound phytosterols (γ-oryzanol) or free phytosterols. β-Sitosterol, campesterol, and stigmasterol (Fig. 2) are three major free phytosterols found in RBO, where the total of these phytosterols were sometimes used to express the total phytosterols in RBO (Mingyai, et al., 2018; Sawadikiat and Hongsprabhas, 2014). The total phytosterol concentration in crude and commercially refined RBO were 848–1034 mg/100 g and 1362–1376 mg/100 g, respectively (Sawadikiat and Hongsprabhas, 2014). More recently, Mingyai et al. (2018) reported that the concentration of phytosterol in chemically refined RBO ranged from 878 to 1143 mg/100 g. They also reported that the total phytosterol concentration was >10 times the total γ-oryzanol.

The effects of RBO and phytosterol on serum lipid and serum cholesterol were well documented (Nicolosi, et al., 1991; Sharma and Rukmini, 1987). In 2000, a study by Vissers et al. (2000) specifically examined the effect of plant sterols from RBO in 70 healthy, normolipidemic volunteers (38 men and 32 women). Subjects who received 2.1 g plant sterols in 29 g margarines each day for a period of 3 weeks showed a 9% reduction in LDL cholesterol, but no effect on HDL-C and triacylglycerol concentration. They postulated that, upon consumption, the phytosterols could bind with dietary cholesterol and reduce its absorptions in the gut or promote attachment of cholesterol to bile acids, which would then be excreted through feces.

Squalene, a precursor of cholesterol biosynthesis, is a natural lipid synthesized in humans and various organisms. Commercially available squalene extract is mainly from shark liver oil and smaller amounts are from olive oil, palm oil, wheat-germ oil, and RBO. The use of plant-based squalene was limited perhaps due to its sensitivity to oxidative deterioration. Nevertheless, plant-based squalene has great potential because shark squalene is not a sustainable source. The extraction method plays an important role on the squalene quality. Sugihara et al. (2010) demonstrated a novel fractionation method for RBO squalene and phytosterols. Using supercritical carbon dioxide extraction produces high purity squalene and phytosterol with oxidation stability. Squalene is well known for its health benefits. It has chemopreventive activity, enhances immune response, and improves oral drug delivery and has a hypocholesterolemic effect (Reddy and Couvreur, 2009). Nevertheless, to this author's knowledge, no research exists on squalene from RBO. It is unsure if the RBO squalene has the same health benefits of squalene from shark or other sources.

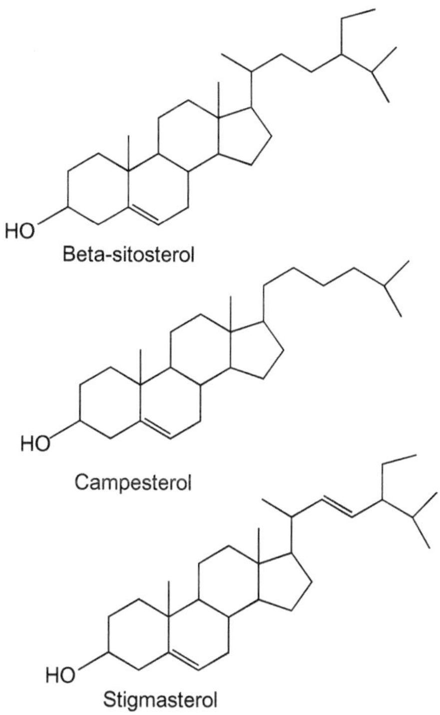

Fig. 2 Structure of the three common phytosterols found in RBO: β-sitosterol, campesterol, and stigmasterol.

4. WAXES AND POLICOSANOL

Rice bran wax (RBW) is a byproduct of the RBO refinery process. It is commonly removed by winterization treatment on crude RBO and treated as waste. Depending on the extraction parameters and rice bran source, RBW ranging from 1% to 7% of crude RBO contained predominately a mixture of saturated ester of fatty acids and policosanol (Liu et al., 2008; Saunders, 1985). Policosanol is a mixture of long-chain aliphatic alcohols (20–36 carbons) found predominantly in sugar cane wax and beeswax, and a significant amount was found in the unsaponifiable fraction of rice bran. Liu et al. (2008) studied the composition of policosanols from high-intensity ultrasonic hydrolyzed rice bran and reported that triacontanol, C30 (26.95%), was the major policosanol followed by octacosanol, C28 (17.04%), and other minor components including dotriacotanol (C32), tetracosanol (C24), hexacosanol (C26), nonacosanol (C29), heptacosanol (C27), and pentacosanol (C25). In 2004, Crovotto et al., using a similar hydrolysis method with high-intensity ultrasonication, reported that the predominate policosanol components in rice bran were octacosanol (46%) followed by triacontanol (32%) and hexacosanol, C26 (15%); whereas in 2005, Vali et al. reported that the major policosanols were triacontanol (24%–27%), docotricotanol (17%–20%), and octacosanol (12%–17%). The major policosanols in RBW are even-numbered aliphatic alcohols (C14–C32), with approximately 5% odd-numbered aliphatic alcohols (Liu et al., 2008).

There are mixed reports on the health benefits of rice policosanol. In 2003, Wang et al. examined the effect of policosanols using hamsters. The hamsters were divided into 5 groups with 10 hamsters each. Group 1 was fed a control diet, and group 2 to group 5 were fed a control diet plus Octa 6 (a policosanol mixture from sugar cane wax, 25 mg/kg BW) (Group 2), ricewax (a policosanol mixture from rice wax with 50% converted into the corresponding acids, 50 mg/kg BW) (Group 3), phytosterol (Cholestatin, 1000 mg/kg BW) (Group 4), and phytosterol (1000 mg/kg BW) and ricewax (50 mg/kg BW). They reported that the use of Octa 6 (Group 2) and ricewax (Group 3) did not show a significant effect on total cholesterol, HDL-C, and non-HDL-C. However, they also reported that the use of Octa 6 and ricewax did not cause toxicity to the hamster's brain, liver, heart, or kidney. A similar observation was reported by Reiner et al. (2005) investigating the effect of rice policosanol in hypercholesterolemic human subjects in an 8-week study supplementing 10 mg/day policosanol. They reported that the rice policosanol supplement showed no significant effect on plasma TG, HDL-C, LDL-C, oxidized-LDL-C, apoproteins B, fibrinogen, homocysteine, and C-reactive protein, but significant reduction in plasma total cholesterol (from 7.37 to 6.99 mmol/L) and Apoprotein AI (from 1.49 to 1.58 mmol/L) were reported. Singh et al. (2006) studied the effect of policosanol in cultured rat hepatoma cells and reported that policasanol treatment downregulated the HMG-CoA reductase (a rate-controlling enzyme in cholesterol biosynthesis) activity.

A more recent study by Wong et al. (2016) working on hyperlipidemic rat platelets found that the crude rice bran policosanol extract (RBE) at low dosage (100 mg/kg body weight), but not high (500 mg/kg) or medium (250 mg/g) dose, inhibited ex vivo platelet aggregation by 42.32%, which is comparable with the effect of 30 mg/kg aspirin treatment. Platelet hyperactivity could cause platelet aggregation, a risk factor of stroke and coronary artery diseases. Platelet activation is also related to hyperlipidemia, diabetes, and hypertension. The authors explained the medium and high dose of RBE did not exhibit the same platelet aggregation effect due to hormesis effect, however the author reported no adverse effects on all RBE treatments. In a follow-up study, they reported RBE inhibition of in vitro rat platelet aggregation in a dose-dependent manner (125–1000 μg/mL) (Wong et al., 2016). The health benefits of RBO policosanol is not conclusive. More thorough investigations are needed to evaluate the physiological activities and health effects of this compound.

5. VITAMIN E

Vitamin E refers to a family of eight molecules having a chromanol ring (chroman ring with an alcoholic hydroxyl group) and a 12-carbon aliphatic side chain containing two methyl groups in the middle and two more methyl groups at the ends. Approximately 1% of the unsaponifiable fraction of RBO is vitamin E (Nicolosi et al., 1993). Among other plant commodities, tocopherol contents in rice bran is relatively high. Vitamin E in RBO is not limited to tocopherol but also includes tocotrienol (Sen et al., 2007). Tocopherols and tocotrienols share similar structures, but the side chain for tocopherols is saturated whereas tocotrienols are unsaturated and possess an isoprenoid side chain (Theriault et al., 2002). Both tocopherols and tocotrienols have alpha, beta, gamma, and delta form-named based on the number and position of methyl groups on the chromanol ring (Juturu, 2008). The alpha form has three methyl groups; the beta and gamma forms have two methyl groups; and the delta has only one methyl group.

Vegetable oil provide the best sources of these vitamin E forms, particularly palm oil and RBO, which contain higher amount of tocotrienols. Other sources of tocotrienols include grape seed oil, oats, hazelnuts, maize, olive oil, buckthorn berry, rye, flax seed oil, poppy seed oil, and sunflower oil. Rice bran or rice germ oil is rich in tocotrienols especially γ-tocotrienol with a potent biological effect against carcinogenesis. In 2000, Qureshi et al. reported that two novel tocotrienols were isolated from rice bran. Their structures were established as desmethyl tocotrienol and didesmethyl tocotrienol (Qureshi et al., 2000) (Fig. 3).

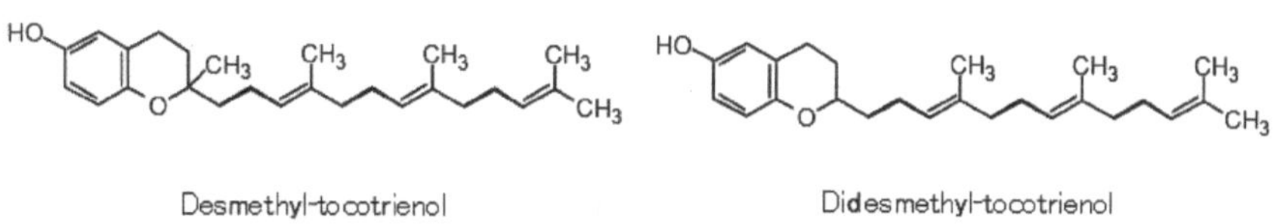

Fig. 3 Structure of desmethyl-tocotrienol and didesmethyl-tocotrienol.

5.1 Tocotrienols

The tocotrienols (T3) are members of the vitamin E family, have excellent antioxidant properties, and are able to prevent the autocatalytic lipid peroxidation process (Seppanen et al., 2010). Potential health benefits of tocotrienols are much less studied compared to tocopherol. The function of tocotrienols can be very specific. Tocotrienol is reported to inhibit cholesterol biosynthesis (Berger, et al., 2005; Qureshi et al., 2002) and have neuroprotective properties (Khanna et al., 2005; Park et al., 2011; Sen et al., 2010). Moreover, α-T3 has outstanding antioxidative activity in liver microsomes, 40–60 times greater than that of α-tocopherol (α-T), and shows special anticarcinogenic properties.

Tocotrienols possess neuroprotective (Khanna et al., 2010; Mangialasche et al., 2012; Sen, et al., 2010), antioxidant (Chotimarkorn, et al., 2008; Lai et al., 2009; Müller et al., 2010), anticancer (Park et al., 2010; Pierpaoli et al., 2010; Shirode and Sylvester, 2010), and cholesterol-lowering properties (Yuen et al., 2011; Zaiden et al., 2010) that often differ from the properties of tocopherols. Micromolar amounts of tocotrienol suppress the activity of HMG-CoA reductase, the hepatic enzyme responsible for the synthesis of cholesterol. Tocotrienols are thought to have more potent antioxidant properties than α-tocopherol. The unsaturated side chain of tocotrienol allows for more efficient penetration into tissues that have saturated fatty layers such as the brain and liver. Experimental research examining the antioxidant, free radical scavenging effects of tocopherol and tocotrienol have found that tocotrienols appear superior due to their better distribution in the lipid layers of the cell membrane (Tiwari et al., 2009). One major conclusion often used to undermine tocotrienol research is the relative inferiority of the bioavailability of orally taken tocotrienols as compared to that of α-tocopherol (Abuasal et al., 2012). Numerous studies indicate that tocotrienols exhibit antioxidant, antiproliferative, antisurvival, proapoptotic, antiangiogenic, and anti-inflammatory activities.

5.2 Antitumor Effects

Tocotrienols have been proven to be effective in inducing apoptosis in a wide variety of tumor cells. The effects of tocotrienols are mediated through the activation of both extrinsic and intrinsic pathways. Induction of death receptors and activation of caspase-8 that leads to the caspase-3 activation is the acting mode of the extrinsic pathway (Lamy et al., 2008). The intrinsic pathway can be conducted by involving mitochondrial depolarization and mediated by upregulation of Bax, cleavage of Bid, release of cytochrome C, and activation of caspase-9, which in turn leads to activation of caspase-3. Because tocotrienol is an unsaturated form of vitamin E, it can mediate apoptosis through DNA fragmentation and upregulation of p53 in certain cells (Kannappan et al., 2010).

In the context of anticancer effects, tocotrienols are proven to be effective in the suppression and apoptosis induction in a wide variety of tumor cells. Breast tumor cell suppression by tocotrienols is widely researched and discussed (Abuasal, et al., 2012; Nesaretnam et al., 2010; Patacsil et al., 2012). Recently, interaction of tocotrienols with estrogen receptors has been reported. γ- and δ-tocotrienols have greater anticancer activity compared to α- or β-tocotrienol (Hsieh and Wu, 2008). Other than that, tocotrienols have exhibited positive results in the suppression of liver (Chan and Chan, 2005; Magosso et al., 2013), lung (Ji et al., 2011), stomach (Sun et al., 2008), skin (Yamada et al., 2008), pancreas (Husain et al., 2011), and prostate cancer cells (Yap et al., 2008).

With the antitumor effects exhibited by tocotrienols, it also showed activity in both prevention and treatment of cancer in animal studies. As early as 1985, a study conducted by Kato el al. showed that tocotrienols have extended the life span of tumor-bearing rats. Antitumor activity was obtained by having tocotrienols administered intraperitoneally to mice with established murine Meth A fibrosarcoma. In most of the studies, tocotrienols were proven to be more effective than α-tocopherol, whereas among the tocotrienols, γ-tocotrienol was more effective than α-tocotrienol as an antitumor agent (Komiyama et al., 1992). Both γ-tocotrienol and δ-tocotrienol managed to delay the growth of tumors in animal studies, and the tumors contained a specific accumulation of tocotrienol analogues when examined.

In cancer prevention models, tocotrienols were found to be effective too. Oils containing tocotrienols, like RBO, are effective in preventing 7,12-dimethylbenz[α]anthracene (DMBA)-induced mammary carcinogenesis in rats, but corn oil and soybean oil, which only contains tocopherol, lack this activity. Long-term administration of tocotrienols have been proven to prevent hepatocarcinogenesis in rat models (Nesaretnam et al., 2010). In the study, tocotrienols managed to weaken the effect of carcinogens. A similar study by others also produced similar findings (Xiong et al., 2016). These studies show that tocotrienols have a high potential to prevent and treat cancer.

5.3 Antiinflammatory Effects

Tocotrienols have been proven to exhibit potent inflammatory activity. Studies have shown that tocotrienols are able to activate the transcription factor NF-κB that is closely related to inflammation (Ralhan et al., 2009). Expression of TNF, IL-1, IL-6, IL-8, inducible nitric oxide synthase, and cyclo-oxygenase 2, all of which mediate inflammation,were also reduced by tocotrienols (Shibata et al., 2010). The STAT3 cell-signaling pathway and the hypoxia-induced factor-1(HIF-1) pathway, which are involved in inflammation, were also shown to be modulated by tocotrienols. Tocotrienols were effective in blocking the activation of NF-κB in several independent lines of evidence (Al Rasheed et al., 2012; Zhang et al., 2015).

5.4 Cardioprotective Effects

Cardiovascular disease remains a leading fatal disease. Tocotrienols exhibit cardioprotective effects by their ability to inhibit a rate-limiting enzymes in cholesterol biosynthesis and their antioxidant and anti-inflammatory activities (Qureshi et al., 2015). Although all the tocotrienol isomers have cardioprotective properties, γ-tocotrienol is the most protective (Qureshi et al., 2011). RBO high in γ-tocotrienol has been studied in rats. In the study, RBO downregulated LDL-cholesterol, hepatic triglyceride concentration, and plasma triglycerides in rats. Meanwhile, the hepatic cholesterol 7-alpha-hydroxylase, hepatic LDL receptor, and HMG-CoA reductase mRNA were increased (Vasanthi et al., 2011). Similarly, RBO was found to improve lipid abnormalities and atherogenic index, and subdued the hyperinsulinemic response in rats with streptozotocin/nicotinamide-induced type 2 diabetes mellitus (Chou et al., 2009; Lai et al., 2012).

Positive effects of γ-tocotrienol were shown in a case studying the lipid peroxidation and total antioxidant status of spontaneously hypertensive rats. Administration of γ-tocotrienol improved the total antioxidant status and superoxide dismutase activity in rats in a 3-month trial (Newaz et al., 2003). Qureshi et al. (2002) reported that a novel tocotrienol-rich fraction, TRF 25, from RBO significantly lowered total cholesterol and LDL-cholesterol by 12% and 16%, respectively. These studies gave an indication that γ-tocotrienol has high potential in reducing lipid peroxides in plasma and blood vessels, and has the ability to enhance total antioxidant status, including superoxide dismutase activity.

5.5 Antidiabetic Effects

Tocotrienol is a potential agent to control the blood glucose level by decreasing levels of plasma and aorta malondialdehyde and 4-hydroxynonenal, and oxidative DNA damage. In a diabetic rat model with a combination of insulin and tocotrienol treatment, the blood glucose level and oxidative stress markers were found to be lowered (Kuhad and Chopra, 2009). The development of diabetes and its complications are mainly contributed by oxidative stress. The crude lipophilic rice bran extract, which includes α-tocopherol, tocotrienols, and phytosterol, was incorporated in the diet of obese diabetic KKAy mice, and it was found that the antioxidative effects of the extract suppressed elevation of plasma malondialdehyde and significantly increased glutathione peroxidase (GPx) mRNA expression at the 0.1% concentration (Aggarwal et al. 2010). In a study done my Fang et al., tocotrienols were proposed to act as peroxisome proliferator-activated receptor (PPAR) modulators (Fang et al. 2010). PPARs are ligand-regulated transcription factors that play essential roles in energy metabolism (Varga et al., 2011). PPARα, PPAR γ, and PPAR δ can be activated by δ-tocotrienol in reporter-based assays. Tocotrienols enhanced the interaction between the purified ligand-binding domain of PPARα and the receptor-interacting motif of coactivator PPAR γ coactivator-1alpha. They also found that TRF improved whole-body glucose utilization and insulin sensitivity of diabetic Db/Db mice by selectively regulating PPAR target genes (Fang et al., 2010). All these results indicate that tocotrienols have antidiabetic potential.

5.6 Bone-Protective Effects

Because nicotine in cigarettes has been identified to be a risk factor that contributes to the development of osteoporosis, some supplements were studied for their role in attenuating the effect of tobacco. Adult male rats treated with γ-tocotrienol (60 mg/kg) were found to have significantly higher trabecular thickness and less eroded surface than the control group. For glucocorticoid-induced or free radical-induced bone loss models, the administration of tocotrienols were found to improve the bone health of adrenalectomized rats (Hermizi et al., 2009; Mehat et al., 2010; Shuid et al., 2010). There is a protective effect posted by tocotrienols against free radical damage in the rat femur bones. Ima-Nirwana et al. showed that treatment with γ-tocotrienol (60 mg/kg body weight/day) reduced body fat mass and increased fourth lumbar vertebra bone calcium content in rats, whereas α-tocopherol was ineffective (Ima-Nirwana and Suhaniza, 2004). Therefore γ-tocotrienol has the potential to be utilized as a prophylactic agent in the prevention of skeletal side effects of long-term glucocorticoid and tobacco use.

5.7 Immunomodulatory Effects

In an experiment conducted by Gu et al. (1999), the administration of tocopherols or tocotrienols enhanced the expression of interferon-γ, IgA, and IgG, but not IgE, and decreased the proportion of $CD4^+$ T cells. Tocotrienols decreased the expression of TNF-α (Gu et al., 1999). These investigators concluded that oral administration of tocopherols and tocotrienols affect the proliferation and function of spleen and mesenteric lymph node lymphocytes.

6. CONCLUDING REMARKS

This chapter reviews the available data on the nutritional effects of RBO and its main components. They have demonstrated abilities in improving the plasma lipid pattern of rodents, rabbits, nonhuman primates, and humans, reducing total plasma cholesterol and triglyceride concentration and increasing the HDL-C level. Other potential properties of RBO and its components, studied both in vitro and in animal models, include modulation of pituitary secretion, inhibition of gastric acid secretion, antioxidant action, and inhibition of platelet aggregation. In spite of having good nutritional composition as well as providing health benefits to humans, RBO is still underutilized. So keeping in mind the benefits, it should be provided opportunity for incorporation into food products to enhance their nutritional value. After all, diet is the first, and sometimes only, therapeutic approach to some of the ailments hitting the 21st century.

REFERENCES

Abuasal, B.S., Lucas, C., Peyton, B., Alayoubi, A., Nazzal, S., Sylvester, P.W., Kaddoumi, A., 2012. Enhancement of intestinal permeability utilizing solid lipid nanoparticles increases γ-tocotrienol oral bioavailability. Lipids 47 (5), 461–469.

Accinni, R., Rosina, M., Bamonti, F., Della Noce, C., Tonini, A., Bernacchi, F., Campolo, J., Caruso, R., Novembrino, C., Ghersi, L., Lonati, S., Grossi, S., Ippolito, S., Lorenzano, E., Ciani, A., Gorini, M., 2006. Effects of combined dietary supplementation on oxidative and inflammatory status in dyslipedemic subjects. Nutr. Metab. Cardiovasc. Dis. 16 (2), 121–127.

Aggarwal, B.B., Sundaram, C., Prasad, S., Kannappan, R., 2010. Tocotrienols, the vitamin E of the 21st century: its potential against cancer and other chronic diseases. Biochem. Pharmacol. 80 (11), 1613–1631.

Al Rasheed, N., Abdelbaky, N.A., Al Rasheed, N., Shebly, W., Ahmed, A., Faddah, L.M., 2012. Effect of vitamin E and-lipoic acid on nano zinc oxide induced renal cytotoxicity in rats. Afr. J. Pharm. Pharmacol 6 (29), 2211–2223.

Aladedunye, F., Przybylski, R., Rudzinska, M., Klensporf-Pawlik, D., 2013. γ-Oryzanols of North American wild rice (*Zizania palustris*). J. Am. Oil Chem. Soc. 90 (8), 1101–1109.

Bakota, E.L., Winkler-Moser, J.K., Hwang, H.-S., 2014. Properties of rice bran oil-derived functional ingredients. Lipid Technol. 26 (8), 179–182.

Barker, R., Herdt, R.W., Rose, B., 2014. The Rice Economy of Asia. Routledge.

Berger, A., Rein, D., Schäfer, A., Monnard, I., Gremaud, G., Lambelet, P., Bertoli, C., 2005. Similar cholesterol—lowering properties of rice bran oil, with varied γ-oryzanol, in mildly hypercholesterolemic men. Eur. J. Nutr. 44 (3), 163–173.

Bhaskaragoud, G., Rajath, S., Mahendra, V.P., Kumar, G.S., Krishna, A.G.G., Kumar, G.S., 2016. Hypolipidemic mechanism of oryzanol components-ferulic acid and phytosterols. Biochem. Biophys. Res. Commun. 476 (2), 82–89.

Callaway, E., 2014. The birth of rice. Nature 514 (7524), S58.

Champagne, E.T., 2004. Rice: Chemistry and Technology. American Association of Cereal Chemists.

Chan, H.H., Chan, K.K., 2005. Effects of tocotrienols on cell viability and apoptosis in normal murine liver cells (BNL CL. 2) and liver cancer cells (BNL 1ME A. 7R. 1), in vitro. Asia Pac. J. Clin. Nutr. 14 (4), 374.

Chia, S.L., Boo, H.C., Muhamad, K., Sulaiman, R., Umanan, F., Chong, G.H., 2015. Effect of subcritical carbon dioxide extraction and bran stabilization methods on rice bran oil. J. Am. Oil Chem. Soc. 92 (3), 393–402.

Cho, J.-Y., Lee, H.J., Kim, G.A., Kim, G.D., Lee, Y.S., Shin, S.C., et al., 2012. Quantitative analyses of individual γ-oryzanol (steryl ferulates) in conventional and organic brown rice (*Oryza sativa* L.). J. Cereal Sci. 55 (3), 337–343.

Cho, S.S., Samuel, P., 2009. Fiber Ingredients: Food Applications and Health Benefits. CRC Press.

Chotimarkorn, C., Benjakul, S., Silalai, N., 2008. Antioxidant components and properties of five long-grained rice bran extracts from commercial available cultivars in Thailand. Food Chem. 111 (3), 636–641.

Chou, T.-W., Ma, C.-Y., Cheng, H.-H., Chen, Y.-Y., Lai, M.-H., 2009. A rice bran oil diet improves lipid abnormalities and suppress hyperinsulinemic responses in rats with streptozotocin/nicotinamide-induced type 2 diabetes. J. Clin. Biochem. Nutr. 45 (1), 29–36.

Cicero, A.F.G., Gaddi, A., 2001. Rice bran oil and γ-oryzanol in the treatment of hyperlipoproteinaemias and other conditions. Phytother. Res. 15 (4), 277–289.

De Deckere, E.A.M., Korver, O., 1996. Minor constituents of rice bran oil as functional foods. Nutr. Rev. 54 (11), S120.

Devarajan, S., Chatterjee, B., Urata, H., Zhang, B., Ali, A., Singh, R., Ganapathy, S., 2016a. A blend of sesame and rice bran oils lowers hyperglycemia and improves the lipids. Am. J. Med. 129 (7), 731–739.

Devarajan, S., Singh, R., Chatterjee, B., Zhang, B., Ali, A., 2016b. A blend of sesame oil and rice bran oil lowers blood pressure and improves the lipid profile in mild-to-moderate hypertensive patients. J. Clin. Lipidol. 10 (2), 339–349.

Fang, F., Kang, Z., Wong, C., 2010. Vitamin E tocotrienols improve insulin sensitivity through activating peroxisome proliferator-activated receptors. Mol. Nutr. Food Res. 54 (3), 345–352.

FAO, 2017. Rice Market Monitor (RMM). Available at http://www.fao.org/economic/est/publications/rice-publications/rice-market-monitor-rmm/en/ (accessed 19 March 2018).

FAOSTAT, 2013. Available at: http://www.fao.org/faostat/en/#data/FBS (accessed 19 March 2018).

Fujii, M., Butler, J.P., Sasaki, H., 2018. Gamma-oryzanol for behavioural and psychological symptoms of dementia. Psychogeriatrics 18 (2), 151–152.

Gertz, C., Klostermann, S., Kochhar, S.P., 2000. Testing and comparing oxidative stability of vegetable oils and fats at frying temperature. Eur. J. Lipid Sci. Technol. 102 (8–9), 543–551.

Ghatak, S.B., Panchal, S.S., 2012. Anti-diabetic activity of oryzanol and its relationship with the anti-oxidant property. Int. J. Diabetes Dev. Countries 32 (4), 185–192.

Gu, J.-Y., Wakizono, Y., Sunada, Y., Hung, P., Nonaka, M., Sugano, M., Yamada, K., 1999. Dietary effect of tocopherols and tocotrienols on the immune function of spleen and mesenteric lymph node lymphocytes in Brown Norway rats. Biosci. Biotechnol. Biochem. 63 (10), 1697–1702.

Gul, K., Yousuf, B., Singh, A.K., Singh, P., Wani, A.A., 2015. Rice bran: nutritional values and its emerging potential for development of functional food—a review. Bioactive Carbohydr. Dietary Fibre 6 (1), 24–30.

Harris, D.R., 2007. Agriculture, cultivation and domestication: exploring the conceptual framework of early food production. Rethink. Agric. Archaeol. Ethnoarchaeol. Perspect. 16–35.

Hegsted, M., Windhauser, M.M., Morris, S.K., Lester, S.B., 1993. Stabilized rice bran and oat bran lower cholesterol in humans. Nutr. Res. 13 (4), 387–398.

Hermizi, H., Faizah, O., Ima-Nirwana, S., Nazrun, S.A., Norazlina, M., 2009. Beneficial effects of tocotrienol and tocopherol on bone histomorphometric parameters in Sprague-Dawley male rats after nicotine cessation. Calcif. Tissue Int. 84 (1), 65–74.

Hiramitsu, T., Armstrong, D., 1991. Preventive effect of antioxidants on lipid peroxidation in the retina. Ophthalmic Res. 23 (4), 196–203.

Hsieh, T.-C., Wu, J.M., 2008. Suppression of cell proliferation and gene expression by combinatorial synergy of EGCG, resveratrol and γ-tocotrienol in estrogen receptor-positive MCF-7 breast cancer cells. Int. J. Oncol. 33 (4), 851–859.

Huang, X., Kurata, N., Wang, Z.-X., Wang, A., Zhao, Q., Zhao, Y., et al., 2012. A map of rice genome variation reveals the origin of cultivated rice. Nature 490 (7421), 497.

Hudson, E.A., Dinh, P.A., Kokubun, T., Simmonds, M.S.J., Gescher, A., 2000. Characterization of potentially chemopreventive phenols in extracts of brown rice that inhibit the growth of human breast and colon cancer cells. Cancer Epidemiol. Prevent. Biomark. 9 (11), 1163–1170.

Husain, K., Francois, R.A., Yamauchi, T., Perez, M., Sebti, S.M., Malafa, M.P., 2011. Vitamin E δ-tocotrienol augments the anti-tumor activity of gemcitabine and suppresses constitutive NF-κB activation in pancreatic cancer. Mol. Cancer Ther. 0424, 2011.

Ima-Nirwana, S., Suhaniza, S., 2004. Effects of tocopherols and tocotrienols on body composition and bone calcium content in adrenalectomized rats replaced with dexamethasone. J. Med. Food 7 (1), 45–51.

Ishihara, M., Ito, Y., Nakakita, T., Maehama, T., Hieda, S., Yamamoto, K., Ueno, N., 1982. Clinical effect of gamma-oryzanol on climacteric disturbance-on serum lipid peroxides (author's transl). Nihon Sanka Fujinka Gakkai Zasshi 34 (2), 243–251.

Islam, M.S., Murata, T., Fujisawa, M., Nagasaka, R., Ushio, H., Bari, A.M., et al., 2008. Anti-inflammatory effects of phytosteryl ferulates in colitis induced by dextran sulphate sodium in mice. Br. J. Pharmacol. 154 (4), 812–824.

Ji, X., Wang, Z., Geamanu, A., Sarkar, F.H., Gupta, S.V., 2011. Inhibition of cell growth and induction of apoptosis in non-small cell lung cancer cells by delta-tocotrienol is associated with notch-1 down-regulation. J. Cell. Biochem. 112 (10), 2773–2783.

Ju, Y.-H., Vali, S.R., 2005. Rice bran oil as a potential resource for biodiesel: a review. J. Sci. Ind. Res. 64, 866–882.

Juturu, V., 2008. Tocopherol and tocotrienols. In: Tocotrienols: Vitamin E Beyond Tocopherols. p. 219.

Kang, H.K., Kim, C.H., 2016. Effects of dietary supplementation with rice bran oil on the growth performance, blood parameters, and immune response of broiler chickens. J. Anim. Sci. Technol. 58 (1), 12.

Kannappan, R., Ravindran, J., Prasad, S., Sung, B., Yadav, V.R., Reuter, S., et al., 2010. γ-Tocotrienol promotes TRAIL-induced apoptosis through reactive oxygen species/extracellular signal-regulated kinase/p53—mediated upregulation of death receptors. Mol. Cancer Ther. 9 (8), 2196–2207.

Kahlon, T.S., 2009. Rice bran: production, composition, functionality and food applications, physiological benefits. Fiber ingredients: food applications and health benefits. CRC Press, Boca Raton, FL. https://doi.org/10.1201/9781420043853-c14.

Khanna, S., Parinandi, N.L., Kotha, S.R., Roy, S., Rink, C., Bibus, D., Sen, C.K., 2010. Nanomolar vitamin E α-tocotrienol inhibits glutamate-induced activation of phospholipase A2 and causes neuroprotection. J. Neurochem. 112 (5), 1249–1260.

Khanna, S., Roy, S., Slivka, A., Craft, T.K.S., Chaki, S., Rink, C., et al., 2005. Neuroprotective properties of the natural vitamin E α-tocotrienol. Stroke 36 (10), e144–e152.

Kim, H.W., Kim, J.B., Cho, S.-M., Cho, I.K., Li, Q.X., Jang, H.-H., et al., 2015. Characterization and quantification of γ-oryzanol in grains of 16 Korean rice varieties. Int. J. Food Sci. Nutr. 66 (2), 166–174.

Koba, K., Yanagita, T., 2014. Health benefits of conjugated linoleic acid (CLA). Obes. Res. Clin. Pract. 8, e525–e532.

Komiyama, K., Hayashi, M., Cha, S., Yamaoka, M., 1992. Antitumor and antioxidant activity of tocotrienols. In: Lipid-Soluble Antioxidants: Biochemistry and Clinical Applications. Springer, pp. 152–159.

Kong, C.K.L., Lam, W.S., Chiu, L.C.M., Ooi, V.E.C., Sun, S.S.M., Wong, Y.-S., 2009. A rice bran polyphenol, cycloartenyl ferulate, elicits apoptosis in human colorectal adenocarcinoma SW480 and sensitizes metastatic SW620 cells to TRAIL-induced apoptosis. Biochem. Pharmacol. 77 (9), 1487–1496.

Kozuka, C., Kaname, T., Shimizu-Okabe, C., Takayama, C., Tsutsui, M., Matsushita, M., et al., 2017. Impact of brown rice-specific γ-oryzanol on epigenetic modulation of dopamine D2 receptors in brain striatum in high-fat-diet-induced obesity in mice. Diabetologia 60 (8), 1502–1511.

Kozuka, C., Yabiku, K., Sunagawa, S., Ueda, R., Taira, S.-i., Ohshiro, H., et al., 2012. Brown rice and its component, γ-oryzanol, attenuate the preference for high-fat diet by decreasing hypothalamic endoplasmic reticulum stress in mice. Diabetes 61 (12), 3084–3093.

Krishna, A.G.G., Khatoon, S., Shiela, P.M., Sarmandal, C.V., Indira, T.N., Mishra, A., 2001. Effect of refining of crude rice bran oil on the retention of oryzanol in the refined oil. J. Am. Oil Chem. Soc. 78 (2), 127–131.

Kuhad, A., Chopra, K., 2009. Attenuation of diabetic nephropathy by tocotrienol: involvement of NFkB signaling pathway. Life Sci. 84 (9–10), 296–301.

Kuriyan, R., Gopinath, N., Vaz, M., Kurpad, A.V., 2005. Use of rice bran oil in patients with hyperlipidaemia. Natl Med. J. India 18 (6), 292.

Lai, M.-H., Chen, Y.-T., Chen, Y.-Y., Chang, J.-H., Cheng, H.-H., 2012. Effects of rice bran oil on the blood lipids profiles and insulin resistance in type 2 diabetes patients. J. Clin. Biochem. Nutr. 51 (1), 15–18.

Lai, P., Li, K.Y., Lu, S., Chen, H.H., 2009. Phytochemicals and antioxidant properties of solvent extracts from Japonica rice bran. Food Chem. 117 (3), 538–544.

Lamy, V., Roussi, S., Chaabi, M., Gossé, F., Lobstein, A., Raul, F., 2008. Lupulone, a hop bitter acid, activates different death pathways involving apoptotic TRAIL-receptors, in

human colon tumor cells and in their derived metastatic cells. Apoptosis 13 (10), 1232–1242.
Lee, S.-C., Tsai, C.-C., Yao, C.-H., Chen, Y.-S., Wu, M.-C., 2013. Ferulic acid enhances peripheral nerve regeneration across long gaps. Evid. Based Complement. Alternat. Med. 2013.
Li, Z.C., Su, Y.B., Bi, X.H., Wang, Q.Y., Wang, J., Zhao, J.B., et al., 2017. Effects of lipid form and source on digestibility of fat and fatty acids in growing pigs. J. Anim. Sci. 95 (7), 3103–3109.
Lichtenstein, A.H., Ausman, L.M., Carrasco, W., Gualtieri, L.J., Jenner, J.L., Ordovas, J.M., et al., 1994. Rice bran oil consumption and plasma lipid levels in moderately hypercholesterolemic humans. Arterioscler. Thromb. Vasc. Biol. 14 (4), 549–556.
Liu, Y., Yu, J., Wang, X., 2008. Extraction of policosanols from hydrolysed rice bran wax by high-intensity ultrasound. Int. J. Food Sci. Technol. 43 (5), 763–769.
Lunsin, R., Wanapat, M., Rowlinson, P., 2012. Effect of cassava hay and rice bran oil supplementation on rumen fermentation, milk yield and milk composition in lactating dairy cows. Asian-Australas. J. Anim. Sci. 25 (10), 1364.
Magosso, E., Ansari, M.A., Gopalan, Y., Shuaib, I.L., Wong, J.-W., Khan, N.A.K., et al., 2013. Tocotrienols for normalisation of hepatic echogenic response in nonalcoholic fatty liver: a randomised placebo-controlled clinical trial. Nutr. J. 12 (1), 166.
Mäkynen, K., Chitchumroonchokchai, C., Adisakwattana, S., Failla, M., Ariyapitipun, T., 2012. Effect of gamma-oryzanol on the bioaccessibility and synthesis of cholesterol. Eur. Rev. Med. Pharmacol. Sci. 16 (1), 49–56.
Malve, H., Kerkar, P., Mishra, N., Loke, S., Rege, N.N., Marwaha-Jaspal, A., Jainani, K.J., 2010. LDL-cholesterol lowering activity of a blend of rice bran oil and safflower oil (8:2) in patients with hyperlipidaemia: a proof of concept, double blind, controlled, randomised parallel group study. J. Indian Med. Assoc. 108 (11), 785–788.
Mangialasche, F., Xu, W., Kivipelto, M., Costanzi, E., Ercolani, S., Pigliautile, M., et al., 2012. Tocopherols and tocotrienols plasma levels are associated with cognitive impairment. Neurobiol. Aging 33 (10), 2282–2290.
Mehat, M.Z., Shuid, A.N., Mohamed, N., Muhammad, N., Soelaiman, I.N., 2010. Beneficial effects of vitamin E isomer supplementation on static and dynamic bone histomorphometry parameters in normal male rats. J. Bone Miner. Metab. 28 (5), 503–509.
Miller, A., Engel, K.-H., 2006. Content of γ-oryzanol and composition of steryl ferulates in brown rice (*Oryza sativa* L.) of European origin. J. Agric. Food Chem. 54 (21), 8127–8133.
Min, B., McClung, A.M., Chen, M.-H., 2011. Phytochemicals and antioxidant capacities in rice brans of different color. J. Food Sci. 76(1).
Minatel, I.O., Francisqueti, F.V., Corrêa, C.R., Lima, G.P.P., 2016. Antioxidant activity of γ-oryzanol: a complex network of interactions. Int. J. Mol. Sci. 17 (8), 1107.
Mingyai, S., Srikaeo, K., Kettawan, A., Singanusong, R., Nakagawa, K., Kimura, F., Ito, J., 2018. Effects of extraction methods on phytochemicals of rice bran oils produced from colored rice. J. Oleo Sci. 67 (2), 135–142.
Molina, J., Sikora, M., Garud, N., Flowers, J.M., Rubinstein, S., Reynolds, A., et al., 2011. Molecular evidence for a single evolutionary origin of domesticated rice. Proc. Natl. Acad. Sci. 108 (20), 8351–8356.
Most, M.M., Tulley, R., Morales, S., Lefevre, M., 2005. Rice bran oil, not fiber, lowers cholesterol in humans. Am. J. Clin. Nutr. 81 (1), 64–68.
Müller, L., Theile, K., Böhm, V., 2010. In vitro antioxidant activity of tocopherols and tocotrienols and comparison of vitamin E concentration and lipophilic antioxidant capacity in human plasma. Mol. Nutr. Food Res. 54 (5), 731–742.
Murase, Y., Iishima, H., 1963. Clinical studies of oral administration of gamma-oryzanol on climacteric complaints and its syndrome. Obstet. Gynecol. Prac. 12, 147–149.

Nam, S.J., Chung, S.I., Ryu, S.N., Kang, M.Y., 2017. Effect of bran extract from pigmented rice Superjami on the lipid and glucose metabolisms in a postmenopause-like model of ovariectomized rats. Cereal Chem. 94 (3), 424–429.

Nesaretnam, K., Selvaduray, K.R., Razak, G.A., Veerasenan, S.D., Gomez, P.A., 2010. Effectiveness of tocotrienol-rich fraction combined with tamoxifen in the management of women with early breast cancer: a pilot clinical trial. Breast Cancer Res. 12 (5), R81.

Newaz, M.A., Yousefipour, Z., Nawal, N.N., Adeeb, N., 2003. Nitric oxide synthase activity in blood vessels of spontaneously hypertensive rats: antioxidant protection by gamma-tocotrienol. J. Physiol. Pharmacol. 54 (3), 319–327.

Nhung, B.T., TUYEN, L.D., Linh, V.A., Nguyen Do Van, Nga, T.T., Thuc, V.T.M., et al., 2016. Rice bran extract reduces the risk of atherosclerosis in post-menopausal Vietnamese women. J. Nutr. Sci. Vitaminol. 62 (5), 295–302.

Nicolosi, R.J., Austrian, L.M., Hegsted, D.M., 1991. Rice bran oil lowers serum total and low density lipoprotein cholesterol and apo B levels in nonhuman primates. Atherosclerosis 88 (2–3), 133–142.

Nicolosi, R.J., Rogers, E.J., Ausman, L.M., Orthoefer, F.T., 1993. Rice bran oil and its heath benefits. Food Sci. Technol. (New York-Marcel Dekker) 421.

Orsavova, J., Misurcova, L., Ambrozova, J.V., Vicha, R., Mlcek, J., 2015. Fatty acids composition of vegetable oils and its contribution to dietary energy intake and dependence of cardiovascular mortality on dietary intake of fatty acids. Int. J. Mol. Sci. 16 (6), 12871–12890.

Orthoefer, F.T., 2005. Rice Bran Oil. Bailey's Industrial Oil and Fat Products.

Pali, V., 2013. Rice bran oil—unique gift of nature: a review. Agric. Rev. 34, 288–294.

Park, H.-A., Kubicki, N., Gnyawali, S., Chan, Y.C., Roy, S., Khanna, S., Sen, C.K., 2011. Natural vitamin E α-tocotrienol protects against ischemic stroke by induction of multidrug resistance-associated protein 1. Stroke STROKEAHA. 110.608547.

Park, S.K., Sanders, B.G., Kline, K., 2010. Tocotrienols induce apoptosis in breast cancer cell lines via an endoplasmic reticulum stress-dependent increase in extrinsic death receptor signaling. Breast Cancer Res. Treat. 124 (2), 361–375.

Patacsil, D., Tran, A.T., Cho, Y.S., Suy, S., Saenz, F., Malyukova, I., et al., 2012. Gamma-tocotrienol induced apoptosis is associated with unfolded protein response in human breast cancer cells. J. Nutr. Biochem. 23 (1), 93–100.

Patel, M., Naik, S.N., 2004. Gamma-oryzanol from rice bran-oil – a review. J. Sci. Res. 63, 569–578.

Pierpaoli, E., Viola, V., Pilolli, F., Piroddi, M., Galli, F., Provinciali, M., 2010. γ- and δ-tocotrienols exert a more potent anticancer effect than α-tocopheryl succinate on breast cancer cell lines irrespective of HER-2/neu expression. Life Sci. 86 (17–18), 668–675.

Piironen, V., Lindsay, D.G., Miettinen, T.A., Toivo, J., Lampi, A.-M., 2000. Plant sterols: biosynthesis, biological function and their importance to human nutrition. J. Sci. Food Agric. 80 (7), 939–966.

Prasanth Kumar, P.K., Manohar, R.S., Indiramma, A.R., Gopala Krishna, A.G., 2014. Stability of oryzanol fortified biscuits on storage. J. Food Sci. Technol. 51 (10), 2552–2559.

Qureshi, A.A., Karpen, C.W., Qureshi, N., Papasian, C.J., Morrison, D.C., Folts, J.D., 2011. Tocotrienols-induced inhibition of platelet thrombus formation and platelet aggregation in stenosed canine coronary arteries. Lipids Health Dis. 10 (1), 58.

Qureshi, A.A., Khan, D.A., Mahjabeen, W., Qureshi, N., 2015. Dose-dependent modulation of lipid parameters, cytokines and RNA by [delta]-tocotrienol in hypercholesterolemic subjects restricted to AHA Step-1 diet. Br. J. Med. Med. Res. 6 (4), 351.

Qureshi, A.A., Mo, H., Packer, L., Peterson, D.M., 2000. Isolation and identification of novel tocotrienols from rice bran with hypocholesterolemic, antioxidant, and antitumor properties. J. Agric. Food Chem. 48 (8), 3130–3140.

Qureshi, A.A., Sami, S.A., Salser, W.A., Khan, F.A., 2002. Dose-dependent suppression of serum cholesterol by tocotrienol-rich fraction (TRF25) of rice bran in hypercholesterolemic humans. Atherosclerosis 161 (1), 199–207.

Raghuram, T.C., Rao, U.B., Rukmini, C., 1989. Studies on Hypolipidemic Effects of Dietary Rice Bran Oil in Human Subjects. Nutrition Reports International, USA.

Ralhan, R., Pandey, M.K., Aggarwal, B.B., 2009. Nuclear factor-kappa B links carcinogenic and chemopreventive agents. Front. Biosci. (Schol. Ed.) 1, 45–60.

Reddy, L.H., Couvreur, P., 2009. Squalene: a natural triterpene for use in disease management and therapy. Adv. Drug Deliv. Rev. 61 (15), 1412–1426.

Reiner, Ž., Tedeschi-Reiner, E., Romić, Ž., 2005. Effects of rice policosanol on serum lipoproteins, homocysteine, fibrinogen and C-reactive protein in hypercholesterolaemic patients. Clin. Drug Invest. 25 (11), 701–707.

Rogers, E.J., Rice, S.M., Nicolosi, R.J., Carpenter, D.R., McClelland, C.A., Romanczyk, L.J., 1993. Identification and quantitation of γ-oryzanol components and simultaneous assessment of tocols in rice bran oil. J. Am. Oil Chem. Soc. 70 (3), 301–307.

Salar, A., Faghih, S., Pishdad, G.R., 2016. Rice bran oil and canola oil improve blood lipids compared to sunflower oil in women with type 2 diabetes: a randomized, single-blind, controlled trial. J. Clin. Lipidol. 10 (2), 299–305.

Saunders, R.M., 1985. Rice bran: composition and potential food uses. Food Rev. Int. 1 (3), 465–495.

Sawadikiat, P., Hongsprabhas, P., 2014. Phytosterols and γ-oryzanol in rice bran oils and distillates from physical refining process. Int. J. Food Sci. Technol. 49 (9), 2030–2036.

Seetharamaiah, G.S., Prabhakar, J.V., 1986. Oryzanol content of Indian rice bran oil and its extraction from soap stock. J. Food Sci. Technol. 23 (5), 270–273.

Sen, C.K., Khanna, S., Roy, S., 2007. Tocotrienols in health and disease: the other half of the natural vitamin E family. Mol. Asp. Med. 28 (5–6), 692–728.

Sen, C.K., Rink, C., Khanna, S., 2010. Palm oil—derived natural vitamin E α-tocotrienol in brain health and disease. J. Am. Coll. Nutr. 29 (sup3), 314S–323S.

Seppanen, C.M., Song, Q., Csallany, A.S., 2010. The antioxidant functions of tocopherol and tocotrienol homologues in oils, fats, and food systems. J. Am. Oil Chem. Soc. 87 (5), 469–481.

Sharma, R.D., Rukmini, C., 1987. Hypocholesterolemic activity of unsaponifiable matter of rice bran oil. Indian J. Med. Res. 85, 278–281.

Shibata, A., Nakagawa, K., Kawakami, Y., Tsuzuki, T., Miyazawa, T., 2010. Suppression of γ-tocotrienol on UVB induced inflammation in HaCaT keratinocytes and HR-1 hairless mice via inflammatory mediators multiple signaling. J. Agric. Food Chem. 58 (11), 7013–7020.

Shimomura, Y., Kobayashi, I., Maruta, S., Ohshima, K., Mori, M., Kamio, N., Fukuda, H., 1980. Effect of γ-oryzanol on serum TSH concentrations in primary hypothyroidism. Endocrinol. Japonica 27 (1), 83–86.

Shirode, A.B., Sylvester, P.W., 2010. Synergistic anticancer effects of combined γ-tocotrienol and celecoxib treatment are associated with suppression in Akt and NFκB signaling. Biomed. Pharmacother. 64 (5), 327–332.

Shuid, A.N., Mehat, Z., Mohamed, N., Muhammad, N., Soelaiman, I.N., 2010. Vitamin E exhibits bone anabolic actions in normal male rats. J. Bone Miner. Metab. 28 (2), 149–156.

Singh, D.K., Li, L., Porter, T.D., 2006. Policosanol inhibits cholesterol synthesis in hepatoma cells by activation of AMP-kinase. J. Pharmacol. Exp. Ther. 318 (3), 1020–1026.

Son, M.J., Rico, C.W., Nam, S.H., Kang, M.Y., 2011. Effect of oryzanol and ferulic acid on the glucose metabolism of mice fed with a high-fat diet. J. Food Sci. 76(1).

Srinivasan, M., Sudheer, A.R., Menon, V.P., 2007. Ferulic acid: therapeutic potential through its antioxidant property. J. Clin. Biochem. Nutr. 40 (2), 92–100.

Sugihara, N., Kanda, A., Nakano, T., Nakamura, T., Igusa, H., Hara, S., 2010. Novel fractionation method for squalene and phytosterols contained in the deodorization distillate of rice bran oil. J. Oleo Sci. 59 (2), 65–70.

Sun, W., Wang, Q., Chen, B., Liu, J., Liu, H., Xu, W., 2008. γ-Tocotrienol-induced apoptosis in human gastric cancer SGC-7901 cells is associated with a suppression in mitogen-activated protein kinase signalling. Br. J. Nutr. 99 (6), 1247–1254.

Suzuki, S., Oshima, S., 1970. Influence of blending of edible fats and oils on human serum cholesterol level (part 1). Jpn. J. Nutr. Diet. 28 (1), 3–6.

Szczesniak, K.A., Ostaszewski, P., Ciecierska, A., Sadkowski, K., 2016. Investigation of nutriactive phytochemical- gamma-oryzanol in experimental animal models. J. Anim. Physiol. Anim. Nutr. 100, 601–617.

Yoshino, G., Kazumi, T., Amano, M., Tateiwa, M., Yamasaki, T., Takashima, S., et al., 1989. Effects of gamma-oryzanol on hyperlipidemic subjects. Curr. Ther. Res. 45 (4), 543–552.

Theriault, A., Chao, J.-T., Gapor, A., 2002. Tocotrienol is the most effective vitamin E for reducing endothelial expression of adhesion molecules and adhesion to monocytes. Atherosclerosis 160 (1), 21–30.

Tiwari, V., Kuhad, A., Bishnoi, M., Chopra, K., 2009. Chronic treatment with tocotrienol, an isoform of vitamin E, prevents intracerebroventricular streptozotocin-induced cognitive impairment and oxidative-nitrosative stress in rats. Pharmacol. Biochem. Behav. 93 (2), 183–189.

Tsuchiya, T., Kaneko, K., Tanaka, A., 1957. Oryzanol content of rice bran oil. Tokyo Kogyo Shikensho Hokoku 52, 1.

Utarwuthipong, T., Komindr, S., Pakpeankitvatana, V., Songchitsomboon, S., Thongmuang, N., 2009. Small dense low-density lipoprotein concentration and oxidative susceptibility changes after consumption of soybean oil, rice bran oil, palm oil and mixed rice bran/palm oil in hypercholesterolaemic women. J. Int. Med. Res. 37 (1), 96–104.

Van Hoed, V., Ayala, J.V., Czarnowska, M., De Greyt, W., Verhé, R., 2010. Optimization of physical refining to produce rice bran oil with light color and high oryzanol content. J. Am. Oil Chem. Soc. 87 (10), 1227–1234.

Varga, T., Czimmerer, Z., Nagy, L., 2011. PPARs are a unique set of fatty acid regulated transcription factors controlling both lipid metabolism and inflammation. Biochim. Biophys. Acta (BBA)—Mol. Basis Dis. 1812 (8), 1007–1022. https://doi.org/10.1016/j.bbadis.2011.02.014.

Vasanthi, H.R., P Parameswari, R.P., Das, D.K., 2011. Tocotrienols and its role in cardiovascular health-a lead for drug design. Curr. Pharm. Des. 17 (21), 2170–2175.

Vaughan, D.A., Lu, B.-R., Tomooka, N., 2008. The evolving story of rice evolution. Plant Sci. 174 (4), 394–408.

Vissers, M.N., Zock, P.L., Meijer, G.W., Katan, M.B., 2000. Effect of plant sterols from rice bran oil and triterpene alcohols from sheanut oil on serum lipoprotein concentrations in humans. Am. J. Clin. Nutr. 72 (6), 1510–1515.

Wang, M., Yu, Y., Haberer, G., Marri, P.R., Fan, C., Goicoechea, J.L., et al., 2014. The genome sequence of African rice (*Oryza glaberrima*) and evidence for independent domestication. Nat. Genet. 46 (9), 982.

Wang, O., Liu, J., Cheng, Q., Guo, X., Wang, Y., Zhao, L., 2015. Effects of ferulic acid and γ-oryzanol on high-fat and high-fructose diet-induced metabolic syndrome in rats. PLoS One 10(2)e0118135.

Wilson, T.A., Ausman, L.M., Lawton, C.W., Hegsted, D.M., Nicolosi, R.J., 2000. Comparative cholesterol lowering properties of vegetable oils: beyond fatty acids. J. Am. Coll. Nutr. 19 (5), 601–607.

Wilson, T.A., Nicolosi, R.J., Woolfrey, B., Kritchevsky, D., 2007. Rice bran oil and oryzanol reduce plasma lipid and lipoprotein cholesterol concentrations and aortic cholesterol ester accumulation to a greater extent than ferulic acid in hypercholesterolemic hamsters. J. Nutr. Biochem. 18 (2), 105–112.

Winkler-Moser, J.K., Hwang, H.-S., Bakota, E.L., Palmquist, D.A., 2015. Synthesis of steryl ferulates with various sterol structures and comparison of their antioxidant activity. Food Chem. 169, 92–101.

Wong, W.-T., Ismail, M., Tohit, E.R.M., Abdullah, R., Zhang, Y.-D., 2016. Attenuation of thrombosis by crude rice (*Oryza sativa*) bran policosanol extract: ex vivo platelet aggregation and serum levels of arachidonic acid metabolites. Evid. Based Complement. Alternat. Med. 2016.

Xiong, A., Yu, W., Liu, Y., Sanders, B.G., Kline, K., 2016. Elimination of $ALDH^+$ breast tumor initiating cells by docosahexanoic acid and/or gamma tocotrienol through SHP-1 inhibition of Stat3 signaling. Mol. Carcinog. 55 (5), 420–430.

Xu, Z., Godber, J.S., 1999. Purification and identification of components of γ-oryzanol in rice bran oil. J. Agric. Food Chem. 47 (7), 2724–2728.

Xu, Z., Godber, J.S., 2001. Antioxidant activities of major components of g-oryzanol from rice bran using a linoleic acid model. J. Am. Oil. Chem. Soc. 78 (6), 645–649.

Xu, Z., Hua, N., Godber, J.S., 2001. Antioxidant activity of tocopherols, tocotrienols, and γ-oryzanol components from rice bran against cholesterol oxidation accelerated by 2,2′-Azobis (2-methylpropionamidine) dihydrochloride. J. Agric. Food Chem. 49 (4), 2077–2081.

Yamada, Y., Obayashi, M., Ishikawa, T., Kiso, Y., Ono, Y., Yamashita, K., 2008. Dietary tocotrienol reduces UVB-induced skin damage and sesamin enhances tocotrienol effects in hairless mice. J. Nutr. Sci. Vitaminol. 54 (2), 117–123.

Yap, W.N., Chang, P.N., Han, H.-Y., Lee, D.T.W., Ling, M.T., Wong, Y.C., Yap, Y.L., 2008. γ-Tocotrienol suppresses prostate cancer cell proliferation and invasion through multiple-signalling pathways. Br. J. Cancer 99 (11), 1832.

Yuen, K.H., Wong, J.W., Lim, A.B., Ng, B.H., Choy, W.P., 2011. Effect of mixed-tocotrienols in hypercholesterolemic subjects. Funct. Foods Health Dis. 1 (3), 106–117.

Zaiden, N., Ong, S., Nesaretnam, K., Shiba, S., 2010. Gamma delta tocotrienols reduce hepatic triglyceride synthesis and VLDL secretion. J. Atheroscler. Thromb. 17 (10), 1019–1032.

Zavoshy, R., Noroozi, M., Jahanihashemi, H., 2012. Effect of low calorie diet with rice bran oil on cardiovascular risk factors in hyperlipidemic patients. J. Res. Med. Sci. 17 (7), 626.

Zhang, L., Ding, Y., Yuan, Z., Liu, J., Sun, J., Lei, F., et al., 2015. MicroRNA-500 sustains nuclear factor-κB activation and induces gastric cancer cell proliferation and resistance to apoptosis. Oncotarget 6 (4), 2483.

Zimmel, D., 2011. Management of Equine Cushing's Disease and Equine Metabolic Syndrome. http://extension.vetmed.ufl.edu/files/2011/10/Cushings-Ds-and-Metabolic-Syndrome.pdf.

Zuo, X., Lu, H., Jiang, L., Zhang, J., Yang, X., Huan, X., et al., 2017. Dating rice remains through phytolith carbon-14 study reveals domestication at the beginning of the Holocene. Proc. Natl. Acad. Sci. 114 (25), 6486–6491.

CHAPTER 3

Processing Technology of Rice Bran Oil

Pradosh Prasad Chakrabarti, Ram Chandra Reddy Jala
Centre for Lipid Science and Technology, CSIR-Indian Institute of Chemical Technology, Hyderabad, India

1. INTRODUCTION

Rice bran is a byproduct of the rice milling industry obtained from milling of brown rice to produce white rice. It contains 15%–20% of oil. Apart from oil, rice bran is a promising source for proteins, carbohydrates, dietary fiber, vitamins, tocopherol, γ-oryzanol, and phospholipids (Tao, 1989; Houston, 1972; Saunders, 1986). Rice bran oil (RBO) is considered to be a superior quality oil as it has a balanced fatty acid profile, and it contains a number of minor components with proven nutritional benefits such as γ-oryzanol, tocotrienol, tocopherol, and squalene (Houston, 1972). RBO has been reported to exhibit serum cholesterol level-lowering properties (Vijayagopalan and Kurup, 1972).

As RBO is not a seed-derived oil, its composition differs from all other common vegetable oils (Table 1). Therefore a slightly different refining protocol is required to obtain very good quality, edible grade RBO. In most seed-derived plant oils, the triglycerides or triacylglycerol (TAG) content are in the range of 95%–99%. Whereas TAG content is much lower in the case of RBO. The lower TAG is due to the hydrolysis triggered by lipases present in the bran, which also results in high mono- (MAG) and diglyceride (DAG) contents in RBO. In the intact grain, lipases are primarily found in the testa-cross layer of the grain, whereas the oils are in the alevrone and subalevrone layers of the grain (Shastry and Raghavendra Rao, 1976; Luh et al., 1991). During rice milling, lipases and oils are brought together resulting in a decrease in TAG and an increase in FFA content. The extent of lipase-catalyzed hydrolysis of RBO depends on the moisture content, temperature, and other environmental conditions. A rise in FFA content of 5%–7% per day and up to 70% in a month during bran storage has been observed (Saunders, 1986). Therefore if a conventional alkali refining process is adopted for refining of RBO, a significantly high oil loss will be

Rice Bran and Rice Bran Oil
https://doi.org/10.1016/B978-0-12-812828-2.00003-2

Table 1 Composition of crude RBO, SBO, and GNO

Component		Percentage		
		RBO	**SBO**	**GNO**
Saponifiable lipids				
	Neutral lipids	88–89	–	–
	Triglycerides	83–86	94.0–95.0 (%)	93.3–95.8 (%)
	Diglycerides	3–4	–	–
	Monoglycerides	6–7	–	–
	Free fatty acids	2–4	0.3–0.7 (%)	0.5–1.5 (%)
	Waxes	6–7	–	–
	Glycolipids	6–7	–	–
	Phospholipids	4–5	3.0–4.0 (%)	0.3–0.7 (%)
Unsaponifiable Matters		4–2	≤15 (g/kg); 1.3–1.6 (%)	≤10 (g/kg)
	Phytosterols	43%	1800–4500 (ppm)	900–4344 (ppm)
	Sterol esters	10%	–	–
	Triterpene alcohols	28%	845	360
	Hydrocarbons	18%	n.d-170 ppm; 0.38 (%)	270–1296 (ppm)
	Tocopherols/ tocotrienols	1%	993–3370/n. d-173 (ppm)	137–934/n. d-474 (ppm)

Fereidoon, S., 2005. Bailey's Industrial Oil and Fat Products, sixth ed., vol. 6, pp. 465–489, 577–653; Maria, E.C., Amalia, A.C., 2010. Eur. J. Lipid Sci. Technol. 112, 697–707. RBO, rice bran oil; SBO, soybean oil; GNO, groundnut oil.

encountered. Therefore the primary objective of RBO refining should be eliminating the high FFA to obtain good quality oil.

Another striking difference is the presence of a higher amount of glycolipids. Glycolipids together with MAG and DAG act as surfactants, which causes new challenges to the refining of RBO. Phospholipid contents (gums) are also higher in RBO as compared to many other oils. Phosphorous-containing glycolipids, which are unfound in any other oils, were also found in RBO (Vali et al., 2005a, b). RBO is very dark in color. It is therefore important for RBO processors to find an economical way to remove RBO color to an extent of aesthetical acceptance by the customers. RBO also contains around 5% of wax, which has to be removed. Improper oil dewaxing will lead to cloudiness and lower customer acceptance. In addition, cloudiness of RBO might also be due to high amounts of saturated fatty acids. Thus winterization is essential to lower the cloud point of RBO.

Besides that, refining of crude RBO has to be designed in such a way that all unwanted materials like gums, waxes, pigments, odoriferous compounds, and oxidation products are removed efficiently with a maximum retention of nutritionally beneficial components such as oryzanol, tocopherol, tocotrienols, phytosterols, sterylesters, squalene, etc. Moreover, the refining process should be economically feasible. It definitely poses a great challenge to the refiner.

2. STABILIZATION OF RICE BRAN

A major problem of using RBO for food purposes is the time lag between bran production and oil extraction. Oil should be extracted immediately as soon as the bran is produced. Otherwise, oil will be hydrolyzed into free fatty acids (FFAs) and glycerol by the action of very active lipase enzymes present in rice bran. Under favorable conditions, the rate of oil degradation and FFA formation could be very high (Houston, 1972). Oxidation of FFA leads to production of various odoriferous compounds such as aldehydes, ketone, etc. In general, RBO with an excess of 10% FFA is unfit for human consumption; bran with >5% FFA will not be feasible for refining by a chemical method due to higher refining loss. Hence, RBO extraction and/or protein separation are needed and proper methodologies have to be adopted to produce healthy and good quality RBO.

Generally, the bran can be stabilized by four different methods: two external heating methods, namely oven drying and steam retorting; one internal method, which is microwave heating; and one nonthermal process, such as irradiation. Oven-dried bran needs to be stored in sealed containers to maintain its stability, which causes poststabilization and transportation problems. In addition, prolonged heating during oven drying can result in significant nutrient losses and color fixation.

Several researchers have investigated microwave-heating methods (Patil et al., 2016; Qian, 2008; Guo et al., 2014; Hu and He, 2015; Huang et al., 2015). Microwave heating has a significant advantage as it causes internal heating of particles within the energy-penetrating depth. This makes it extremely suitable for thermal processing of rice bran that has low thermal conductivity. Unlike oven drying and steam retorting, microwave energy results in disruption of weak hydrogen bonds as it causes dipole rotation of free water molecules. The dipole moment causes molecular friction and heats the bran uniformly throughout the penetrating depth of microwave (Patil et al., 2016; Qian, 2008; Guo et al., 2014; Hu and He, 2015; Huang et al., 2015).

Rice bran stabilized by microwave heating was as good as steam-retorted bran in terms of FFA. Moisture content of rice bran played an important role in the process by microwave heating for stabilization of rice bran. Increasing the power of the microwave up to a certain limit may help in stabilizing bran with an increased moisture level. Wang et al. (2017) and Su et al. (2012) conducted studies using infrared radiation heating. It was observed that the acid value of microwave-heated rice bran did not raise much compared to that of raw rice bran during 8-week storage at 25°C. In recent years, the cost of microwave-heating equipment has also reduced significantly because of the technical advances made in this area. Results of physical, thermal, and nutritional studies have shown that microwave heating is a technically feasible method for inactivation of lipase responsible for rice bran degradation and rise in FFA. To date, no detailed information was given in the reported literature about the effect of microwave heating on color, odor, and physicochemical properties of the oil and, more importantly, effects on micronutrients.

Various heating methods such as dry-heating, wet-heating (Geng et al., 2006), and ohmic heating (Lakkakula et al., 2004) were also investigated. The other kind of cooking and drying treatments such as extrusion cooking (Randall et al., 1985; Sharma et al., 2004; Xu, 2007; Rafe and Sadeghian, 2017), fluidized bed drying (Fernando and Hewavitharana, 1993), and drying of rice bran using an extruder, pellet cooker, pellet mill, and expander cooker (Riaz et al., 2010) were examined for better results. Rodchuajeen et al. (2016) carried out selected moving-bed drying methods, that is, hot-air fluidized bed drying, superheated-steam fluidized bed drying, and IR vibrated bed drying.

Another alternative strategy adopted by most of the industries is to extract the oil from bran within 2–4 h of bran production. Most of the solvent extraction plants are surrounded by rice mills, and all the bran produced in these mills are collected immediately and brought to solvent extraction plants. In this way, the FFA contents of the oil can be controlled within 3%–5%.

Apart from the physical methods, some researchers tried different variations of bran stabilization using chemical methods. Prabhakar and Venkatesh (1986) proposed a simple chemical method for stabilization of rice bran. The principle behind this process is that lipase activity is low at low pH. Therefore it is suggested to use hydrochloric acid at 40 L/ton of bran for lowering the pH of rice bran from 6.9 to 4.0. They claimed that the acid can be efficiently applied by mechanical mixing. Besides that, sodium sulfite (Ma et al., 2008), sodium metabisulfite, and sodium bisulfate (Wells and Belcher, 1999), and hand-sprinkling of concentrated hydrochloric acid (Hossain et al., 1997; Haque et al., 2006) were also used to stabilize bran.

3. PREFERRED PROCESS FOR REFINING OF RICE BRAN OIL

3.1 Physical Refining Versus Chemical Refining

Fig. 1 shows the various stages of the physical and chemical refining processes. Physical refining of RBO includes four basic operations, namely degumming, bleaching, dewaxing, and deodorization/deacidification. All the unit operations have their own complexities and importance in

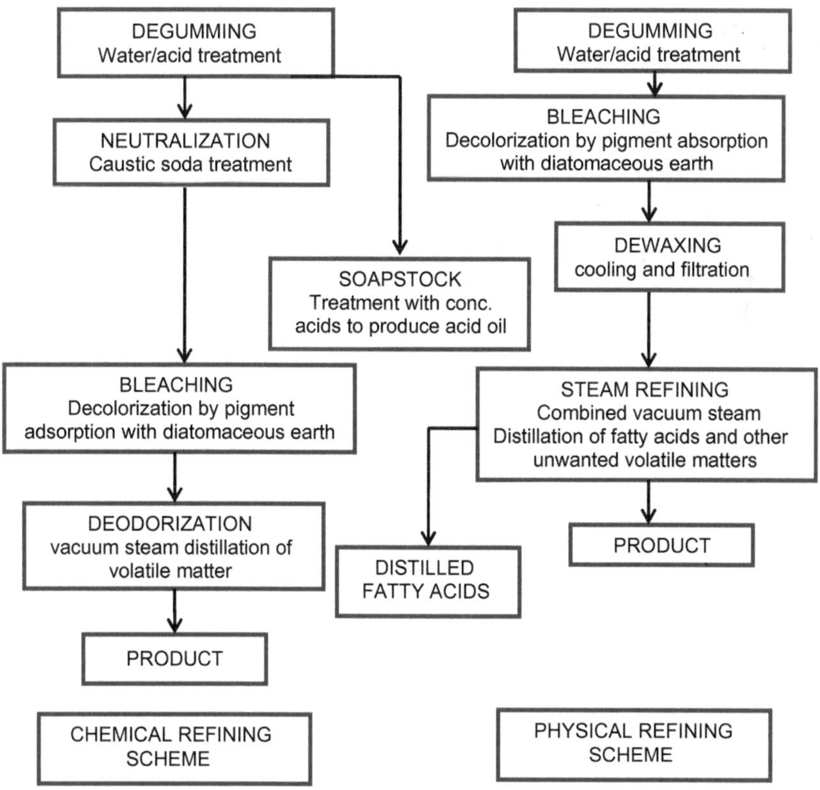

Fig. 1 Various stages of chemical and physical refining.

achieving a good quality edible oil. Alkali refining is regarded as the most effective method to remove impurities present in oils and fats. The methods and technical details are well documented (Norris, 1982; Hoffman, 1989). Alkali refining process neutralizes FFA, removes gums (partially removed by water or acid degumming in previous step), and partially removes pigments and metals. Bleaching generally follows and further removes the pigments, peroxides, and some metal contaminants and residual phosphatides. For oils like RBO or sunflower oil, an additional step of dewaxing is required to remove the waxes present in oils. The last step in refining is deodorization. Deodorization removes the constituents responsible for malodor and reduces FFA down to the levels of 0.01%–0.03%.

The most commonly used alkaline is caustic soda (sodium hydroxide), which not only neutralizes the FFA but also effectively decolorizes the oil. In the industry, the common practice is to use an excess of alkali over the stoichiometric requirement. This ensures total removal of FFA and also most of the coloring bodies. However, there are enormous disadvantages. Alkali, when it reacts with FFA, produces soap, which is an emulsifying agent. Particularly for high FFA oils, emulsification causes higher refining loss. Moreover, the excess alkali used may saponify some of the triacylglycerols leading to neutral oil loss. In the case of RBO, it contains a higher amount of mono- and diglycerides and also a higher amount of phospholipids and glycolipids. All these compounds are surfactant by nature. These will all have a huge impact on refining loss. On the other hand, the soap stock produced in the process will have serious implications on the environment. Soap stock is a sticky product, which has little commercial value. Soap stocks are converted to acid oil by acidulation with strong mineral acids. This is a tedious operation and, for environmental reasons, the acid wash water cannot be discharged as such in the sewage system and must be neutralized beforehand. This increases the total dissolved solids (TDS) content of the effluent, which is also undesirable. Another major disadvantage of alkali refining of RBO is alkali destroys most of the antioxidants and nutritional components present in oil. It was observed that, during alkali neutralization, oryzanol, the most important antioxidant present in RBO, is reduced drastically (from around 1% to 1.2% to as low as 0.1%).

Physical refining process is an alternative strategy. This process was long recognized for its ability to remove FFA from oils and fats. This method, known as physical (steam) refining, involves processing of oils and fats where FFA are removed by steam stripping during deodorization, as opposed to chemical (alkali) refining where they are removed in the form of soaps after neutralization. Because RBO contains more FFA, some other methods were also tried to refine RBO. One of the major processes tried by many scientists is miscella refining. Although it has multiple advantages, this was not practiced by industries except some in Japan due to higher plant and machinery costs (Ghosh, 2007; Canavag, 1976). Nogima (1950) and Bhattacharyya et al. (1986) have described various methods for refining of RBO using a miscella refining technique (Nogima, 1950; Bhattacharyya et al., 1986). For RBO with high FFA, hexane and other solvents having more polarity, such as ethanol or isopropanol, was also used for selective extraction of FFA, color, and other relatively polar compounds (Bhattacharyya and Bhattacharyya, 1983, 1985). Because generally these methods are not used in the industry, the discussion here is restricted to chemical and physical refining only.

Removal of FFA, unsaponifiable matters, and odoriferous compounds using steam stripping at higher temperature under high vacuum eliminates the use of alkali and formation of objectionable soap stock associated with chemical refining. Consequently, oil losses are reduced, the quality of distilled FFA is improved, and the operation is simplified. This process requires less operating costs as it needs less steam, water, and power as compared to a conventional alkali refining method. Due to its environment-friendly nature, there have been interests in physical refining of vegetable oils for quite some time (Posschelle, 1981; Forster and Harper, 1983; Tandy and Mcpherson, 1984), and it is now preferred as the most likely alternative to the conventional alkali refining process.

The possibilities of deacidifying oils by distillation with superheated steam were suggested in the early 20th century. This process was initially utilized to reduce the level of FFA in oils containing a high level of FFA prior to chemical refining. In the 1950 s, the physical refining process was found to be very useful for palm oil. With a refining factor of about 2, the loss from a 5% FFA content oil was about 10%, but a physical refining method resulted in only 6.5% loss (Young, 1990). An oil containing 4% FFA would have a refining loss of 8% by conventional methods, but the loss can be reduced to about 4.4% if a physical refining method is employed. The higher temperature of the physical refining process also breaks down the carotene pigments and reduces the color of the oil resulting in higher consumer acceptance. Most of the minor constituents having health benefits would also be retained in the oil refined by using physical refining process.

4. PROBLEMS WITH PHYSICAL REFINING

Initially, only oils with high FFA levels such as tallow, palm oil, palm kernel oil, and coconut oil were refined by physical method. These are also oils that have low phosphatide content, thus pretreatment with bleaching earth can adequately serve the purpose. However, in processing of oil with high phosphatides such as soybean oil, it was observed that processing by water degumming, activated earth bleaching, and steam refining-deodorization resulted in oil that lacked satisfactory quality and customer acceptance (James, 1958). It was understood that removal of phosphatides and metals are of utmost importance for successful implementation of physical refining process as it involves exposure to higher temperature.

Drawbacks of physical refining process have been previously presented (Norris, 1985). It is very difficult to remove phospholipids completely from oil as nonhydratable phosphatides remain after water degumming and are resistant to many other types of degumming. If phosphoric acid is used as a degumming agent, the gums produced may not be commercially attractive because of their dark color. More bleaching earth may be necessary to efficiently remove phosphatides, adding to the cost and disposal problems associated with spent earth. In this case, acid wash water is to be disposed of, although the load is much less than when soap stock is acidulated. The deodorizer must be designed to handle higher concentrations of fatty acids present and requires more corrosion-resistant materials of construction.

5. QUALITY REQUIREMENT OF OILS MEANT FOR PHYSICAL REFINING

Practical experience with physical refining shows desirable results can only be achieved when a very good quality of starting material is used. For poor quality oils or oils with high phosphatides, efficient pretreatment steps are needed to get a good quality finished product. Minor constituents that need to be removed for successful physical refining are phosphorous, iron, other trace metals, pigments, carbohydrates, sulfur, etc. The major difficulty in physical refining is the darkening of oil due to presence of phospholipids.

Soybean oil quality, which is suitable for physical refining, has been specified in Table 2. It has been claimed that, due to advancement of process controls and process reagents, E 230 value can be extended to 2.5 and the iron content to 2.0 mg/kg (Young, 1990). It was further suggested that higher values can be accepted for rapeseed and sunflower oils.

Table 2 Specification for crude degummed soybean oil for physical refining

Characteristics	Value
Acid value (mg KOH/g)	$\leq$1.5
Red[1]	$\leq$5.0
Yellow[1]	$\leq$50.0
Anisidine value	$\leq$1.5
Peroxide value, m.equiv./kg Extinction, 1% solution in 1 cm cell	$\leq$2.0
E 230 nm	$\leq$1.5
E 270 nm	$\leq$0.2
Iron content, mg/kg	$\leq$1.0
Phosphorous content, mg/kg	$\leq$20

[1]Lovibond color, 1″ cell (Ong, 1980).

Leysen (1981) stated that, for successful operation of physical refining, the phosphorous content of the oil should be in the range of 5 ppm.

Although, initially it was felt that the iron content should be <0.1 ppm (Ong, 1980; Grothues, 1981), a more systematic study revealed that, with the iron content of >0.1 ppm and phosphorous content of as high as 16 ppm, an oil of good keeping quality could easily be achieved (Cleenewerck and Dijkstra, 1992). It was also reported that the risk of getting a bad quality of oil started only when these standards are not maintained. In several other communications, it was found that an oil of very good quality could be obtained by physical refining if these specifications are strictly maintained (Posschelle, 1981; Segers, 1982; Cvengros, 1995). Other impurities such as pigments, sugars and carbohydrates, trace metals like calcium and magnesium, sulfur, etc. can be easily removed by treatments like activated earth bleaching, heat bleaching, etc.

6. PRETREATMENT—THE KEY FACTOR

There was no single physical refining process that can handle oil of different qualities (Forster and Harper, 1983). Partial removal of undesirable components from the oil in pretreatment steps has to be compensated for increased bleaching earth requirements (Ohlson, 1992). Minimal oil losses attained by steam refining compared to conventional alkali refining do not offset the oil losses due to the increased volumes of bleaching earth needed if proper pretreatments are not done (Balicer et al., 1985).

In general, quality of the final product is always determined by quality of crude oils. Even caustic soda refining would not be able to give high yields of stable and superior quality of oils if the crude oil contains high FFA, peroxide value, metal, and chlorophyll (Cvengros, 1995). The major emphasis, therefore, has to be placed on preliminary processing of crude oil prior to steam refining, which is aimed at removal of any component of the oil that may, during the high temperature operation, darken the color or undergo other adverse alterations and thus decrease the quality of the final product. It is noteworthy that the technology of physical refining is more about how to remove gums and other impurities in upstream processing, that is, in degumming and bleaching (Norris, 1985). Therefore it is concluded that development of physical refining technique is more dependent on the development of the pretreatment methods, degumming method in particular. It is important for an oil that is dark in color such as RBO. Hence, the degumming of RBO is discussed in detail in this chapter.

7. DEGUMMING OF RICE BRAN OIL

The most important pretreatment step is the degumming of crude oil. Oil degumming is carried out to remove undesired phospholipids, metal ions, and other impurities from oil. These impurities, if not removed efficiently, may eventually create difficulties in the subsequent refining steps and may result in poor quality of refined oil. Phospholipids present in oils are broadly classified as hydratable and nonhydratable types. Hydratable phosphatides can be removed by a simple water degumming step, whereas nonhydratable phospholipids require special treatment.

Phospholipids are the phosphoglyceride molecules where two fatty acids are esterified to the first and the second hydroxyl group of the glycerol. The third hydroxyl group of glycerol is esterified with phosphoric acid. Some phosphoglycerides also contain a second alcohol, which is again esterified with the phosphoric acid. This alcohol is situated on the polar head of the phosphoglyceride molecule. The parent compound of phosphoglycerides is phosphatidic acid without any head alcohol. The other phosphoglycerides present in oils are phosphatidyl ethanolamine, phosphatidyl choline, phosphatidyl inositol, and to a lesser extent phosphatidyl serine. Phosphoglycerides from oils and fats are often called phosphatides. Structure of phosphatidic acid and other phosphatides are given in Fig. 2. Phosphatides thus have two very different kinds of groups, a polar hydrophilic head and a nonpolar hydrophobic tail.

All phosphatides have a negative charge on the phosphoric group at pH 7. The head alcohol group will also contribute electric charges at pH near 7. It is, therefore, evident that phosphatidic acid and other phosphatides

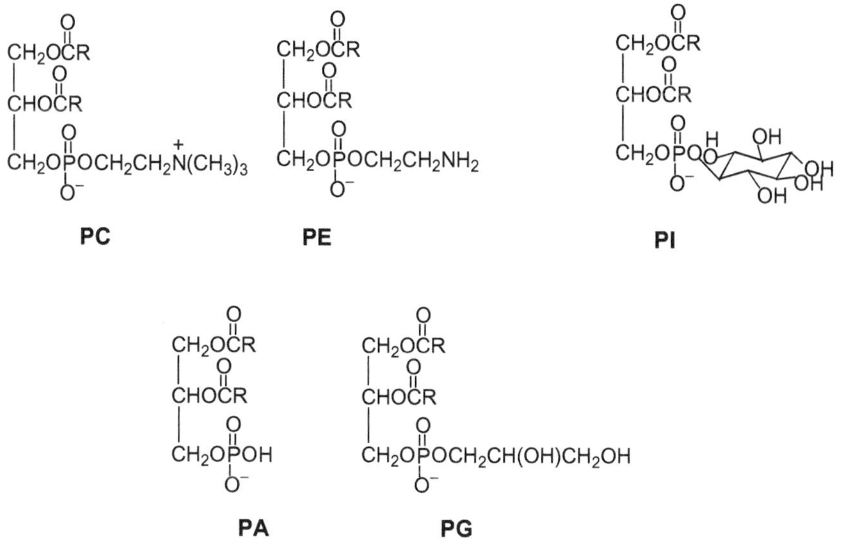

PC, phosphatidyl choline; PE, phosphatidyl ethanolamine; PI, phosphatidyl inositol; PA, phosphatidic acid; PG, phosphatidyl glycerol

Fig. 2 Phospholipids present in rice bran gums.

have cation (metal ion) binding capacity. If phosphatides are complexed with monovalent ions like sodium or potassium, they are hydratable. If they are bound with bivalent metal ions like calcium or magnesium they become nonhydratable (Blak, 1990). These salts are strong emulsifying agents and, if not removed, are responsible for higher losses during further steps of refining. On the other hand, the water degumming step alone is not sufficient for preparing the oil ready for physical refining. Phosphatidyl choline is hydratable. Due to the presence of highly polar inositol group, phosphatidyl inositol irrespective of forming complex of monovalent or divalent cation is hydratable. Phosphatidic acid and phosphatidyl ethanolamine are hydratable when complexed with monovalent cations yet nonhydratable if complexed with divalent ions.

At pH 7, complex formation of phosphatidic acid and phosphatidyl ethanolamine with Ca^{2+} and Mg^{2+} is more favorable. Binding of bivalent ions is 1000 times stronger than the binding to sodium and potassium ions at pH 6 and pH 7 (Hvolvy, 1971). As a result, under the conditions of water degumming, these phospholipids are not hydratable. Addition of an acid, sufficiently strong, liberates free phosphatidic acids and phosphatidyl ethanolamine, and thus renders them hydratable. Bivalent metal ions generally bind with the added acid and are removed with water.

8. COMMONLY USED DEGUMMING TECHNIQUES

8.1 Water Degumming

In water degumming, oil is mixed with 1%–3% water at 60–80°C. The gums became hydrated and insoluble in oil. After 20–30 min, the agglomerated gums can be efficiently separated by centrifugation. The detailed reviews on water degumming process are well documented in the literature (Braae, 1976; Carr, 1976; List et al., 1981; Moulton and Mounts, 1990). This process cannot remove nonhydratable phosphatides and not suitable as a pretreatment for physical refining. RBO has considerable amount of nonhydratable phospholipids. Hence, this is not sufficient for degumming of RBO. Water degumming can be used for preparation of rice bran lecithin, a value added product from RBO.

8.2 Acid Degumming

One of the common methods for degumming of RBO is acid degumming. Ca^{2+} and Mg^{2+} salts of the phosphatidic acids are the major components of nonhydratable phospholipids. Addition of phosphoric/citric acids causes formation of calcium and magnesium salts of those acids, and the process sets free the corresponding phosphatidic acids (Hvolvy, 1971). Acid degumming can be considered as an extension of the water degumming process as it uses water and acid simultaneously for effective removal of both hydratable and nonhydratable phosphatides. Acid degumming leads to a lower residual phosphorous content than water degumming. If this method is properly executed, this can be a pretreatment step for physical refining (Dijkstra, 1993). Another major advantage of this process is that the metal content of the oil is also reduced. However, due to the presence of degumming acid, the gums produced cannot be utilized for lecithin recovery. Proper mixing of acid is a crucial point for the success of this process (Mag and Reid, 1980) and ultrashear mixing is suggested (Carlson, 1986). Soybean oil degummed by this method is reported to have <5 ppm phosphorous (Carlson, 1986). Decomposition of phosphatide complexes with divalent ions reduces the divalent ion content of the oil resulting in lesser bleaching earth consumption. In the case of RBO, acid degumming could not bring down the phosphorous content of the oil to <5 ppm even after bleaching and dewaxing in many cases. This requires uses of more bleaching earth and may create problems during dewaxing. If acid degumming is performed for RBO, further refining steps should be modified accordingly after monitoring the phosphorous content of the oil at various stages.

8.3 Super Degumming

Super degumming process was promoted by Unilever (Segers, 1982). In this process, oil is heated to 70°C with citric acid and for some specific cases with modified lecithin, reacted for 5–15 min, cooled to 25°C, mixed with water, and held in a vessel for 3 h for formation of liquid crystals of phosphatides. The oil is then again heated to 65°C, and gums are separated by centrifugation. In another variation, Cambrian Engineering Group carried out the acid treatment at below 40°C (Kaji, 1988). The super-degumming process leads to lower residual phosphorous than standard acid degumming, but these levels are quite dependent on crude oil quality (Segers and Van de Sande, 1992). In the case of RBO, no report is available on industrial scale application of super degumming. Similarly, other variations of degumming such as uni degumming, special degumming, etc. may be used by the industry. However, no data is available.

8.4 Total Degumming Process (TOP)

This is a modified two-stage acid degumming process developed in Vandermoortele (Dijkstra and Van Opstal, 1987). Modification of this process suggested a reduction in wash water requirement (Cleenewerck and Dijkstra, 1992). It is claimed that this process is very much useful to reduce the phosphorous and iron level to attain oil quality that may be suitable for physical refining (Dijkstra, 1993). This may be one useful method for RBO degumming. Unfortunately, no result is reported for degumming of RBO using this process.

8.5 Enzymatic Degumming

Degumming of ricebran oil using phospholipase enzymes (PLA_1 and PLA_2) is a well-established process. In this process, phospholipids are converted to lysophospholipids and, because of its hydrophilic nature, can be separated easily from the bulk oil (Fig. 3). Aalrust et al. (1992) developed a process for soybean oil using PLA_2 enzyme. The phosphorous content and iron content were found to be very low (Dijkstra, 1993).

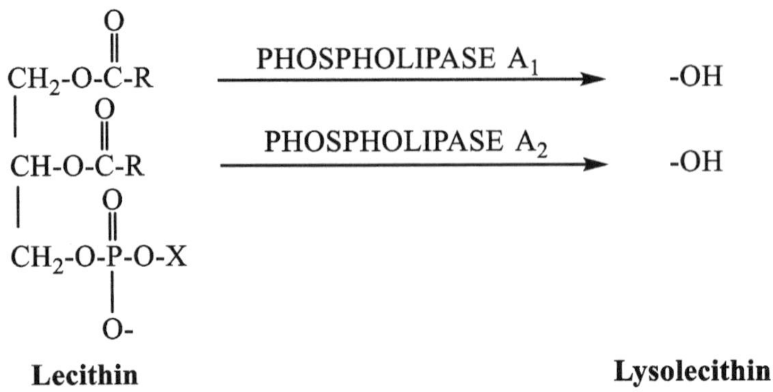

Fig. 3 Enzymatic degumming employing PLA_1 & PLA_2.

Chakrabarti et al. have developed another process for degumming of RBO using commercially available phospholipase A_1 (Chakrabarti et al., 2009, US Patent No. 7494676). This process was applied in commercial production of physically refined RBO in India. This results in good quality RBO keeping most of the nutraceuticals intact in the oil. The phosphorous content of the degummed, bleached, and dewaxed RBO processed by this method was found to be <5 ppm.

8.6 Membrane Degumming

Membrane separation technology has achieved a level of maturity during the last three decades. This method is less complicated, ecofriendly, and capable of removing most of the undesirable materials in a single step making it a suitable method for pretreatment of edible oil refining. This method of edible oil purification was firstly suggested by Sengupta (1977, 1978) for complete removal of phospholipids. In another modification, calculated amount of liquid ammonia was added to remove phosphatides and also FFAs by membrane filtration in a single step (Sengupta, 1985). Many researchers have worked on membrane degumming of soybean oil and RBO, and found this process can remove all undesirable components like gums and mucilage, polar materials, coloring pigments, metal contaminants, etc. in a single step (Koseoglu et al., 1990a; Subrahmanyam and Bhowmick, 1999).

Phospholipids are amphiphilic molecules (Fig. 4). In a nonpolar environment, phospholipids form inverse micelles (Fig. 4) having hydrophilic core (Sengupta, 1988; Koseoglu et al., 1990b; Bhowmick, 1992). The principles of membrane degumming can be observed in Figs. 5 and 6. The hydrophilic core retains all impurities such as trace metals, carbohydrates, sugars, etc. and

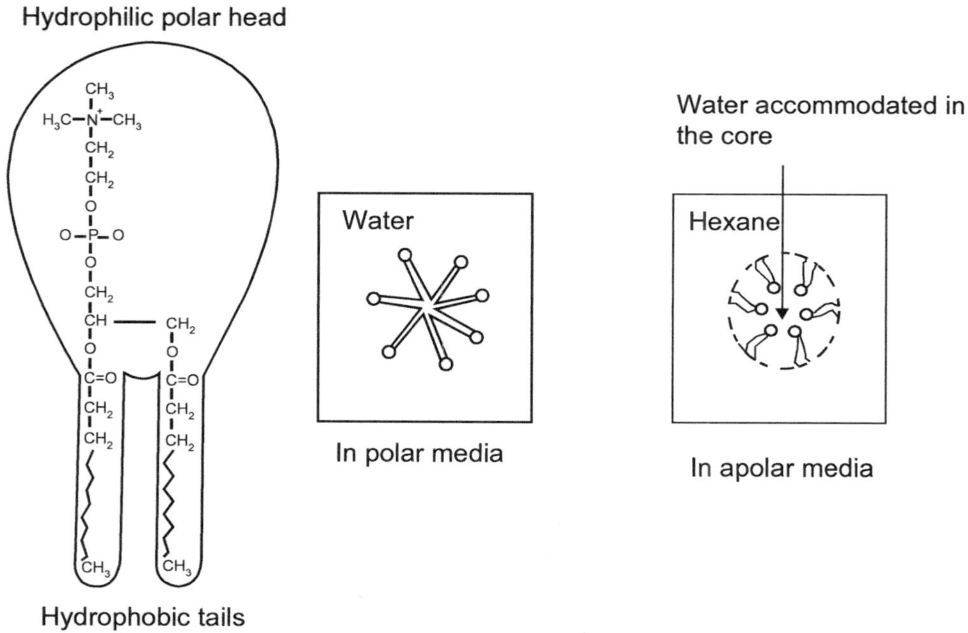

Fig. 4 Amphiphilic structure of phospholipids and miscelle formation.

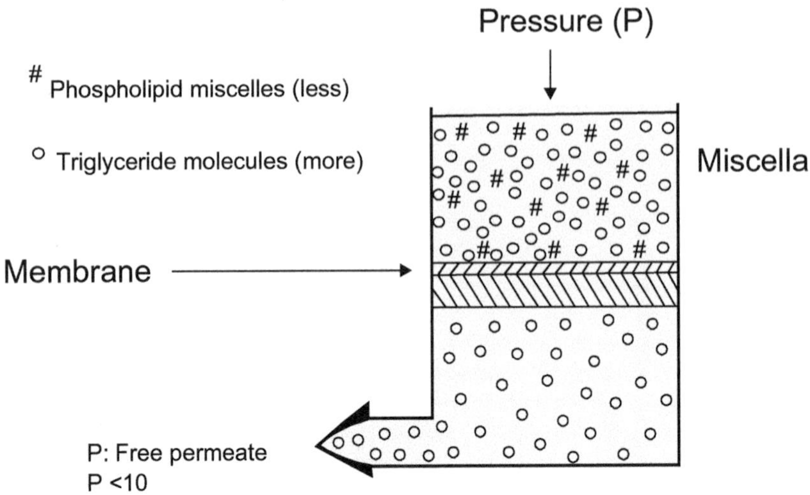

Fig. 5 Principles of membrane degumming.

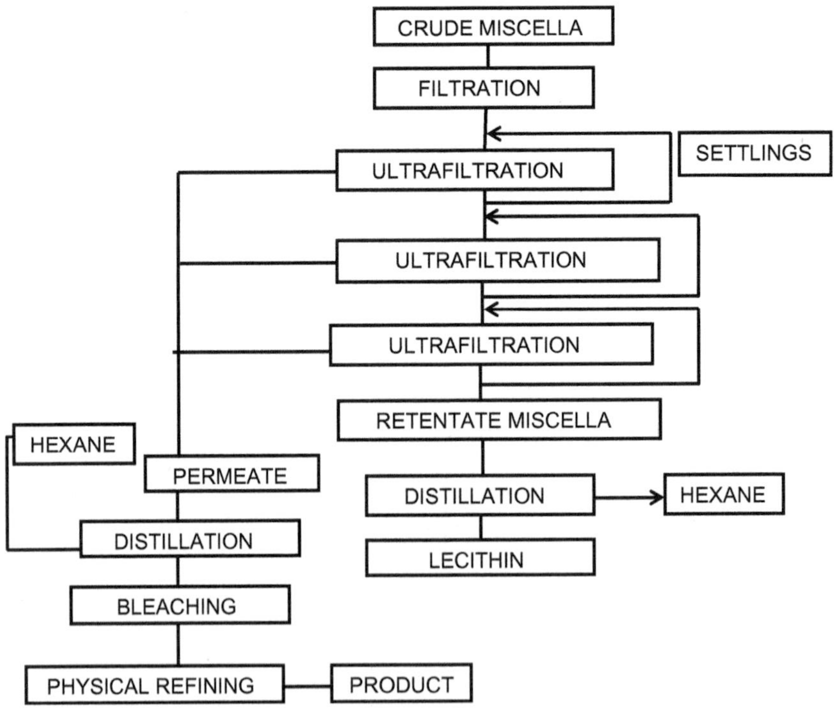

Fig. 6 Introduction of ultrafiltration step in edible oil refining scheme.

are removed along with the inverse micelles during ultrafiltration with a suitable membrane having a pore size less than that of the inverse micelle. In vegetable oil-hexane miscella, phospholipids are present as inverse micelles with a size between 18 and 200 nm, which is at least 10-fold larger than the size of the triacylglycerols (1.5 nm) (Sengupta, 1987; Segers and Van de Sande, 1980). The ultrafiltered miscella gives an oil of such quality that may go for physical refining after removal of hexane. The flow diagram of the processing of oils through membranes and then by physical refining are presented in Fig. 8. The details of the physical chemistry behind separation of phospholipids by ultrafiltration and the results of studies with different oils and various membranes were presented in few scientific communications (Sengupta, 1987, 1988).

However, the major drawback of this process is the nonavailability of economically viable solvent-resistant membranes. With the advent of ceramic membranes, the possibility of commercialization of membrane degumming process increased manifold. Due to the presence of wax, RBO requires two stage processes. Initially, the waxes are to be removed using microfiltration membrane passing the crude oil miscella at cold condition. Permeate will then pass through an ultrafiltration membrane. However, until now, no commercial data are available for degumming using a membrane separation technique. Although the quality of oil produced may be much superior, the limitations of higher processing cost, less throughput of membranes, and fouling of membranes restricts the industrial use of this method at present.

9. BLEACHING OF RICE BRAN OIL

Bleaching is one of the most important parts of RBO refining due to its dark color. Bleaching process is done for removal of coloring bodies like pheophytins, carotenoids, chlorophyll, xanthophylls, residual soaps (if alkali is added), phosphatides, trace metals, and high molecular weight oxidative products. In addition, added chemicals during planting, processing, contaminants from the equipment, pesticides, solvents, hydrocarbons (PAHs), acids (phosphoric acid, citric acid), dimers, and polymers are also impurities to be removed for edible applications (Proctor and Brooks, 2005; Dennis, 2005; Werner, 2000). Natural coloring bodies such as carotenoid pigments produce carotenes and xanthophylls, which are responsible for red and yellow colors, respectively. Oil processed from damaged seeds impart brown and blue color due to presence of degraded proteins, phospholipids, carbohydrates, etc. Some oils have green color contributed by chlorophyll pigment, frost damage, etc. (Proctor and Brooks, 2005; Dennis, 2005; Werner, 2000).

There are generally five types of bleaching methods available, namely chemical bleaching, light bleaching, heat and air bleaching, decolorizing with solvent, and adsorptive purification with bleaching earth and activated carbon (Patterson, 1992). In adsorptive purification, there exists a weak type of interaction between the molecules to be adsorbed and the active sites on the surface of the adsorbents. This is generally guided by van der Waal forces. Industries throughout the world adopted this adsorptive purification method for reducing color in oils and fats including for RBO. Oils or fats are treated with adsorbents, such as powdered bleaching earth or clays, and sometimes powdered activated carbons. Adsorbents must have substantial strength and hardness to withstand the bleaching processes. They usually contain a large surface area within the internal pores of particles. The bleaching earth is also known as bleaching clay or Fullers earth. Two types of natural clays are currently being used for bleaching; calcium montmorillonite (calcium bentonite) and attapulgite (hormite). Adsorption efficacy depends on the specific affinity between pigments and bleaching earth surfaces. Bleaching earth also removes noncolored constituents such as soap, phospholipids, and peroxides. These impurities are removed by molecular entrapment in the porous structure of bleaching earth. The treatment serves to replace aluminum ions in the structure of the bentonite with hydrogen ions so that the surface area and porosity of earth are greatly increased and that can act as ion exchange medium (Patterson, 1992; Roberto, 2006).

Natural coloring matters such as carotenoid pigments and xanthophyll impart yellow to red color in various crude and refined vegetable oils. Carotenoids can be readily adsorbed by activated bleaching earth and/or activated carbon. Carotenoids can also be destroyed by heat and oxidation. Oxidation converts unsaturated fatty acids into colored compounds and increases red and yellow color. Bleaching earth adsorb the yellow color pigments more effectively. Activated carbon generally adsorbs red pigments. A combination of earth and carbon in optimum proportion yields desirable color of the end product. However, brown and blue colors can be reduced appreciably by employing alkali or acid refining. Some greenish oils are bleached effectively by subjecting them to alkali treatment followed by bleaching.

RBO generally contains a high amount of red and yellow pigments, oxidized products, and metallic salts of fatty acids. In the physical refining process, bleaching of RBO is carried out immediately after the degumming step. It is important to note that bleaching should be carried out before alkali treatment as chlorophyll tends to be stabilized by alkali and heat, which made it very difficult to be removed (Cowan, 1976; Ghosh, 2007). Carotenoids and chlorophyll are the major pigments present in crude RBO along with porphyrin and brown pigments (Vandana et al., 2003). It is also worth noting that processing conditions are very important to have an efficient bleaching process. The solvent extraction process has to be performed in such a way that the color fixation is minimized. A highly efficient miscellar filtration unit has to be in the solvent extraction plant for efficient removal of fines and bran particles. Apart from proper miscella filtration, oil exposure to atmosphere at higher temperature has to be minimized. All these factors will impart more colors to the oil. Efficient degumming process is another important step. Presence of gums and mucilage in degummed oil will definitely reduce the efficiency of bleaching process.

In general, bleaching is carried out either by a batch or continuous process under vacuum. When activated earth is used under high vacuum and 110°C temperature, not only are the coloring bodies removed but oxidized products are also reduced drastically. At this temperature, the catalytic activity of activated bleaching earth decomposes hydroperoxides. Earth is more active on yellow pigments, and the ion exchange properties of bleaching earth helps in metal removal. Activated carbon helps to reduce red pigment in RBO. Activated carbon not only adsorbs the color of the oil but also gives shine to oil. Conventional earth bleaching can readily produces oils with Lovibond colors of 2–3 red units, which is an acceptable level for high grade cooking oil and salad oils. Some of the bleaching earth and activated carbon manufacturers produce products specifically developed for reduction of colors in RBO (Fig. 7).

Bleaching equipment also underwent many changes from batch to continuous process in combination with most efficient adsorbents. To date, plant manufacturers are offering countercurrent bleaching and steam-agitated bleaching plants for RBO. Industrial grade RBO is obtained through conventional chlorate bleaching and is being used in soap production. Several researchers (Kaimal et al., 2002, 2005; Chakrabarti et al., 2003; Ram, 2004; Ramaswamy et al., 1980; De et al., 1998) reported various methods on refining of RBO. However, all of them used either a mixture of bleaching earth and activated carbon or bleaching earth alone for their bleaching process. Selection of absorbents depends largely on the quality and characteristics of crude oil. Krishna (1992) reported the decoloration

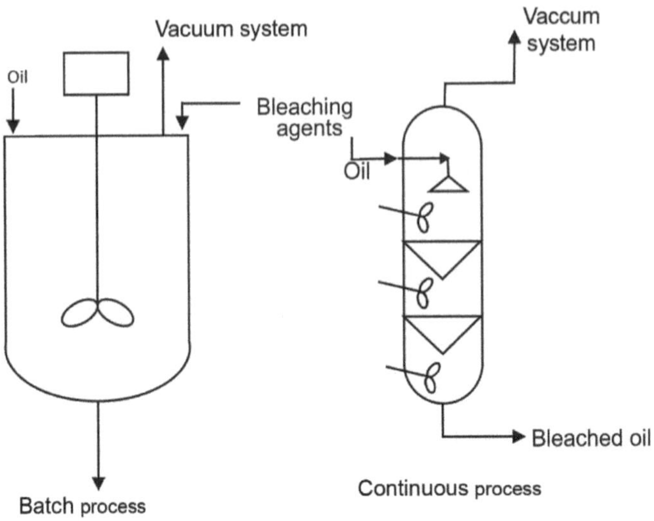

Fig. 7 Bleaching process.

of RBO by silica gel treatment with either column percolation or by merely shaking the miscella with gel followed by decantation. However, this method was not examined at an industrial scale. Based on the available information either in literature or industrial experience, the following bleaching parameters are generally well accepted: bleaching earth, 1%–2%; activated carbon, 0.2%–0.5%; temperature, 100–110°C; vacuum, at least around 700 mm Hg. In some industries, even more activated earth and 1%–2% activated carbon are also used. In many cases, citric acid dosing (0.1% max) is used to get rid of trace metals and phosphatides and to give shine to the oil.

In general, bleaching earth activation improves the qualities such as adsorption capacity, catalytic properties, ion exchange capacity, acid properties, and particle size distribution. Temperature, time, pressure, bleaching earth quantity, and filterability of bleaching earth play significant roles in the bleaching process. Filterability depends on particle size distribution, water content, and impurities (Werner, 2000).

9.1 Bleaching of RBO in a Batch Process

Neutralized RBO samples (in the case of chemical refining) or degummed RBO (in the case of physical refining) should be heated until reaching a temperature of around 100–120°C under constant stirring at an absolute pressure of around 700 mmHg. The moisture present in the oil has to be removed first. Inefficient degumming will also cause reasonable foaming in the bleacher, and the degummed oil quality has to be monitored to avoid this problem. Then, adsorbent should be added to the oil samples and stirring is conducted under high vacuum for about 20 min. Afterward, the bleached oil will be separated from the adsorbent by filtration using a prelayer of diatomaceous earth. Plate, frame, and pressure leaf filters are used by the industry. Propeller design has to be adequate enough to ensure proper mixing. However, most of the industries are switching over to a continuous bleaching processes.

9.2 Challenges During Bleaching and Recommended Precautions

The presence of high foaming and high moisture content should be avoided during the bleaching process. Vacuum should be maintained properly and filtration overloading should be avoided to facilitate efficient filtration. Higher soap content also should be avoided. For efficient bleaching, effective and suitable adsorbents should be used in bleaching. It is preferable to choose adsorbents based on pigment composition and amount of degradation compounds present in the oil.

9.3 Industrial Bleaching Processes

Several adsorption-based bleaching processes are in practice in vegetable oil industries, namely dry bleaching, wet bleaching, two-stage bleaching, two-stage filtration, and multiple dosing, etc.

9.4 Dry Bleaching Process

The dry bleaching process has been practiced in the oil industry for several years and is well known to industry experts throughout the world. In this process, bleaching reactor operates at about 70 Torr vacuum, which greatly reduces the humidity of the oil. The oil should be heated first and then mixed with a citric acid solution to bind trace metals and decompose any residual soap present in the oil. It may reduce the color produced due to presence of chlorophyll. Following the aforementioned pretreatment, the oil should be mixed with bleaching earth, activated carbon, or a mixture of these two in the presence of sparging steam under vacuum to prevent oxidation. After bleaching, filtration should be carried out using pressure leaf filters in which the bleaching agent and other impurity traces will be removed. Filtered oil can be collected at a buffer tank, which operates under a vacuum. Oil collected at the buffer tank will be further taken through a polishing filter to the next process step or storage tank. The dry bleaching set-up requires lower initial investment compared with other types of bleaching.

9.5 Wet Bleaching Process

In this process, water is added to the oil while it is in contact with the bleaching agent in the bleaching reactor. The idea behind the addition of small amounts of water is to utilize expensive bleaching earth in an efficient way, which results in lower bleaching earth consumption and oil losses. Despite the higher initial investment costs, wet bleaching appears to be an attractive solution in terms of cost savings from bleaching earth consumption and lower oil losses. Water will be added in the form of citric acid solution or by taking wet oil from the separate line. Optimum water content in the bleaching reactor will be maintained by careful control of the operating vacuum as >0.5% water content in the reactor will cause filtration problems. In short, the basic difference between dry and wet bleaching installation is whether or not a dryer is located between the bleaching reactor and the filtration system.

9.6 Two-Stage Bleaching Process

The primary objective of this process is to ensure that dosing and filtration of important adsorbents takes place between each stage of oil treatment. It is a combination of wet and dry bleaching coupled with highly efficient two-step filtration in which the adsorption process can be made more efficiently. In this process, adsorption takes place on a continuous basis until there is equilibrium between concentration of pigments, other impurities on the surface of adsorbent, and the concentration remaining in the oil. As soon as this equilibrium is reached, the impure adsorbent is filtered out of the oil and fresh adsorbent (as part of second stage) will be added to the treated oil until equilibrium is again reached. Again, the impure adsorbent is filtered out, and the resulting oil will be passed on either to the next step or storage tank. Two-stage bleaching has several advantages including lower operating costs, reduced bleaching agent loading, and high degree of flexibility.

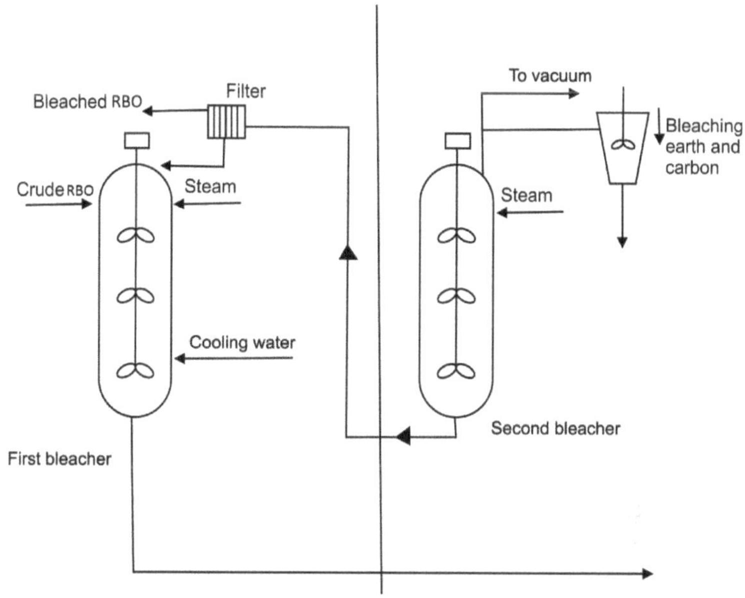

Fig. 8 Flow diagram of two-stage bleaching process.

Multiple dosing systems and control systems: Results from industrial bleaching installations showed efficient bleaching can be achieved by using a combination of bleaching adsorbents. The composition of additives in adsorbents is totally dependent on the type of oil, oil quality, and adsorbents used. Today, bleaching reactors with maximum dosing systems are preferred in the industry,as this enables usage of a substantial range of different adsorbents with different characteristics. The addition of bleaching agent is also crucial; therefore dosing systems should be designed to ensure smooth, proportional flow in relation to the oil flow. Based on the requirement, either simple manually operated local control loops or fully automated plants can be used.

9.7 Countercurrent Bleaching

Compared to the conventional process, 40% less bleaching earth is required by countercurrent bleaching. Presently in a majority of vegetable oil refineries, countercurrent bleaching process is being practiced. Minimizing bleaching earth inputs is in the interest of vegetable oil industry. It would reduce costs for bleaching earth and the amount of oil retained in the bleaching earth. This is more applicable to RBO bleaching, as it requires higher quantities of adsorbents for bleaching.

10. DEWAXING OF RICE BRAN OIL

Most of the vegetable oils do not contain wax. However, few oils like RBO and sunflower oil contain wax ranging from several hundred to several thousand ppm. These are generally considered nonedible, which creates cloudiness of the oil as it has limited solubility in the oil, particularly at lower temperature. Chemically, waxes are esters of high molecular weight fatty alcohols and fatty acids. Waxes are extracted with solvent along with the oil from the bran. It is observed that wax content of expelled oil is always lesser than solvent extracted oil. Waxes present in RBO must be removed to preferably at the level of <10 ppm to pass a cold test. Otherwise, it will have a lower consumer acceptance, particularly in cold countries. Because this process involves formation and maturation of wax crystals and very slow filtration, dewaxing is the rate-determining step of the whole RBO refining process. Another very important factor is presence of molecular species having trisaturates or higher amounts of saturated fatty acids. Lowering of temperature to pass a cold test may result in higher amount of triglycerides losses along with waxes. It is therefore suggested to have dewaxing at a comparatively higher temperature and winterization at a later stage at a lower temperature to get a clear oil that can pass a cold test.

10.1 Factors Affecting Dewaxing Process

Dewaxing is done after degumming and bleaching. It involves crystallization of wax molecules: (1) formation of nuclei; (2) maturation of these nuclei over some time; and (3) filtration of wax crystals. Selection of process variables is important in the dewaxing process. The success rate depends on the formation of nuclei and crystal growth to a large size while cooling continues. Formation of large numbers of small nuclei resulted in difficulties during the filtration process. Initially the oil has to be heated to a certain temperature (~80–90°C) to destroy all the existing crystals. In most of the industrial operations, degummed and bleached oil is taken into specially designed crystallizers. As the oil comes directly from the bleacher after filtration, the temperature of the oil would be around 75–80°C.

The rate of cooling plays a significant role. Supersaturated solution of wax in oil has to be obtained with slow and systematic cooling. There are two to three crystal forms for waxes—alpha (α), beta (β), and beta prime (β^1). Rapid crystallization forms α crystals. These are smaller in size and create problems during filtration. On the other hand, slow and controlled supercooling results in more stable β and β^1 crystals. In addition, viscosity of oil is also kept low enough for a considerable time period to allow the condition for growth of crystals. On the contrary, rapid cooling leads to very little or no crystal growth. After maturation over a certain time period, β and β^1 crystals can be separated by plate and frame type of filtration. During the crystallization process, heat of crystallization is generated, which increases the temperature of oil. Close monitoring of temperature is very essential.

In common practice, hot oil at around 75–80°C is taken into the crystallizer. Initially, oil is cooled to 35–38°C using a cooling tower water within 2–3 h. It is then cooled to crystallization temperature by using chilled water at a rate of 2–3°C/h. Water temperature has to be closely monitored to 4–5°C lower than the oil temperature. This operation on the average takes 10–12 h. Then a maturation time is given by keeping the oil at crystallization temperature under mild stirring. Then the oil is filtered.

Crystallization rate affects formation of proper nuclei and growth of crystals. Slow agitation at a rate of 10–12 rpm with a specially designed agitator is recommended for formation of preferred crystals, minimizing heat and mass transfer gradient, providing a slow motion to the crystals for enhanced nucleation and growth, and most importantly maintaining uniform composition and temperature throughout the bulk of the oil. Higher rate of agitation may lead to fragmentation of crystals creating problems during filtration and inefficient dewaxing. The time of crystallization is also another important factor. Very well designed crystallizers having 12–15 tons or more capacity are available, and if all the parameters are kept under control, preferred crystals can grow within 8–12 h.

The temperature of crystallization is of utmost importance for efficient dewaxing of RBO. However, it depends on the characteristics of the rice paddy, its processing history, and also the geographical location of the place. It was observed that oils obtained from parboiled rice bran have much easier operation compared to raw bran. The temperature of the place also has an influence on dewaxing parameters. Another important factor is the moisture content of the oil. If the oil is not properly dried, the crystal formation and filtration rate alters, creating a huge impact on the dewaxing process. However, the most important factor is efficient degumming before dewaxing of oil. The phosphorous content of the RBO subjected to dewaxing should be as low as possible and preferably in the range of 10–20 ppm. Inefficient degumming will lead to improper wax crystal formation, and gums will block the poresof filter cloth, resulting in very slow filtration. In most of the industries, the filtration during dewaxing is done in an open system, and the oil/wax spillages makes the whole area unhygienic. Proper maintenance is required which is labor intensive. In recent times, closed and automatic controlled filtration systems with very high filtration capacity have come in the market. These systems are highly efficient and reduce the wax content of RBO to a very low ppm level in hygienic conditions. Different filter aids are also used to improve filtration rate and efficiency. Generally, the crystallization is done in the range of 12–16°C. In some places, it is done even at 18°C. The dewaxing parameters, as described earlier, vary from plant to plant and as per the geographical locations. Many variations are reported in the literature.

10.2 Miscella Dewaxing

Several researchers proved that the dewaxing can be performed not only in the oil phase but also in the miscella phase by cooling the oil either in the presence or absence of suitable additives. In Rancher's process (Cavanagh, 1974), RBO miscella is gradually chilled in a compartment fitted with a stirrer with speed 1–10 rpm to speed up crystal formation, and the wax crystals thus formed were separated by centrifugation. This process (Cavanagh, 1974) separated >90% of rice bran *Waxes miscella* phase dewaxing is a well-known process in Japan as the solvent plants are quite large. In addition, miscella dewaxing also depends on the oil content in the miscella, operating temperature, and inclusion of additives (calcium chloride, etc.). According to Marumo (1983a, b), subsequent centrifugation of the oil could remove around 60% of its wax content. A study conducted by Ghosh and Bandyopadhyay (2005) reveals that, in the usual concentration of wax in RBO miscella in a hexane medium, wax crystallizes optimally at 12–15°C after 1 h of incubation. It is further reported by Rao (1983) that the best mixture is generally 1:1 of degummed RBO with hexane. The final crystallization temperature is 5°C, and the crystallization cycle is standardized at 8–10 h per charge. In one patented process, Marumo used calcium chloride to remove waxes from RBO (Marumo, 1983a, b). The use of some surfactants such as soap or sodium lauryl sulfate was also suggested (Haraldsson, 1983). Rajam et al. (2005) has suggested a novel process as pretreatment to physical refining of RBO where they have used higher amount of water and $CaCl_2$ solution followed by crystallization at 20°C. In another process, the use of water is also suggested (Ramakrishna and Manohar, 1995). De and Bhattacharyya (1998) also suggested degumming and dewaxing at a single step at lower temperature.

Miscella dewaxing followed by membrane filtration is also suggested by many investigators. In the mid-1980s, Mutoh et al. described processes to remove wax, gums, FFAs, etc. from vegetable oils including RBO (Mutoh et al., 1985). Later, many researchers had worked on membrane dewaxing of RBO (Asbec and Segers, 1991; Koseoglu and Engelgau, 1990; Tanahashi et al., 1988). The initial studies were done using polymeric membranes. Most of the polymeric membranes were not resistant to hexane. Later on, people have tried dewaxing with ceramic membrane (Muraldihara et al., 1996; De et al., 1998; Subrahmanyam and Bhowmick, 1999). However, due to very poor flux and very high fouling, this technology was not adopted by the industry until now. With the development of newer membranes suitable for this particular application, industrial scale operation is awaited. Furthermore, although it was advised by many researchers that simultaneous degumming and dewaxing either in oil phase or miscella state using membrane would be a better process, until now commercial plants are still using two-stage degumming and dewaxing processes described earlier because of commercial reasons and ease of operation. To develop an efficient dewaxing process of RBO, Yan (2014) invented an apparatus that works under the action of pressure. This prevents oil material from easily bonding on the membrane pore plate, and it has rapid filtration speed and high yield.

11. WINTERIZATION OF RICE BRAN OIL

Winterization of oil is a process of separating the higher melting point components from oil such as waxes or triglycerides by slowly cooling vegetable oils. RBO is the most common oil that is winterized due to its high content of waxes and also due to the presence of a higher amount of saturated fatty acid containing triglyceride molecular species. Around 20% of high temperature melting saturated fatty acids in RBO makes it difficult to dewax this oil at lower temperatures such as 8°C or 10°C (Ghosh, 2007). To get clear oil, RBO is subjected to winterization at lower temperatures compared to dewaxing temperature. Strieder et al. (2017) carried out a study on winterization of RBO in which the highest separation efficiency (89.1%) was obtained using a cooling rate of 2°C per h (30–20°C) and 0.5°C per h (20–10°C) and using the centrifugation operation. However, in this process RBO was obtained with the highest percentage of oleic/linoleic acids with lowest melting point. This clearly demonstrates that during winterization the saturated fatty acid triglycerides were removed from the oil. In an earlier communication, Zheng and his co-researchers (2012) carried out winterization process by introducing the oil into a winterization tank and then preheating to 50°C. It was cooled at 1.5°C/h to 5°C under slow stirring for 24–40h. Later, the contents were filtered to obtain fine quality RBO. The authors claimed that, in this method, no additive was used. The dewaxed and winterized RBO has widespread consumer acceptance across geographic variations including the cold areas.

12. DEODORIZATION/DEACIDIFICATION OF RICE BRAN OIL

Deodorization is a process where all the odoriferous compounds present in the bleached and dewaxed oil are removed by passing a given amount of stripping agent, most commonly the steam, at higher temperature (Ïnsert ">"200°C) under high vacuum. It also removes other volatile components such as oxidative breakdown products, for example, aldehydes and ketones, pesticide residues, some coloring bodies, and other volatile compounds. In the case of RBO, where FFA content is very high, the process is also used for removal of FFAs by carefully choosing the temperature of operation. FFAs are more volatile compared to the triacylglycerol or oil molecules. A little bit higher temperature and strict monitoring of high vacuum help in the removal of FFAs present in the oil along with the other volatile compounds. Hence, this method is known as simultaneous deodorization/deacidification. This is popularly known as physical refining as it does not involve addition of alkali for removal of FFAs. The time of exposure of the oil at this temperature is also a crucial factor and has to be monitored very closely. Some processes go up to 270°C and a time limit of around 45–90min. There exists a possibility of formation of *trans* fatty acids at higher temperatures. A part of tocopherol, tocotrienols, and squalene type of antioxidants may also get distilled at that temperature. This may reduce the nutritional value of the oil. As it is totally dependent on the quality of the oil processed, the time, temperature, and vacuum are to be monitored under an expert eye and by regularly assessing the quality of the final oil.

12.1 Various Studies on Rice Bran Oil Deodorizing Conditions

Deodorization of degummed, dewaxed, and deacidified RBO can be done in the typical way used for other vegetable oils. According to Ghosh (2007), general conditions of deodorization include a temperature between 200°C and 220°C and a pressure of 6–10mmHg. In the case of steam refining of high FFA RBO, the deodorization can be performed simultaneously with deacidification. The temperature and vacuum used in steam refining is rather high (~250°C, under vacuum at 1–3 mmHg) (Ghosh, 2007). The oil has to be given an enough residence time for complete deodorization and the destruction of heat labile pigments. In the initial days, people used to like the typical odor of the oils (Lee and King, 1937). Later on, when products like margarines were made from vegetable oils, the development of a deodorization process was carried out. Initially, batch deodorizers were developed that operated at atmospheric pressure. During the beginning of 20th century, a deodorizer operating under vacuum was first reported (AOCS Lipid Library). With time, there was huge progress in designing of deodorizer and physical refining units. In a very recent patent, Desmet Ballestra Engineering NV SA described a three-stage process for physical refining of vegetable oils (Bart et al., 2015). In this process, 60% of the volatiles are evaporated in the first stage in a flash vessel. The second stage removes a part of the residual volatile matter by countercurrent steam stripping, and further residual volatile matters are removed using a cross-flow tray system in the third stage. Modern deodorizer/deacidification units are using more or less these distillation/evaporation techniques. Generally, the deodorized oil is countercurrently cooled initially by the incoming oil and then by water to around 50°C temperature. The cooled, deodorized oil is then passed through a polishing filter to get better customer acceptability.

Chen et al. (2016) deodorized RBO in vacuum at 160–170°C for 5 min and 180–190°C for 1–1.5 h and then ultrafiltered the oil. Recently Chen et al. (2016) reported a RBO refining system that comprises a water circulation system, a deacidification and deodorization system, and a decoloration system. This method comprises three-step refining, that is, deacidification and deodorization, decoloration, and defatting by winterization. Cheng (2014) deodorized RBO by treating at vacuum of 5–7 kPa and temperature of 220–240°C under stirring. In this study, the authors claimed low energy consumption, high oil yield, good decolorization effect, and less impurity content.

He (2014) reported the deacidification and deodorization process of RBO, which includes decoloration treatment and oil-oil heat exchange, then feeding into a deacidification tower and heating to 250–260°C and then again feeding into the deodorization tower, performing the deodorization process. This oil was further allowed to return into the deacidification tower for circular heating and finally discharged the oil when the temperature of upper layer oil is 250°C and lower layer oil is 230°C. Tsujiwaki and Hamaguchi (2014) reported a deodorization process of RBO with improved flavor and color and reduced *trans* fatty acids. They claimed that deodorization at low temperature condition inhibits the generation of *trans* fatty acids.

Yao et al. (2009) reported industrial refining method and apparatus of RBO in which heating was carried out up to 170–180°C under 6 mmHg pressure and introduced dry steam to remove FFA, ketones, aldehydes, and odoriferous compounds. FFA were recovered by cooling deodorized RBO to below 150°C, and then further cooling to 80°C, which was followed by filtration and storage. The authors claimed that this invention has the advantages of low cost, high efficiency, and good effect on deodorization. Yu et al. (2011) reported the method for preparation of low acid value RBO. In this method, the deodorization step was carried out at vacuum of 5–7 kPa and 200–250°C and 60–80 revolutions/min for 20–30 min to give the product a 1.0–3.0 Acid Value. De and Patel (2010) reported deodorization of RBO at 220°C temperature and 2–4 mmHg pressure.

Prasanna and Kale (2004) reported thin film deacidification and deodorization as an efficient method to remove fatty acids and odoriferous compounds from crude RBO. In their study, a thin film deacidification and deodorization unit was fabricated and used for physical refining of RBO, which basically is a packed column with structured packing. The influence of process conditions, namely temperature, vacuum, and throughput, were studied. They found that better results were obtained for lower flow rates of RBO and steam, and higher vacuum and temperature. The optimal parameters for maximal reduction in FFA were 5 mL/min oil flow, 0.5 mL/min steam flow, 230°C temperature, and 1 mmHg absolute pressure.

Nasirullah and Ramanatham reported a unistep deodorization and deacidification technique for physical refining of RBO. Silicic acid-based adsorbent column comprising of hydrated silicic acid of 100–200 mesh (up to 5% per silica gel) as a stationary phase, nonpolar solvents such as hexane or petroleum ether (b.p. 40–60°C), and polar solvent such as diethyl ether as mobile phase were used for deodorization and deacidification of pretreated refined (water degummed, adsorbent bleached) RBO. As per the TLC analysis, first fraction was pure triglyceride and second fraction contained partial glycerides, FFA, odor factors, and polar components. Physicochemical characteristics of these fractions and room odor test in a closed door room indicated that first fraction of physically refined RBOs gave no odor or negligible odor at a delayed time compared to control and conventionally refined RBO samples.

12.2 Different Types of Deodorizers and Problems Encountered in Deodorization

Various types of deodorizers currently in use are batch deodorizers, semi-continuous deodorizers, continuous deodorizers, packed column deodorizers, and soft column deodorizers. The following problems generally arise in deodorizer sections, for example, air leaks appear through flanges, pump seals, and barometric leg. Steam coil leaks and live steam valve leaks also take place during deodorization. Sometimes the cooling water temperature becomes high, and it may be >32°C. The low steam pressure is another problem faced at deodorization. Once in a while, the ejector/booster nozzles wear out. If a vacuum is not properly maintained, darkening of oils may occur. Another major problem is polymerization and *trans* fat formation in a deodorizer due to high temperatures in the deodorization step. A frequent power failure may aggravate these problems.

Cheng (2016a, b, c) invented three utility models for deodorization unit operation. Song and Fang (2016) invented another refining apparatus useful in RBO degumming, deodorization, and drying, which comprise a degumming tank, a deodorization tank, and a drying tank. The utility model has a reasonable structure design, easy to operate, and it integrates degumming, deodorization, and drying in one. Further, this model can greatly improve the RBO refining efficiency and reduce intermediate links of the degumming, holding, and drying processes, and moreover it reduces operating costs significantly. Li and Xiong (2015) also invented a crude RBO deodorizing apparatus. The utility model has the reasonable design, simple structure, low cost and fully functional, and is able to improve the productivity and achieve better deodorizing effect.

12.3 Effect of Deodorization Conditions on the Minor Components/Micronutrients

Because RBO contains significant amounts of several bioactive minor components, such as γ-oryzanol, tocotrienols, and phytosterols, the refining is crucial. Alkali treatment or neutralization results in a significant loss of oryzanol as well as bringing significant change in the individual phytosterol composition. In the physical refining process, these losses are avoided as no alkali is added in the process. However, in the deodorization/deacidification step, due to relatively high volatility, phytosterols and tocotrienols are stripped from the RBO and are concentrated in the deodorizer distillate. Silkeberg and Kochhar (1999) reported the removal of odoriferous compounds either replacing the deodorization process by the use of inert gas or making a short deodorization with substantially lower heat, that is, 120–150°C instead of the usual 210–270°C. However, this process is not used commercially.

12.3.1 Effect of Deodorization Temperature and Time on the Oryzanol Content

The effect of deodorization temperature and time on the oryzanol content in RBO was studied by Liang et al. (2008). Deodorization temperature and time aggravated the degradation and loss of the oryzanol in RBO, and oryzanol degradation and steam loss were increased with the increasing deodorization temperature and time. At 210°C deodorization temperature and 20 min time, the highest obtained oryzanol content was 1.73%. On the other hand, at 270°C and 120 min time, the oryzanol content in the RBO was found to be 1.28%. Exposure of oil at that high temperature will have a significant effect on quality of oil. However, if the deodorization process is too short, some precursors may remain in the oil resulting in development of unwanted off-flavors during storage.

12.4 Effect of Deodorizing Conditions on the Content of *Trans* Fatty Acid

The effect of deodorizing temperature and time on the *trans* fatty acid content in RBO was studied by Liang (2007). They found that deodorizing temperature and time had a minor effect on the *trans* oleic acid content. However, during their study they observed that deodorizing temperature had a significant effect, and deodorizing time had a secondary effect on *trans* linoleic acid content in RBO. At the particular deodorization time, the highest total *trans* fatty acid content was about 30–40 times of the least one at the different deodorizing temperatures. When the temperature is below 250°C, the total *trans* fatty acids are formed slowly and its content was found to be relatively low. Under the certain temperature, the highest *trans* fatty acid content was about eight times of the least at the different deodorizing times. A very judicious decision has to be taken to fix the temperature and time of operation based on the quality of the feed oil to obtain the refined oil having nutritional values intact and also for removal of undesirable components. A typical commercial deodorization/deacidification unit works at 250–260°C at an exposure time of 30–50 min.

13. VALUE ADDITION TO RICE BRAN OIL REFINING BYPRODUCTS

The refiners always look forward to utilizing the byproducts produced during refining. Value-addition to byproducts would certainly have an impact on economics of the refining process. RBO refining produces gums (in the case of water/acid degumming) or lyso gums (for enzymatic degumming), spent bleaching earth, waxes, and deodorizer distillates as byproducts. It also produces soap stocks, if processed by alkali refining technique. Vandana et al. (2003) studied bleaching and phospholipid enrichment of rice bran gums to produce good quality rice bran lecithin. Reddy et al. (2008) investigated the characterization of rice bran lyso lecithin and its evaluation for surfactant properties. Further, rice bran lyso lecithin was critically evaluated as an energy source in a broiler chicken diet by Raju et al. (2011, 2017a, b) and found that it is suitable as an energy source in a poultry diet and no negative effects on birds were observed. The highest phytochemical concentrations were found in the precipitated soap for γ-oryzanol (14.2mg per g, representing 95.3% of total γ-oryzanol in crude RBO). Rao et al. (2002) developed a process for isolation of γ-oryzanol from RBO soap stock. As a common practice, spent bleaching earth can be used for earth filling. Some industrial manufacturers also used it for oil recovery. However, the recovered oil was utilized for nonedible applications. In addition, a few researchers (Prasad et al., 2008, 2009) developed a process for the upgradation and bleaching of crude rice bran wax. Further, Kaimal et al. (2005) claimed a two-step method for the production of food grade wax, which includes bleaching of the wax with sodium borohydride.

RBO deodorization distillate was found to be the best source of tocopherols (576mg per 100g, representing 6.7% of total tocopherols in crude RBO). Chen et al. (2010) studied the composition of deodorizer distillate. They found that the saturated fatty acid content was 23.8% and the unsaturated fatty acid content was 74.5%, including oleic acid 41.4, linoleic acid 31.7, and linolenic acid 1.4%. The fatty acid composition of deodorizer distillate was in accordance with the RBO. According to them, the sterols content of the deodorizer distillate was 1.4%, and stigmasterol and β-sitosterol content was 0.23% and 0.25%, respectively. Liu et al. (2010) also investigated the contents of tocopherols and tocotrienols in RBO and its deodorizer distillate. According to them, tocotrienols accounted for 60%–70% of the total vitamin E in crude RBO, and 25%–40% of the total vitamin E was present in RBO deodorizer distillate. Stigmasterol, campesterol, and sito sterols are the main sterols found in the RBO deodorizer distillate. However, the total sterol content was only 0.3%. The distillate also contained traces of germanicol, lanosterol, and stigma sterols. Tsuno et al. (2005) studied rice bran unsaponified matter condensate, which contains high amounts of phytosterols, triterpenes, squalene, tocopherol, and tocotrienol. According to them, the condensate is suitable for use in a food composition for prevention of hyperlipidemia in diabetic patients. Sugihara et al. (2010) studied enrichment of squalene (~8%) and phytosterols (~4%) in deodorization distillate, which is essential for use in cosmetics and food industries. Supercritical fluid extraction (SFE) employing supercritical carbon dioxide was investigated at 30°C temperature, 100 kg/cm^2 pressure and 7 mL/min flow rate of carbon dioxide. They found that squalene was effectively concentrated to 25% with nearly quantitative recovery and then a more highly concentrated squalene (approximately 50% purity) was obtained by employing supercritical fluid chromatography (SFC) with silica gel packed into the extraction vessel. Moreover, squalene with approximately 68% purity was obtained by repeating the SFC twice. When deodorization distillate was successively saponified and fractionated with hexane, highly purified phytosterols (97% purity) were obtained. Furthermore, from the unsaponifiable materials, highly purified squalene (81% and 100% purity) was also obtained by using SFC combined with the solvent fractionation technique. These processes may be implemented in RBO refineries to have valorization of the refining byproducts. This will definitely improve the overall economy of the refining industry.

14. CONCLUSION

It is quite evident that RBO is one of the most difficult oils for refining. RBO contains a high amount of FFA due to high lipase activity in the bran. If FFA content is <3%–4%, conventional chemical refining using alkali can be used. This will produce light-colored oil in an economically feasible way. However, a majority of the minor constituents present in the oil will be removed by alkali. Hence the nutritional benefits will be lost. In comparison, physical refining processes that can remove FFA and also retain most nutritionally important compounds are preferable to produce nutritionally enriched oil. A well-planned pretreatment process, particularly degumming process, will be the most important factor for the success of this technique. Close monitoring of the process parameters at each and every unit operation is the key to get a refined RBO having maximum health benefits.

REFERENCES

Aalrust E.; Beyer, W.; Ottofrikenstein, H.; Penk, G.; Plainer, H.; Reiner, R. European Patent E P 0 513 709 (1992).

Asbec, L., Segers, J.C., 1991. Dewaxing of dried oil. USP 5066504.

Balicer, Z.; Leibovitz, Z.; Ruckenstein, C.. In: J. Hollo, Proceedings of the 16th ISF Congress, Held During October 4–7, 1983, Akadémini Kiadó, Budapest, 1985, 393.

Bart, S., Marcelo, U., Marc, K., 2015. Physical refining of triglyceride oils and fats. US Patent US8951592B2.

Bhattacharyya, A.C., Bhattacharyya, D.K., 1983. Refining of high FFA ricebran oil by alcohol extraction and alkali neutralization. J. Oil Tech. Assoc. India 15, 36–38.

Bhattacharyya, A.C., Bhattacharyya, D.K., 1985. Refining of high FFA rice bran oil by mixed solvent-alkali neutralization process. J. Oil Tech. Assoc. India 17, 31–32.

Bhattacharyya, A.C., Majumdar, S., Bhattacharyya, D.K., 1986. Edible quality rice bran oil by Miscella refining. J. Am. Oil Chem. Soc. 63, 1189–1191.

Bhowmick, D.N. In Proceedings of Indo-EC workshop, organized by Department of Science & Technology, Govt. of India, Oxford & IBH Publishing Company Pvt. Ltd., 1992, 157.

Blak, H.W., 1990. In: Erickson, D.R. (Ed.), Proceedings of the World Conference on Edible Fats and Oils Processing: Basic Principles and Modern Practices, Held in Maastricht, the Netherlands in 1989. American Oil Chemists' Society, Champaign, IL, p. 94.

Braae, B., 1976. Degumming and refining practices in Europe. J. Am. Oil Chem. Soc. 53, 353–357.

Canavag, G.C., 1976. Miscella refining. J. Am. Oil Chem. Soc. 53, 361–363.

Carlson, K.F., 1986. Hui, Y.H. (Ed.), Bailey's Industrial Oil and Fat Products, fifth ed. In: vol. 4. John Wiley & Sons Inc., New York.

Carr, R.A., 1976. Degumming and refining practices in the U.S. J. Am. Oil Chem. Soc. 53, 347–352.

Cavanagh, G.C., 1974. Removal of wax from rice bran oil. USP 789120.

Chakrabarti, P.P., Rao, B.V.S.K., Roy, S.K., Bethala, L.A.P.D., Karna, N.P.R., Vemulapalli, V., Chelimi, K., Karthika, G., Kale, V., Prasad, R.B.N., 2003. Process for the pre-treatment of vegetable oils for physical refining. PCT Int. Appl. WO 2003102118 A2 20031211.

Chakrabarti, P.P., Rao, B.V.S.K., Roy, S.K., Prabhavathi Devi, B.L.A., Prasanna Rani, K.N., Vandana, V., Kalyani, C., Gaddam, K., Kale, V., Prasad, R.B.N., 2009. Process for the pretreatment of vegetable oils for physical refining. US 7494676.

Chen, H., Xi, F., Li, J., Jin, W., Jiang, K., 2016. Refining purifying method of rice bran crude oil. Faming Zhuanli Shenqing, CN 105713722 A 20160629.

Chen, L., Liu, Y., Zhang, X., 2010. Determination of sterols in rice bran oil deodorizer distillate. Zhongguo Youzhi 35 (4), 57–62.

Cheng, F., 2016a. One kind of crude rice bran oil deodorizer. Shiyong Xinxing Zhuanli Shuomingshu, CN 205258410 U 20160525.

Cheng, F., 2016b. One kind of crude rice bran oil efficient deodorization tank. Shiyong Xinxing Zhuanli Shuomingshu, CN 205258408 U 20160525.

Cheng, F., 2016c. One kind of crude rice bran oil refining equipment. Shiyong Xinxing Zhuanli Shuomingshu, CN 205258407 U 20160525.

Cheng, J., 2014. Refining production method for rice bran oil. Faming Zhuanli Shenqing, CN 103981022 A 20140813.

Cleenewerck, B., Dijkstra, A.J., 1992. The total degumming process—theory and industrial application in refining and hydrogenation. Fat Sci. Technol. 94, 317–322.

Cowan, C.J., 1976. Degumming, refining, bleaching, and deodorisation theory. J. Am. Oil Chem. Soc. 53, 344–346.

Cvengros, O.J., 1995. Physical refining of edible oils. J. Am. Oil Chem. Soc. 72, 1193–1196.
De, B.K., Bhattacharyya, D.K., 1998. Physical refining of rice bran oil in relation to degumming and dewaxing. J. Am. Oil Chem. Soc. 75, 1683–1686.
De, B.K., Das, R., Dutta, B.K., Bhattacharyya, D.K., 1998. Membrane degumming and dewaxing of Rice bran oil and its refining. Fett-Lipid 100 (9), 416–421.
De, B.K., Patel, J.D., 2010. Effect of different degumming processes and some non-traditional neutralizing agent on refining of RBO. J. Oleo Sci. 59 (3), 121–125.
Dennis, R.T., 2005. Bailey's Industrial Oil and Fat Products, sixth ed. vol. 5 A Wiley- Interscience Publication, pp. 285–328.
Dijkstra, A.J., 1993. In: Applewhite, T.H. (Ed.), Proceedings of the World Conference on Oilseed Technology and Utilization, Held in Budapest Hungary in 1992. American Oil Chemists Society, Champaign, IL, p. 138.
Dijkstra, A.J., Van Opstal, M., 1987. Process for producing degummed vegetable oils and gums of high phosphatidic acid content. US Patent 4,698,185.
Fernando, W.J.N., Hewavitharana, L.G., 1993. Effect of fluidized bed drying on stabilization of Rice bran. Dry. Technol. 11 (5), 1115–1125.
Forster, A., Harper, A.J., 1983. Physical refining. J. Am. Oil Chem. Soc. 60, 265–271.
Geng, R., Zhou, H., Hou, C., 2006. Different approaches of rice bran stabilization. Shipin Keji 126 (2), 120–122.
Ghosh, M., 2007. Review on recent trends in rice bran oil processing. J. Am. Oil Chem. Soc. 84, 315–324.
Ghosh, M., Bandyopadhyay, S., 2005. Studies on crystal growth of rice bran wax in a hexane medium. J. Am. Oil Chem. Soc. 82, 229–231.
Grothues, B. In: Proceedings of the Second American Soyabean Association Symposium on Soybean Processing, 1981.
Guo, J.J., Liu, L., Zhang, M.W., Wei, Z.C., Tang, X.J., 2014. Study on technology of stabilized storage of rice bran by continuous microwave processing. Guangdong Nongye Kexue 41 (13), 77–82.
Haque, M.Z., Ahsan, M.A., Jalil, M.A., 2006. Studies on the stabilization of rice bran and its oil (part II). J. Bangladesh Chem. Soc. 19 (1 & 2), 45–50.
Haraldsson, G., 1983. Degumming, dewaxing and refining. J. Am. Oil Chem. Soc. 60, 251–256.
He, X., 2014. Deacidification and deodorization process of rice oil with golden yellow color and high nutritive value. Faming Zhuanli Shenqing CN 103509649 A 20140115.
Hoffman, G., 1989. The Chemistry and Technology of Edible Oils and Fats and Their High Fat Products. Academic Press Ltd., London.
Hossain, M.G., Chowdhury, E.E.R., Ali, M.M., Jahan, M.S., Ahsan, M.A., Jalil, M.A., Chowdhury, M.F.K., 1997. Stabilization of Rice bran by chemical method. Bangladesh J. Sci. Ind. Res. 32 (4), 638–646.
Houston, D.F., 1972. Rice bran and polish. In: Houston, D.F. (Ed.), Rice: Chemistry and Technology. Am. Assoc. of Cereal Chem., St. Paul, MN, p. 272.
Hu, J., He, D., 2015. Continuous stabilization method of rice bran by tunnel microwave. Faming Zhuanli Shenqing, CN 104304444 A 20150128.
Huang, H., Liu, L., Hu, J., 2015. Comparison of rice bran stabilization with different treatment methods. Zhongguo Youzhi 40 (8), 50–53.
Hvolvy, A., 1971. Removal of nonhydratable phospholipids from soybean oil. J. Am. Oil Chem. Soc. 48, 503–509.
James, E.M., 1958. The refining and bleaching of vegetable oils. J. Am. Oil Chem. Soc. 35, 76–83.
Kaimal, T.N.B., Vali, S.R., Chakrabarti, P.P., 2005. A process for preparation of improved rice bran oil. IN 195177 A1 20050128.

Kaimal, T.N.B., Vali, S.R., Rao, B.V.S.K., Chakrabarti, P.P., Vijayalakshmi, P., Kale, V., Prasanna Rani, K.N., Rajamma, O., Bhaskar, P.S., Rao, T.C., 2002. Origin of problems encountered in rice bran oil processing. Eur. J. Lipid Sci. Technol. 104, 203–211.

Kaji, T., 1988. Process for degumming triglyceride oils. European Patent EP 0 269 277.

Koseoglu, S.S., Engelgau, D.E., 1990. Membrane application and research in the edible oil industry: an assessment. J. Am. Oil Chem. Soc. 67 (4), 239–249.

Koseoglu, S.S., Lawhon, J.T., Lusas, E.W., 1990a. Membrane processing of crude vegetable oils: pilot plant scale removal of solvent from oil Miscellas. J. Am. Oil Chem. Soc. 67, 315–322.

Koseoglu, S.S., Rhee, K.C., Lusas, E.W., 1990b. In: Erickson, D.R. (Ed.), Proceedings of the World Conference on Edible Oils and Fat Processing: Basic Principles and Modern Practices (1989): Maastrict, the Netherlands. American Oil Chemists Society, Champaign, IL, p. 182.

Krishna, A.G.G., 1992. A method for bleaching rice bran oil with silica gel. J. Am. Oil Chem. Soc. 69, 1257–1259.

Lakkakula, N.R., Lima, M., Walker, T., 2004. Rice bran stabilization and rice bran oil extraction using Ohmic heating. Bioresour. Technol. 92 (2), 157–161.

Lee, A.P., King, W.G., 1937. Edible oil deodorizing equipment and methods: a short historical sketch. Oil Soap 14, 263–269.

Leysen, R. Proceedings of the Second American Soyabean Association Symposium on Soybean Processing, 1981.

Li, Y., Xiong, W., 2015. Crude rice bran oil deodorizing apparatus. Shiyong Xinxing Zhuanli Shuomingshu, CN 204385163 U 20150610.

Liang, S., 2007. Effect of deodorizing conditions on the content of *Trans* fatty acid in rice bran oil. Zhongguo Youzhi 32 (2), 37–40.

Liang, S., Lin, M., Zhao, X., Yang, X., 2008. Effects of deodorization on oryzanol content of rice bran oil in physical refining process. Zhongguo Liangyou Xuebao 23 (2), 77–80.

List, G.A., Avellaneda, J.M., Mounts, T.L., 1981. Effect of degumming conditions on removal and quality of soybean lecithin. J. Am. Oil Chem. Soc. 58, 892–898.

Liu, Y., Peng, T., Ma, Y., 2010. Determination of tocopherols and tocotrienols in rice bran oil and the byproduct deodorizer distillate. Zhongguo Youzhi 35 (3), 70–74.

Luh, B.S., Barber, S., deBarger, C.B., 1991. In: Luh, B.S. (Ed.), Rice utilization. vol. 2. Avi Book, Van Nostrand Reinhold, New York.

Ma, T., Zhou, D., Feng, X., Leng, J., 2008. Stabilization of rice bran by composite of Na_2SO_3 and extrusion. Shipin Gongye Keji 29 (5), 158–160.

Mag, T.K., Reid, M.P., 1980. Continuous process for contacting of triglyceride oils with an acid. US Patent 4, 240, 972.

Marumo, H., 1983a. Method for removal of wax from rice bran oil. JP 6188 (1951) and CA 47, 3588d.

Marumo, H., 1983b. Method for separation of wax from rice bran oil. IN 175738.

Moulton, K.J., Mounts, T.L., 1990. Continuous ultrasonic degumming of crude soybean oil. J. Am. Oil Chem. Soc. 67, 33–38.

Muraldihara, H.S., Jirjis, B.F., Seymour, G.F., 1996. Process for removing vegetable waxes by fast cooling vegetable oil and using a porous non-metallic inorganic filter. USP 5,482,633.

Mutoh, Y., Matsuda, K., Ohshima, M., Ohruchi, H., 1985. Method of dewaxing a vegetable oil. USP 4545940.

Nogima, S., 1950. Miscella refining of rice bran oil. JP 1819.

Norris, F.A., 1982. In: Swern, D. (Ed.), Bailey's Industrial Oil and Fat Products, fourth ed. vol. 2.John Wiley & Sons, New York.

Norris, F.A., 1985. In: Applewhite, T.H. (Ed.), Bailey's Industrial Oil and Fat Products. vol. 3. John Wiley & Sons, New York.

Ohlson, R., 1992. Modern processing of rapeseed. J. Am. Oil Chem. Soc. 69, 195–198.
Ong, V.T.L., 1980. Die Notwendigkeit, neue Qualitätsanforderungen für rohes Sojaöl einzuführen. Fette Seifen Anstrich. 82, 169–171.
Patil, S.S., Kar, A., Mohapatra, D., 2016. Stabilization of Rice bran using microwave: process optimization and storage studies. Food Bioprod. Process. 99, 204–211.
Patterson, H.B.W., 1992. Bleaching and Purifying Fats and Oils, Theory and Practice. AOCS Press.
Posschelle, G.L., 1981. De Smet process for physical refining of Soybean oil. J. Am. Oil Chem. Soc. 58, 203–205.
Prabhakar, J.V., Venkatesh, K.V.L., 1986. A simple chemical method for stabilization of rice bran. J. Am. Oil Chem. Soc. 63 (5), 644–646.
Prasad, R.B.N., Rao, B.V.S.K., Prabhavathi Devi, B.L.A., Roy, S.K., Kaimal, T.N.B., 2008. A process for the upgradation and bleaching of crude rice bran wax. JP 4125532.
Prasad, R.B.N., Rao, B.V.S.K., Prabhavathi Devi, B.L.A., Roy, S.K., Kaimal, T.N.B., 2009. A process for the upgradation and bleaching of crude rice bran wax. Indian Patent 228674.
Prasanna, R.K.N., Kale, V., 2004. Thin film deacidification and deodorization of high FFA Rice bran oil and castor oil. J. Oil Technol. Assoc. India (Kanpur, India) 36 (3), 99–104.
Proctor, A., Brooks, D.D., 2005. Baileys's Industrial Oil and Fat Products, sixth ed. vol. 5 A Wiley-Interscience Publication, pp. 267–282.
Qian, J., 2008. Method for processing brown rice to stabilize rice bran with microwave. Faming Zhuanli Shenqing, CN 101301006 A 20081112.
Rafe, A., Sadeghian, A., 2017. Stabilization of Tarom and Domesiah cultivars rice bran: physico-chemical, functional and nutritional properties. J. Cereal Sci. 74, 64–71.
Rajam, L.D.R., Soban, K., Sundaresan, A., Arumughan, C., 2005. A novel process for physically refining of ricebran oil through simultaneous degumming and dewaxing. J. Am. Oil Chem. Soc. 82, 213–220.
Raju, M.V.L.N., Rao, S.V.R., Chakrabarti, P.P., Rao, B.V.S.K., Panda, A.K., Prabhavathi Devi, B.L.A., Sujatha, V., Reddy, J.R.C., Shyam, S.G., Prasad, R.B.N., 2011. Rice bran lyso lecithin as a source of energy in broiler chicken diet. Br. Poult. Sci. 52, 769–774.
Raju, M.V.L.N., Rao, S.V.R., Chakrabarti, P.P., Rao, B.V.S.K., Panda, A.K., Prabhavathi Devi, B.L.A., Sujatha, V., Reddy, J.R.C., Shyam, S.G., Prasad, R.B.N., 2017a. Effects of dietary inclusion of lysolecithin from rice bran oil on performance, serum biochemical profile, organ weights, immune response and nutrient digestibility in broiler chicken. Indian J. Anim. Res. 51 (4), 700–705.
Raju, M.V.L.N., Rao, S.V.R., Chakrabarti, P.P., Rao, B.V.S.K., Panda, A.K., Prabhavathi Devi, B.L.A., Sujatha, V., Reddy, J.R.C., Shyam, S.G., Prasad, R.B.N., 2017b. Comparative evaluation of lysolecithin from rice bran oil vis-a-vis lipotropic agents in broiler chicken diet. Indian J. Anim. Sci. 87 (2), 241–244.
Ram, S.A., 2004. Production of refined rice bran oil. IN 194862 A1 20041204.
Ramakrishna, P., Manohar, B., 1995. A method for enhanced dewaxing of crude rice bran oil or other edible oils. IN175738.
Ramaswamy, K.G., Gopalakrishna, A.G., Sen, D.P., 1980. Refining of Rice bran oil. J. Oil Technol. Assoc. India (Mumbai, India) 12 (1), 16–19.
Randall, J.M., Sayre, R.N., Schultz, W.G., Fong, R.Y., Mossman, A.P., Tribelhorn, R.E., Saunders, R.M., 1985. Rice bran stabilization by extrusion cooking for extraction of edible oil. J. Food Sci. 50 (2), 361–364 (368).
Rao, K.V.S.A., Rao, B.V.S.K., Kaimal, T.N.B., 2002. An improved process for the enrichment of oryzanol from rice bran oil soap stock. US Pat No. 6,410,762.
Rao, V.V., 1983. Dewaxing of Rice Bran Oil. Proceedings of the Seminar on Rice Bran Oil, Status and Prospects. Oil Technologist Association of India (SZ), RRL, Hyderabad, pp. 47–53.

Reddy, J.R.C., Rao, B.V.S.K., Chakrabarti, P.P., Karuna, M.S.L., Prabhavati Devi, B.L.A., Prasad, R.B.N., 2008. Characterization of rice bran Lyso lecithin and its evaluation for surfactant properties. J. Lipid Sci. Technol. 40, 10–15.

Riaz, M.N., Asif, M., Plattner, B., Rokey, G., 2010. Comparison of different methods for rice bran stabilization and their impact on oil extraction and nutrient destruction. Cereal Foods World 55 (1), 35–40.

Roberto, B., 2006. Achieving optimal bleaching performance. Oil Mill Gazetteer 112, 2–6.

Rodchuajeen, K., Niamnuy, C., Charunuch, C., Soponronnarit, S., Devahastin, S., 2016. Stabilization of rice bran *via* different moving-bed drying methods. Dry. Technol. 34 (15), 1854–1867.

Saunders, R.M., 1986. Rice bran: composition and potential food uses. Food Rev. Intl. 1 (3), 465.

Segers, J.C., 1982. Super degumming, a new degumming process and its effect on the effluent problems of edible oil refining. Fette Seifen Anstrichm. 84, 543–546.

Segers, J.C., Van de Sande, R.L.K.M., 1980. In: Erickson, D.R. (Ed.), Proceedings of the World Conference on Edible Fats and Oils Processings: Basic Principles and Modern Practices, Held at Maastricht, the Netherlands in 1989. American Oil Chemists Society, Champaign, IL, p. 86.

Segers, J. C. and Van de Sande, R. L. K. M. Paper Presented at the 83rd Annual Meetings of American Oil Chemists' Society, Toronto, 1992.

Sengupta, A.K., 1977. Process for refining crude glyceride oils by membrane filtration. US Patent, 4,062,882.

Sengupta, A.K., 1978. Purification process. US Patent, 4, 093, 540.

Sengupta, A.K., 1985. Refining. US Patent, 4, 533, 501.

Sengupta, A.K. Paper Presented at ISF Congress, Brighton, September, 1987.

Sengupta, A.K., 1988. Micellar structures and their implication in the chemistry and technology of fats and other lipids. Fat Sci. Technol. 90, 251–256.

Sharma, H.R., Chauhan, G.S., Agrawal, K., 2004. Physico-chemical characteristics of rice bran processed by dry heating and extrusion cooking. Int. J. Food Prop. 7 (3), 603–614.

Shastry, B.S., Raghavendra Rao, M.R., 1976. Chemical studies on ricebran lipase. Cereal Chem. 53 (2), 190–200.

Silkeberg, A., Kochhar, P.S., 1999. Refining of edible oil retaining maximum antioxidative potency. Eur. Pat. Appl., EP 936266 A1 19990818.

Song, D., Fang, H., 2016. Rice bran oil degumming, deodorization and drying apparatus. Shiyong Xinxing Zhuanli Shuomingshu, CN 204939431 U 20160106.

Strieder, M.M., Pinheiro, C.P., Borba, V.S., Pohndorf, R.S., Cadaval, T.R.S.J., Pinto, L.A.A., 2017. Bleaching optimization and winterization step evaluation in the refinement of rice bran oil. Sep. Purif. Technol. 175, 72–78.

Su, D., Li, S., Zhao, F., Cao, Y., Liu, W., 2012. Infrared heating of stabilization of rice bran. Shipin Keji 37 (4), 142–146.

Subrahmanyam, C.V., Bhowmick, D.N., 1999. Membrane degumming of rice bran oil. J. Oil Technol. Assoc. India 31, 193–196.

Sugihara, N., Kanda, A., Nakano, T., Nakamura, T., Igusa, H., Hara, S., 2010. Novel fractionation method for squalene and phytosterols contained in the deodorization distillate of rice bran oil. J. Oleo Sci. 59 (2), 65–70.

Tanahashi, S., Nagano, K., Kasai, M., Tsubone, F., Iwama, A., Kazuse, Y., Tasaka, K., Isooka, Y., 1988. Process for purification of crude glyceride oil compositions. USP 4, 787, 981.

Tandy, D.C., Mcpherson, W.J., 1984. Physical refining of edible oil. J. Am. Oil Chem. Soc. 61, 1253–1258.

Tao, J., 1989. Rice bran stabilization by improved internal and external heating methods. LSU Historical Dissertations and Theses, p. 4812. http://digitalcommons.lsu.edu/gradschool_disstheses/4812.

Tsujiwaki, Y., Hamaguchi, T., 2014. Manufacturing method of edible fats and oils. Jpn. Kokai Tokkyo Koho, JP 2014000012 A 20140109.

Tsuno, T., Kato, H., Tsuno, T., Karitani, T., Kanetani, Y., 2005. Functional rice bran oil deodorized scum-derived unsaponified matter condensate. Jpn. Kokai Tokkyo Koho, JP 2005255563 A 20050922.

Vali, S.R., Chakrabarti, P.P., Kaimal, T.N.B., 2005a. Process for the isolation of glycolipids. US Patent No. 6,953,849 B2.

Vali, S.R., Ju, Y.H., Kaimal, T.N.B., Chem, Y.T., 2005b. A process for the preparation of food-grade rice bran wax and the determination of its composition. J. Am. Oil Chem. Soc. 82, 57–64.

Vandana, V., Karuna, M.S.L., Reddy, J.R.C., Vijeeta, T., Rao, B.V.S.K., Prabhavathi Devi, B.L.A., Prasad, R.B.N., 2003. Bleaching and enrichment of phospholipid content in commercial rice bran lecithin. J. Oil Technol. Assoc. India 35, 139–143.

Vijayagopalan, P., Kurup, P.A., 1972. Hypolipidaemic activity of whole paddy in rats fed a high-fat-highcholesterol diet. Isolation of an active fraction from the husk and bran. Atherosclerosis 15, 215.

Wang, T., Khir, R., Pan, Z., Yuan, Q., 2017. Simultaneous rough rice drying and rice bran stabilization using infrared radiation heating. LWT—Food Sci. Technol. 78, 281–288.

Wells, D.M., Belcher, J.M., 1999. Process for stabilization of oil from plant materials. PCT Int. Appl., WO 9910456 A1 19990304.

Werner, Z., 2000. Modern Technology in the Oils & Fats Industry, first ed. Oil Technologist's Association of India (International Coordination Council), New Delhi, pp. 119–131 (Chapter 7).

Xu, S., 2007. Rice bran: extrusion processing and nutrition. Zhongguo Liangyou Xuebao 22 (3), 12–16.

Yan, G., 2014. Static-pressure filtration-type winterization dewaxing apparatus during rice bran oil refining. Shiyong Xinxing Zhuanli Shuomingshu, CN 203545973 U 20140416.

Yao, J., Cheng, K., Yao, Z., 2009. Industrialized refining method and apparatus of rice bran oil. Faming Zhuanli Shenqing, CN 101473873 A 20090708.

Young, F.V.K., 1990. In: Erickson, D.R. (Ed.), Proceedings of the World Conference on Edible Fats and Oils Processing: Basic Principles and Modern Practices, Held in Maastricht, the Netherlands in 1989. American Oil Chemists Society, Champaign, IL, p. 124.

Yu, F., Lin, Q., Wu, Y., Yang, T., Wu, W., Li, L., Wang, Q., Xiao, H., 2011. Method for preparation of low acid value rice bran oil. Faming Zhuanli Shenqing, CN 102295986 A 20111228.

FURTHER READING

Chen, H., Zhu, X., 2016. Refining system for rice bran oil and fat. PCT Int. Appl., WO 2016034105 A1 20160310.

Fereidoon, S., 2005. Bailey's Industrial Oil and Fat Products, sixth ed. vol. 6. John Wiley & Sons, Inc. Publication, pp. 319–359. 577–653.

Kasturi, V.S.A.R., Bhamidipati, V.S.K.R., Narayana, B.K.T., 2002. Process for the isolation of oryzanols from rice bran oil soap stock. US Patent US6410762B1.

Maria, E.C., Amalia, A.C., 2010. Peanut oil: compositional data. Eur. J. Lipid Sci. Technol. 112, 697–707.

Nasirullah, R.G., 2002. A unistep deodorization and deacidification technique for physical refining of heat sensitive vegetable oils. J. Oil Technol. Assoc. India (Mumbai, India) 34 (2), 57–62.

Pestana-Bauer, V.R., Zambiazi, R.C., Mendonca, C.R.B., Beneito-Cambra, M., Ramis-Ramos, G., 2012. Γ-Oryzanol and tocopherol contents in residues of rice bran oil refining. Food Chem. 134 (3), 1479–1483.
Schurz, K., 2008. Simple method for bleaching fats and oils. PCT Int. Appl. WO 2008012072 A1 20080131.
Shick, L.J., Hoo, S.Y., 1981a. Process for separation of wax from rice bran oil. KR 8100413.
Shick, L.J., Hoo, S.Y., 1981b. Process for separation of wax from rice bran oil. KR 8100413.
Sool, P.S., 1981. Method for dewaxing of rice bran oil. KR 8100007.
Thengumpillil, N.B.K., Ongole, R., Potzla, B.S., 2002. Preparation of low-phosphorus rice bran oil using degumming method. Jpn. Kokai Tokkyo Koho, JP 2002309282 A 20021023.
Xia, C., Bai, F.N., Song, Y., Chen, J., 2015. Study on rice bran stabilization by enzymatic method. Shipin Gongye (Shanghai, China) 36 (11), 152–155.
Zeng, X., 2012. Method for producing grade one edible rice bran oil. Faming Zhuanli Shenqing, CN 102690723 A 20120926.

CHAPTER 4

Bioprocessing Technology of Rice Bran Oil

Yuanrong Jiang
Wilmar Biotechnology Research & Development Center, Shanghai, China

1. INTRODUCTION

Enzymes are versatile biocatalysts produced by animals, plants, and microorganisms. Like other catalysts, enzymes catalyze reactions to its equilibrium position more quickly through lowering its free energy. However, as compared with other catalysts, enzymes have the unique advantages of high specificity, high efficiency, and mild reaction conditions. During the past decades, enzymes have been widely used in food industry such as baking, fruit juice production and processing, brewing, and dairy, to improve texture, flavor, nutritional value, and shelf life, or to improve process yield and safety, through catalysis of different chemical reactions. The commercial application of enzymes in oil and fat processing has developed rapidly due to growing interest in milder processes and availability of new commercial enzymes. Special interests are focused on enzymatic degumming, which has become increasingly popular in recent years to replace traditional degumming processes, followed by enzymatic deacidification, enzymatic interesterification (EIE), and enzymatic stabilizing (Dayton and Galhardo, 2012).

2. ENZYMATIC STABILIZING OF RICE BRAN

2.1 Introduction

Rice bran is a nutrient-rich byproduct from rice milling. Lipids extracted from rice bran are liable to hydrolysis by endogenous enzymes during the milling process, which decreases its value for further application. Reports show that free fatty acid (FFA) content of rice bran oil (RBO) from untreated raw bran would dramatically increase to 47.5% after 30 days of storage (Prabhakar and Venkatesh, 1986).

In comparison to popular methods of chemical treatment (solvent extraction, pH adjustment, et al.), heat treatment (extrusion, visible light

Rice Bran and Rice Bran Oil
https://doi.org/10.1016/B978-0-12-812828-2.00004-4

irradiation, microwave, et al.) and deep freezing, biochemical methods (i.e., enzymatic stabilizing) are seldom studied due to its limitation on industrialization. However, enzymatic processes are becoming increasingly popular due to their ability to preserve nutrient composition and bioactive components in rice bran due to their mild reaction conditions.

2.2 Enzymatic Stabilizing

In 1994, RIBUS® successfully applied its patented enzymatic technology on rice bran stabilizing by irreversibly inactivating its lipolytic enzyme using an antilipase enzyme with intact functional nutrients through stabilizing treatment (Hammond and Cordova, 1994a, b, 1996, 1998). A protease of ENZECO Fungal Protease brand concentrate, an enzyme fermented from various fungi, was involved as the antilipase enzyme. As a result of stability treatment, consequent oxidation and rancidity of lipids were strongly prohibited. Huge improvement can be observed in terms of extended shelf life from a few days to more than a year. Stable, functional, rich lipids from rice bran unlocked value and possibility on lipids posttreatment for food or personal care products. The enzymatic stabilizing process mainly includes the following steps:

(1) First, a mixture of antilipase enzyme, rice bran, and water is formed.
(2) Then, the mixture was kept at a certain temperature for a period of time to deactivate the lipolytic enzyme.
(3) Finally, the obtained oil was dried to a certain moisture content.

2.3 Section Summary

Although RIBUS®-implemented enzymatic stabilization process is already industrialized, this process is seldom used due to scattering distribution of bran source in small amounts, such as in China. Additionally, except RIBUS®, no other applications on enzymatic stabilizing have been reported, and the dominating approach is extrusion.

3. ENZYMATIC DEGUMMING OF RICE BRAN OIL

3.1 Introduction

Traditional degumming processes like water degumming, acid degumming, and alkaline degumming have either high oil losses or high residue phospholipids. Enzymatic degumming is a relatively new technology that emerged in the past 20 years. In comparison with water/acid degumming or alkali

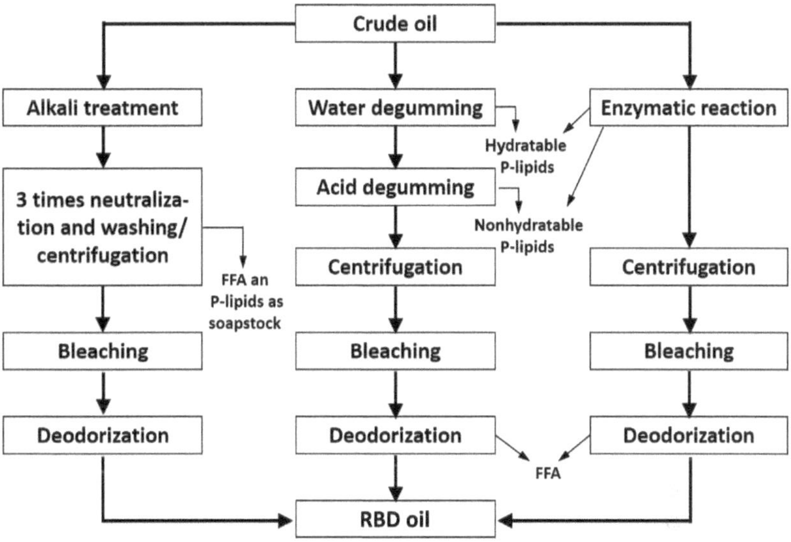

Fig. 1 Flow chart comparing chemical refining to physical refining, including enzymatic degumming. RBD, refined, bleached, deodorized; FFA, free fatty acid (List, 2009).

treatment, the benefit of using enzymes as auxiliary agents in degumming vegetable oils is that it reduces total oil loss and waste water, as well as operating costs, yet enabling physical refining by lowering the residue phosphorus (P) to <10 mg/kg (Fig. 1) (List, 2009). There are already several review articles elaborating all aspect of enzymatic degumming (De Maria et al., 2007; Dijkstra, 2010, 2011), meanwhile the key advantage for the enzymatic degumming process is that the process converts almost all the gums to their hydratable form, therefore enzymatic degumming has the lowest yield loss amongst all the different degumming methods.

3.2 Phospholipase

Enzymes specific for phospholipids are known as phospholipases. Enzymatic degumming refers to usage of phospholipases to hydrolyze part of phospholipids, converting nonhydratable phospholipids to hydratable lysophospholipids, and making it easier to eliminate from the water phase by subsequent centrifugation. There are several types of phospholipases, including phospholipase A1 (PLA1), phospholipase A2 (PLA2), phospholipase B (PLB), phospholipase C (PLC), and phospholipase D (PLD). Fig. 2 demonstrates the acting sites of different phospholipases (List, 2009) (Dijkstra, 2011).

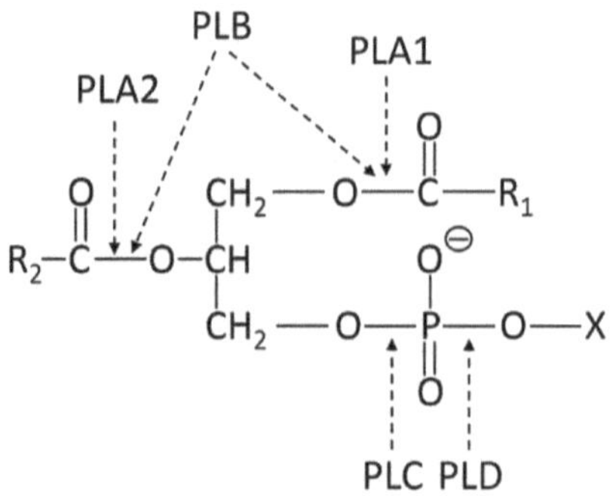

Fig. 2 The acting sites of different phospholipases.

PLA1 and PLA2 hydrolyze the ester bond in the *sn*1 and *sn*2 sites, respectively. Then the hydroxyl group makes the molecule affinitive to the water and easily separated from oils. PLB can hydrolyze both *sn*1 and *sn*2 sites, generating two hydroxyl groups, which have much more affinity to water. PLC is a novel phospholipase. It only hydrolyzes the undesired phosphate moiety, leaving diacylglycerol (DAG) that can be actually treated as oil. Thus PLC not only excludes the undesired part, but also increases oil yield of the process. PLD is present in seeds and is responsible for destruction of phosphatidyl choline (PC) and formation of phosphatidic acid (PA) during storage and processing. PLD is of little use in the edible oil degumming process. Various phospholipases used in enzymatic degumming processes are listed in Table 1 (Dijkstra, 2010, 2011).

3.3 Enzymatic Degumming Processes

The first industrial enzymatic degumming process was Lurgi's EnzyMax® process launched in 1992, which utilized a porcine phospholipase A2 (Dijkstra, 2010). This classical process consists of several important steps (see Fig. 3):

- adjusting the temperature for enzyme reaction;
- adding citric acid and retention for a period of time;
- adding sodium hydroxide to the expected optimal pH for the enzyme;
- adding enzyme and water to the oil and mixing vigorously;
- conducting the enzyme reaction under sufficient mixing;
- heating the oil to optimal temperature for separation in the centrifuge;
- separating the gum from the oil in the centrifuge.

Application of porcine phospholipase A2 did not last long due to its limited availability. Then, Novozymes developed Lecitase® Novo, which is a PLA1

Table 1 Various phospholipases used in enzymatic degumming processes

Type	Name	Supplier, year	Notes
PLA2	Rohalase® MPL	AB Enzymes, 2004	Microbial PLA2 from *Trichoderma reesei*
PLA2	Lecitase® 10L	Novozymes, 2006	Porcine PLA2, firstly used in EnzyMax® process
PLA1	Lecitase® Novo	Novozymes, 2006	Microbial PLA1 from *Fusarium oxysporum*
PLA2	LysoMax®	Dupont/ Danisco, 2007	2-specific lipid acyl transferase (LAT) from *Streptomyces violaceoruber* with PLA2 activity
PLA1	Lecitase® Ultra	Novozymes, 2008	Thermostable microbial PLA1 from *Thermomyces lanuginosa/Fusarium oxysporum*
PLC	Purifine®	DSM/ Verenium, 2008	Microbial PLC from *Pichia pastoris*, specific for PC, PE
PLA2	GumZyme™	DSM, 2011	Microbial PLA2 from *Aspergillus niger*
PLB	–	–	Commercial unavailable

phospholipase obtained from *F. oxysporum*. This new enzyme, obtained through fermentation, opened the possibility for higher dosages and lower costs for enzymes. Later, Lecitase® Ultra, an updated version of Lecitase® Novo with improved thermal and pH stability, was developed. Purifine®, a PLC-type phospholipase, was also developed at the same time. Although PLC has the unique advantage of oil yield improvement, the result demonstrated that only PC and PE could be split off by Purifine®. This indicated that oil that has been treated with Purifine® requires further degumming treatment (Dijkstra, 2010).

Another enzymatic process is the Danisco process, which utilizes lysophospholipid acyl transferase (LAT) that acts on all phosphatides by catalyzing the transfer of an acyl moiety from the sn-2 position in a phosphatide to a nonesterified sterol or stanol (Dijkstra, 2010). Because the transferred acyl moiety is not lost as FFA, esterified sterol is not lost during vacuum steam stripping, and the lysophosphatides retain less oil during degumming processing, the LAT-catalyzed transesterification increases the final oil yield. However, the degree of oil yield increment depends on the type of oil being treated.

Researchers have also tried to use PLB to degum edible oils considering the specific hydrolysis activity of PLB. PLBs from *Pyrococcus furiosus*, *Thermotoga*

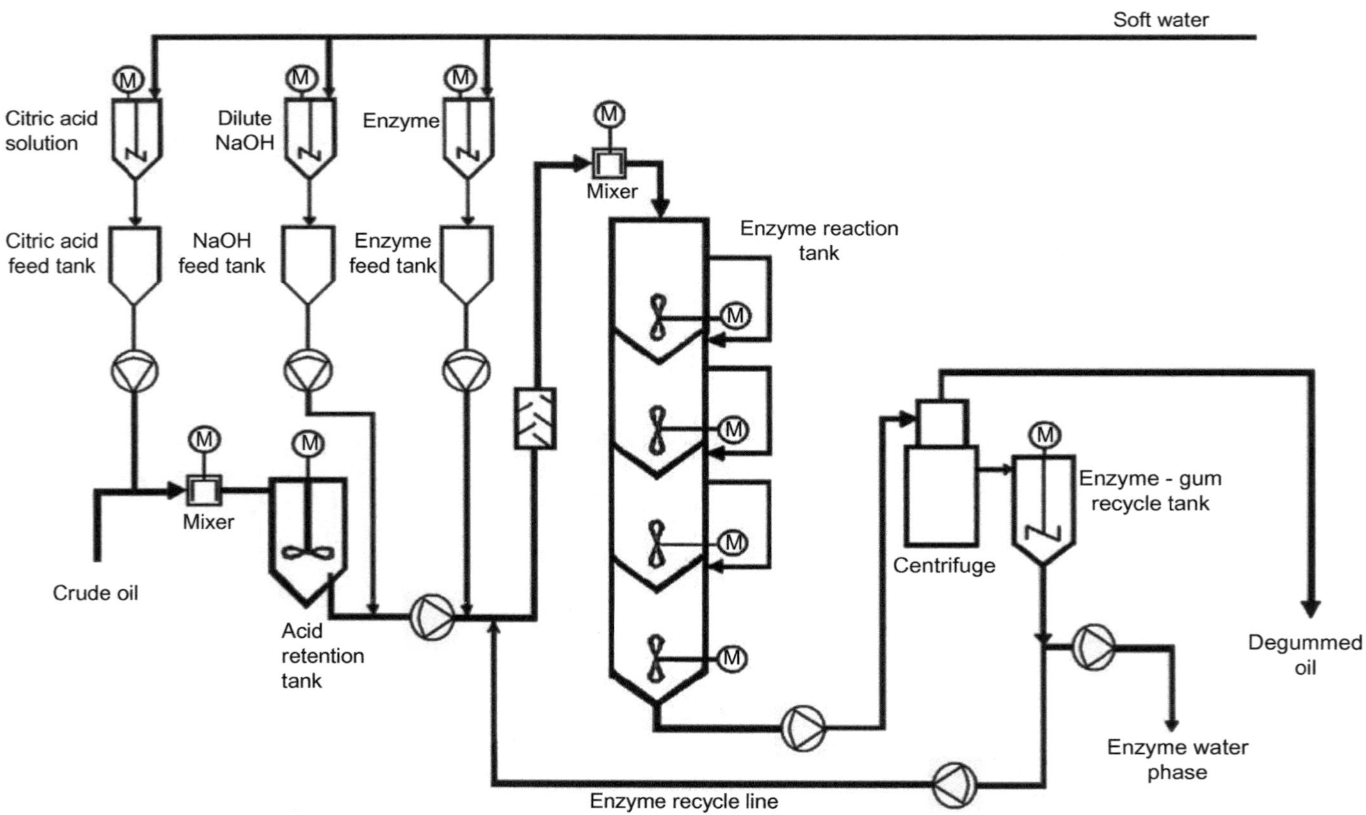

Fig. 3 EnzyMax® degumming process flow diagram (Dijkstra, 2010).

lettingae, and *Pseudomonas fluorescens* were expressed in *Escherichia coli* or *P. pastoris* systems in the lab by various researchers (Wei et al., 2015; Huang et al., 2014; Jiang et al., 2011; Su et al., 2017). Although some PLBs exhibited effective degumming effects on various edible oils, industrialization of PLB is difficult due to low expression level during fermentation.

Given that PLA, PLB, and PLC have respective weak points, it is natural to propose the idea of combining several phospholipases together. Dayton and Galhardo (2015) (US08956853) described the use of a combination of PLC and PLA enzymes for complete hydrolysis of phospholipids in a degumming process. A synergy between the two different types of enzymes was discovered, even when reaction conditions were not optimal for both. The reaction time was reduced to 30 min to 1 h through use of a combination of enzymes.

3.4 Enzymatic Degumming of RBO

RBO differs from other vegetable oils because of its higher FFA content along with its unusually high contents of wax, unsaponifiable constituents, polar lipids, and pigments (Kaimal et al., 2002). To retain as much nutrients such as oryzanol as possible, physical deacidification rather than chemical neutralization is needed, as oryzanol will be excluded during chemical neutralization due to its weak acid nature. Traditional water-based degumming methods cannot reduce the P content below 10 mg/kg, which is disadvantageous as P-containing components cause color fixation in the final oil during exposure to higher temperatures of physical refining. Enzymatic degumming turned out to be the best process for reducing the phosphorous content of crude RBO < 5 mg/kg. In comparison to a traditional degumming process, enzymatic degumming of RBO has advantages such as (Sengar et al., 2014):

- Better utilization of byproducts such as gums and waxes;
- Decreased oil losses due to elimination of chemical neutralization;
- Decreased oil loss in gum;
- Well-preserved oryzanol;
- Environmental-friendly process with less waste water;
- Less oil loss through water washing.

Although RBO is different from other oils in various aspects, the general process for enzymatic degumming of RBO is similar to other oils. Both liquid and immobilized enzymes have been investigated for RBO degumming in past years, however only liquid enzymes are now commercially available for industrial application.

3.4.1 Liquid Enzyme

In 2002, Roy et al. (Panesar et al., 2011) made the first attempt of enzymatic degumming with crude RBO by applying Lecitase® Novo and EnzyMax® process. The P levels of crude RBO (403 mg/kg) and water-degummed oil (60 mg/kg) were only reduced to 15–18 mg/kg after enzymatic degumming. But after bleaching of the enzymatic degummed RBO, the P levels was reduced to <5 mg/kg, making it suitable for physical refining.

To simplify the enzymatic degumming process, a new enzymatic degumming process for pretreatment of RBO was published in a patent in 2004, where citric acid and sodium hydroxide were added together with phospholipase A1; although sometimes no citric acid or sodium hydroxide was added (US20040005399, 2004). The enzymatic degumming process could reduce P content to <5 mg/kg after bleaching and dewaxing.

In 2008, Jahani et al. (2008) optimized the process conditions for enzymatic degumming of RBO by employing Lecitase® Ultra. Four processing factors affecting residual P and FFA content were identified: hydrolysis time, temperature, added water, and enzyme dosage. Amongst the individual variables, reaction time had the maximum effect on P content, which decreased by increasing reaction time. Hydrolysis temperature showed significant quadratic effect on residual P content because higher temperature increased enzyme hydrolytic activity, but at temperatures over 50°C the enzyme lost activity because of denaturation. The added water had a negative effect on residual P content in the degummed oil due to the hydrolase nature of the enzyme and direct contribution of water in the hydrolysis of phospholipids. Moreover, the linear effect of reaction time and temperature, and the quadratic effect of the enzyme dosage on FFA content, dominated the other variables whereas reaction time had the maximum influence on FFA content and process temperature had a negative effect on FFA content. At optimum operating conditions of 4.07 h reaction time, 50 mg/kg enzyme dosage, 1.5 mL/100 g water and 49.2°C reaction temperature, phosphorous and FFAs contents of degummed oil were found to be 8.86 mg/kg (crude oil, 196.25 mg/kg) and 2.01 g/100 g as oleic acid (crude oil, 1.6 g/100 g as oleic acid), respectively.

Manjula et al. (2011) assessed the efficacy of enzymatic degumming by using third generation PLA1, Lecitase® Ultra from *Thermomyces lanuginose*/*F. oxysporum* with different qualities of crude RBO in 2011. The P content in the oil was reduced from 390 to ~10 mg/kg and 128 to 9 mg/kg, respectively, for crude RBO and water-degummed RBO after enzymatic degumming. However, P content in the acid-degummed RBO could only be reduced to the extent of 31 mg/kg. The residual excess added phosphoric acid trapped in nonhydratable phospholipids (NHP) present in the acid-degummed oil could be an effective inhibitor for phospholipase, reducing the substrate binding and hence the extent of hydrolysis of substrate. Attempts were also made on enzymatic degumming of RBO in a hexane medium because hexane is the common solvent used for the extraction of oil from oilseeds and pressed cakes. However, the degumming efficiency was not comparable to oil-phase degumming possibly due to inhabitation of organic solvent and poor contact between the miscella and aqueous phase. To reuse the enzymes, the authors proposed an integrated approach by combining enzymatic treatment and membrane separation to explore the possibility of recycling/reusing the enzyme in the process.

As previously described, the use of a combination of PLC and PLA enzymes could lead to complete hydrolysis of phospholipids as well as increased oil yield. Jiang et al. (2015) systematically investigated the applications of PLC and PLA1 in the degumming process of different kinds of crude oils including RBO. Results showed that a citric acid pretreatment followed by double-enzyme (PLC + PLA1) degumming process efficiently reduced the P content and maintained the maximal oil yield of most oil samples. Considering different effects of PLC, PLA1, and the quality of different crude oils, it was also suggested that specific enzymatic degumming process should be selected. In the case of RBO, a citric acid pretreatment followed by PLA1 hydrolysis generated degummed RBO with <10 mg/kg P, whereas the addition of PLC besides PLA1 helped to increase the oil yield by 0.4% (see Table 2).

Table 2 Residual phosphorus, oil yield, and DAG content after degumming with different agents

Degumming agent	Residual P (mg/kg)	Oil yield (%)	DAG (%)
Water	68.3 ± 5.7	97.04 ± 0.08	5.21 ± 0.38
Citric acid	27.9 ± 0.9	96.84 ± 0.05	5.19 ± 0.12
PLC	56.6 ± 5.4	97.72 ± 0.07	5.60 ± 0.27
PLA1	45.6 ± 2.9	97.31 ± 0.04	5.15 ± 0.30
PLC + PLA1	36.9 ± 1.9	97.55 ± 0.06	5.40 ± 0.22
Citric acid-PLC	25.1 ± 0.2	97.74 ± 0.04	5.55 ± 0.11
Citric acid-PLA1	5.9 ± 1.7	97.22 ± 0.08	5.17 ± 0.25
Citric acid-PLC + PLA1	0.9 ± 0.3	97.62 ± 0.06	5.32 ± 0.18

The initial P and DAG of crude RBO was 392.6 ± 0.9 and 5.12 ± 0.26 respectively (Jiang et al., 2015).

In 2017, Li et al. (2017a, b) demonstrated that the P content of a high-acid crude RBO could be decreased from 229.7 mg/kg to 7.9 mg/kg using the degumming method reported by Yang et al. (2008), which was similar to EnzyMax® process employing PLA1.

3.4.2 Immobilized Enzymes

At present, most enzymes used in industrialized enzymatic degumming processes are disposed directly without possibility of reuse. As enzyme is the main cost in degumming process, many researchers have tried to develop various techniques to reuse enzymes and reduce cost. The general strategy to reuse enzymes is to immobilize them with substrates for ease of recoverability and reusability.

In 2008, Sheelu et al. (2008) investigated the immobilization of phospholipase A1 in gelatin hydrogel as well as its stability for RBO degumming. The thermostability of PLA1 was improved by immobilization, and the activity of the immobilized phospholipase A1 became more pH-sensitive. The immobilized enzyme was used in a spinning basket bioreactor for 6 recycles of RBO degumming without loss of initial enzyme activity. P content of the degummed RBO could be reduced from 400 to 50–70 mg/kg in each cycle. After charcoal treatment and dewaxing, a second enzymatic treatment was needed to bring down the P content to below 5 mg/kg.

The enzyme carrier has also expanded into nanomaterials such as nanopolymers, as the nanoparticle could provide a larger surface area for attachment of enzymes, leading to higher enzyme loading and could be easily and rapidly separated from the reaction medium when placed in a magnetic field. The magnetic nanoparticles could also improve the stability of the immobilized biomolecules. In 2013, Yu et al. (2013) immobilized PLA1 onto magnetic nanoparticles and achieved complete enzyme reusability through the unique separating capability of these magnetic nanoparticles. Compared to the free PLA1 process, the immobilized PLA1 had a broader pH profile as well as improved thermostability. After 10 cycles through the soybean oil degumming process at 55°C, the immobilized PLA1 still possessed >80% of its initial activity.

The enzymatic degumming process involves interface reactions, thus the blending effect of oil and enzyme has strong influence in degumming effects. Singhania et al. (2015) investigated the immobilization of the PLA1 from *Thermomyces lanuginosus* and evaluated the degumming efficiency of the immobilized enzyme. The calcium alginate immobilized PLA1 showed ability to conquer mass transfer resistance at the lipid-water interface and increase the enzyme catalytic efficiency. Although the immobilized PLA1 was superior in removing P content compared to its free form, the residue P content was quite high (57.2 mg/kg for immobilized PLA1 and 266.7 mg/kg for free PLA1), probably due to no acid pretreatment before enzymatic degumming. Furthermore, the removal of P content from oil was also accompanied by a significant reduction in color and restoration of iodine value to the desired level.

3.5 Section Summary

Enzymatic degumming is a new way of degumming edible oils and has now become an accepted tool in various vegetable oil refining due to its benefits in P reduction and oil recovery. The availability of commercial phospholipases such as PLA1 and PLC from Novozymes and DSM/Verenium, respectively, enabled industrial application of enzymatic degumming. The general enzymatic degumming process first includes an acid pretreatment followed by addition of alkali for pH adjustment. Then, the enzymes are added to the oil together with a proper amount of water and mixed vigorously before pumping into the enzyme reactor. The enzyme reaction takes place under sufficient mixing to maintain the emulsion. The mixture is then heated and separated through centrifugation to harvest degummed oil. Nowadays, only free liquid enzymes are commercially available, and different methods have been investigated for enzyme immobilization.

Main process parameters such as enzyme dosage, reaction temperature, reaction time, and water dosage are reported to affect the residue P content and FFA differently depending on the quality of different crude oils. For RBO or water-degummed RBO, which is especially suitable for enzymatic degumming due to high FFA and valuable nutrients, slightly modified Enzy-Max® process could be used to generate degummed RBO with <10 mg/kg P content. In the case of high P content in the crude oil, a combination of PLC and PLA1 might also help increase the oil yield.

Overall, enzymatic degumming has a reduced environmental impact compared with chemical refining by using lower amounts of chemicals, water, and energy. Therefore enzymatic degumming of RBO processing will be applied more and more widely in the future as the development of RBO industry.

4. ENZYMATIC DEACIDIFICATION OF RICE BRAN OIL

4.1 Introduction

To our knowledge, enzymatic deacidification is a contribution to green RBO refining process without generation of any hazardous content generation through the whole refining process. It facilitates RBO deacidifying under mild operating conditions with the aid of enzyme. Existing literatures showed crude RBO generally retains 5–50 wt% of FFA (Ajay, 1992). It is well known that the quality of RBO is highly affected by FFA content because of its undesirable rancid flavor and easier oxidization of FFA. Until now, commercial RBO are mainly produced by alkali deacidification, physical refining, and miscella deacidification with high neutral oil and bioactive compound loss during saponification and large waste water generated during posttreatment of soap stock. In the past decades, efforts on alternative approaches to improve the deacidification process, including enzyme reesterification, biological deacidification, solvent extraction, supercritical fluid extraction (SFE), and membrane technology, are basically overcoming the drawbacks of conventional processes for ecofriendly and neutral oil losses minimality (Bhosle and Subramanian, 2005). However, biological deacidification cannot be applied for linoleic acid and short-chain FFAs (C no. <12) and reesterification are high destruction on functional components caused by high operating temperature; the uncertainty safety of the catalysts also increased the difficulty in subsequent decolorization. As for the extraction practices, solvent extraction will lead to incomplete deacidification because TAG solubility varies with FFA content whereas SFE is a costly processing technique. As for membrane technology, the membrane used is not only difficult to clean but also susceptible to pollution, resulting in low quality and membrane life expectancy. In summary, enzymatic deacidifying is the best promising method for production. Nowadays, enzymatic deacidification has already achieved promising results on a laboratory scale. Literature shows that the FFA can be reacted with acyl acceptor to reduce FFA to lower than 2 wt% (De, 2006; Zheng et al., 2015) after enzyme modification, high quality oil with bioactive ester-rich substances and well reservation of γ-oryzanol with outstanding features in unsaponifiable substances remaining. Then the treated oil was transferred downstream for further purifying to meet edible oil standards (FFA content $<1.5\%$, Chinese national standard of the RBO: GB19112-2003). Fig. 4 shows the typical treatment route for edible RBO processing (Ghosh, 2007). Generally, the raw RBO should be pretreated to have phosphorous and wax reduction, and it also extends the

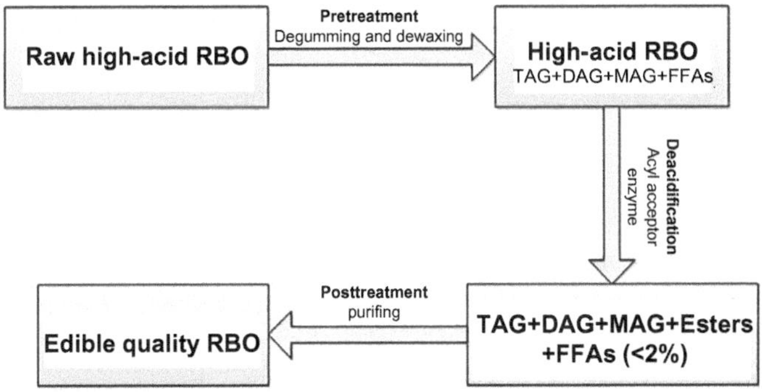

Fig. 4 Typical enzymatic refining (Ghosh, 2007).

enzyme life span during following deacidification. After the FFA reduces to a certain low level (usually <2%), it is delivered for further purifying, such as deodorization.

4.2 Lipase for Deacidification

Esterification of acyl acceptors and FFAs is the main reaction of enzymatic deacidification, which is promoted by lipase under proper temperature, pressure, enzyme loading, substrate/oil ratio, water content, and reaction time. Recently, few works involved amidation to reduce FFA contents under similar operating conditions used in esterification (Wang et al., 2017). Regardless of proposed esterification or amidation, compared to immobilized enzymes, liquid enzyme has a limitation on the application of nutritive lipid production as well as their difficulty in after-use recovery, therefore immobilized enzymes are the most popular catalyst in bioprocessing technology.

Reports presented a greener process of FFA removal by esterification between FFA and glycerol by using 1,3-specific lipase under proper reaction conditions (Ajay, 1992; Bhattacharyya and Bhattacharyya, 1989). Bhattacharyya et al. reduced the FFA content from 30% to 3.6% based on Mucor miehei catalytic esterification (Bhattacharyya and Bhattacharyya, 1989). Ajay Lakshmanan and B.K. De et al. also employed Mucor miehei lipase (Ajay, 1992) in enzymatic esterification, which reduced the FFA content from 45.5% to 15.7%. The FFA content could be lowered to 4% with step-wise addition of fresh enzyme and acyl acceptor. For an improvement, B.K. De successfully reduced the FFA content from 16.2% to 0.8%–1% at 6 h and 0.3%–0.8% after 7 h by Mucor miehei catalytic esterification with stoichiometric amount addition of DAG or 25% excess.

If monoacylglycerol (MAG) is used, FFA was reduced to 1.2% and 0.6%–0.8% after 6 and 7h with stoichiometric amount MAG or 50% excess, respectively (De, 2006). Similar studies have been carried out with excellent results. Song et al. adopted Novozym 435, Lipozyme RM IM, and Lipase TL IM to facilitate FFA-MAG esterification (Song et al., 2012), which reduced the FFAs down to 0.28%.

Recently, immobilized enzymes were developed for FFA deacidifying with a good application. Zheng et al. immobilized *Candida rugosa* lipase (CRL) on the octyl-functionalized ordered mesoporous silica (C8-OMMs) to catalyze esterification of phytosterol ester-enrich functional oil, which resulted in FFA content below 2.3%. (Zheng et al., 2015). Li et al. immobilized crude fermentation broth of SMG1-F278N onto the epoxy-functionalized resin ECR8285 by covalent binding, resulting in favorable stability in both hydrolysis of partial glycerides (MAG, DAG) to FFA, which reduced the FFA to 0.03%–0.08% (Zhao et al., 2017; Li et al., 2016, 2017a, b).

Several studies have been conducted to investigate the selectivity of enzymes in oil deacidification. For example, Wang et al. reported that Lipozyme 435 (food grade) is better than Novozyme 435 (industrial grade) in deacidification of Lipozyme 435 and Novozyme 435 applications, which resulted in FFA reduction from 21.5 to 7.7 and 7.4 mg/g after 10h reaction, respectively (Wang et al., 2017). Also, the units of activity of Lipozyme 435 and Novozym 435 (1.63 and 1.71 U/g, respectively) (Song et al., 2012) show comparative high enzyme activity, but the former is cheaper. Hence, Lipozyme 435 is a more suitable enzyme in enzymatic deacidification. Another research spot is DAG-rich oil producing. Song et al. reported the capability of commercial lipases (Novozym 435, Lipase RM IM, Lipase TL IM) was in the order of Novozym 435 > Lipase RM IM > Lipase TL IM (Song et al., 2012), and the efficiency on deacidifying are a positive correlation in the units of activity of Novozym 435, Lipozyme TL IM, and Lipozyme RM IM (1.71, 0.76, and 1.82 U/g, respectively). But it is worth noting that the 1,3-positional DAG product, when Lipase RM IM and Lipase TL IM were taken as catalysts, was beneficial on lipid metabolism. Above all, compared with Lipase TL IM and Novozym 435, Lipase RM IM was a better catalyst for enzymatic deacidification due to its good reusability, low price, and relatively high capacity on esterification (Song et al., 2012).

In conclusion, more researchers have tried to utilize immobilized enzymes in enzymatic deacidification processes due to their unique characteristics such as thermal stability, excellent performance on FFA esterification of its highly operational activity, stability, and easier implementation on enzyme separation and regeneration.

4.3 Acyl Acceptor for Enzymatic Deacidification

4.3.1 Glycerol and its Ester Derivatives as Acyl Acceptor

Generally, the selection of acyl acceptor mainly depends on the final product where the esters formed are either excellent in terms of functional properties or easily removed. In the past decades, a wide range of interest and attention have been directed toward involving glycerol and its ester derivatives as acyl acceptors due to their remarkable capacities on the inclination of TAG formation.

Few works introduced glycerol as an acyl acceptor in RBO deacidification. For example, several reports took glycerol (Ajay, 1992; Bhattacharyya and Bhattacharyya, 1989; Lu et al., 2016; Kosugi et al., 1994) and its ester derivatives (MAG, DAG) (De, 2006; Song et al., 2012) as an acyl acceptor, which demonstrated strong capability for FFA reduction and the formation of acylglycerols (MAG, DAG, and TAG). After deacidifying, S. Bhattacharyya and D.K. Bhattacharyya increased TAG content from 53.3% to 80.3% without significant change in MAG or DAG contents (Bhattacharyya and Bhattacharyya, 1989). For the same purpose, Song et al. (2012) increased DAG content from 0.31% to 27.98% by introducing MAG (Song et al., 2012).

For an acyl acceptor of glycerol and its ester derivative-based deacidifying, the required glycerol estimation, residual glycerol content, water produced estimation of apparent reaction rate, and estimation of dehydration rate are similar, and they can be estimated by the following equations:

(1) Estimation of required glycerol amount (Kosugi et al., 1994):

Assumption 1 The stoichiometric amount of glycerol necessary for converting all FFA, MAG, and DAG into TAG in 1 g of the high FFA RBO (Eq. 1).

$$\begin{aligned}\text{Weight of glycerol required (g)} &= (92/3)\,[(\text{the molar amount of FFA})\\ &\quad - 2\,(\text{the molar amount of MAG}) - (\text{the molar amount of DAG})]\\ &\quad - \text{glycerol in 1 g of the high FFA RBO} \qquad (1)\end{aligned}$$

When the stoichiometric amount of substrate was used, the stoichiometric amount of product was produced by this enzymatic conversion (Kosugi et al., 1994). Thus the residual glycerol content can be calculated (Eq. 2):

$$\text{Glycerol}\left(\frac{\text{mol}}{\text{g}}\right)=\frac{1}{3}\times\left[\frac{\text{FFA content }(\%wt.)}{277}-2\times\frac{\text{MAG content}}{351}-\frac{\text{DAG content}}{610}\right]\times 10.2 \quad (2)$$

where the average molecular weights of fatty acid, MAG, DAG, and TAG were 276, 351, 610, and 869, respectively.

Assumption 2 TAG is synthesized from the main component of the FFA in 1 g of the high FFA RBO (Eq. 3).

$$\text{Glycerol (g)}=\frac{92}{3}\times\frac{\text{AV}}{1000}\times\frac{1}{56.1} \quad (3)$$

(2) Estimation of water amount (delta W): (Eq. 4)

$$\Delta W\text{ (g)}=18\times\text{substrate amount}\left(\Delta[\text{FFA}]+\Delta[\text{H}_2\text{O}]\right) \quad (4)$$

where Δ[FFA] and $\Delta[H_2O]$ are differences of concentration (mol/g) from product to substrates.

(3) Estimation of apparent reaction rate:

The apparent average TAG production rate (μmol/(g^{-1} h^{-1})) was estimated by using the following equation (Eq. 5):

$$\frac{\Delta\text{TAG}}{\Delta t}=10^6\times\frac{\text{TAG}_1-\text{TAG}_2}{869}\times 100\times(t_1-t_2) \quad (5)$$

where TAG_1 and TAG_2 are the TAG content (%, *w*/w) at t_1 and t_2, respectively.

(4) Estimation of dehydration rate: (Eq. 6)

$$\text{Dehydration rate}=\Delta(\text{H}_2\text{O}+\text{FFA})/\Delta t \quad (6)$$

4.3.2 Other Acyl Acceptors

As glycerol and its ester derivatives as acyl acceptors lead to formation of partial glycerides that might increase risk for glycidyl ester (GEs) formation during deodorization process (Craft et al., 2012), monoalcohols as acyl acceptors, such as ethanol (Li et al., 2016, 2017a, b) and methanol (Li et al., 2017a, b), have been selected for enzymatic deacidification to reduce both FFA and partial glyceride content. As a result, final DAG and FFAs can be reduced to 0.05% and 0.09%, respectively, and the

TAG content increased to 99.86% before oil processing. Additionally, the esters (fatty acid methyl ester or ethyl esters) produced can then be easily removed by molecular distillation under 120°C for its comparative lower boiling point.

On the other hand, the quest for bioactive compounds from natural oil has been implemented by scientific research of these targeted lipids against a vast array of diseases in their food application. In view of the need for desirable bioactive lipids production, phytosterol (Wang et al., 2016; Zheng et al., 2015) was proposed in the last few years to be utilized in the FFA-reducing process to obtain phytosterol ester-rich oil (>84.7 wt% by Zheng or 29.3 wt% by Wang) at a relatively high reaction rate. The final oil not only has higher nutritional value but also has an extended shelf life. However, the high melting point of phytosterol and its low solubility in both water and edible oils makes it inconvenient in application. Recently, the improved studies on ethanolamine-based enzymatic amidation reaction were driven by Xiaosan Wang et al. that gives an alternative way to get better deacidifying in much shorter time under optimal reaction conditions; 2% of Lipozyme 435, about 1:1 mass ratio of oil to solvent, and at 76°C, acid value was reduced from 21.5 to 1.6 mg/g after 4 h reaction, and rich fatty acid ethanolamides (11.9 wt%) were obtained (Li et al., 2017a, b).

4.4 Enzymatic Deacidification Process Development

It is well known that batch operation and continuous operation are the most commonly used process operations. For enzymatic deacidification, researchers analyzed the process based on batch operation for easy control. Under batch operation, less complicated operation control is involved, and regardless of the process, the focused parameters were similar: temperature range of high enzyme activity, certain vacuum, 2–5 wt% on oil enzyme loading, stoichiometric or excess substrate loading, and reaction time usually lasting longer than 4 h. Meanwhile, to maximize esterification under enzyme catalysis, the water produced was drawn off with various measures online, such as spontaneous evaporation, molecular sieves, vacuum, and dry air bubbling. For further variable optimization, a mathematical model—the response surface technique (RSM)—is usually proposed for optimum variable determination. Bhattacharyya and Bhattacharyya (1989) studied the effects of enzymes and glycerol concentrations using the classical method, and found that the optimum time and temperature were 4 h and 70°C, respectively (Ajay, 1992).

Compared to batch operation, continuous operation refers to a sophisticated design and plays a key role on process industrialization. In 1994, Kosugi et al. raised a set of continuous processes, making continuous and consecutive conversion of FFA in RBO to TAG (Kosugi et al., 1994). The continuous process is mainly comprised of two paths, one for esterification and the other for a dehydration-dedicated nitrogen line. When the process starts, glycerol and high FFA RBO are pumped from the glycerol vessel and high FFA RBO vessel separately. After dry nitrogen dehydration, the glycerol and RBO are further charged to take esterification in fixed-bed reactors, which are filled with immobilized lipase. During the reaction, water is removed mainly by nitrogen bubbling in dehydrators, and the used hydrated nitrogen can be regenerated in a molecular-sieve column. In this process design, the controller parameter is a two-moisture level for DAG and TAG formation purpose. It is at 110–160 ppm in the first loop (the first fixed-bed reactor and dehydrator), mainly for DAG formation, and the second loop (the second reaction and dehydrator) at 40–70 ppm, mainly for TAG formation. The product flow is 4.31 g/h with the balance of total glycerol 0.228 and RBO 4.2 g/h, and the mean residence time takes the first loop 5.7 h and second loop 22.6 h, respectively. Under this process, the FFA can be reduced from 41.4% to 11.8% (24 h) and 12.8% (48 h) or from 27.6% to 5.8%–9.7%.

4.5 Enzyme Reusability for Industrialization Purpose

It is widely recognized that the comparatively high cost of enzymes is the chief barrier to industrialization, and the application of immobilized enzyme reusability is the usual approach in the design of an enzyme biocatalyst to solve the issue concerning enzyme cost-lowering in industrialization.

From this purpose, a great effort has been devoted to enzyme reusability. Xiaosan Wang et al. reported that Lipozyme RM IM showed excellent operational stability in the esterification of phytosterol with FFA; 80% of its original activity was retained after being used for 10 batches, and FFA conversion was 92.5% in batch 1, then reduced to 76.8% after 11 batches (Wang et al., 2016). Other literature also gives excellent results: Zhihua Song et al. found Lipozyme RM IM could be used nine times under its optimum condition, with 90% of its original catalytic activity still retained after repeated batches (optimum condition: MG/RBO, 0.25; enzyme load, 4.77%; temperature, 56°C; and reaction time, 5.75 h). Under the suggested conditions, the respective content of FFA and DAG was 0.28% and 27.98%, respectively (Song et al., 2012). Recently, Daoming Li et al. proposed immobilized SMG1-F278N as a highly stable enzyme that could be used for 20 consecutive cycles without significant loss in activity (reaction conditions: ethanol/Fas, 1.5 (mole ratio); enzyme load, 40U/g [U/w, with respect to oil mass]; and temperature, 30°C). A final product with 99.86% TAG, 0.05% DAG, and 0.09% FFA was obtained (Li et al., 2017a, 2017b). Lu et al. set up a process of continuous packed-bed reactor (PBRs) system that, when combined with continuous deacidification and dehydration, the lipase used had good stability in 12 continuous operations with 83.22% of its original ability (Lu et al., 2016). Mingming Zheng et al. developed a lipase of CRL immobilized on C8-OMSs arrays; it retained 80.7% of its original catalytic activity after use 11 times (Zheng et al., 2015). In spite of the factors on enzyme activity of unfavorable conditions, such as water content, operating temperature, and agitation causing structure deformation, lowering its activity, and after-use recovery, enzyme reusability strategy has a promising potentiality for industrial applications.

4.6 Section Summary

In summary, enzyme deacidification preserved functional compound at high efficiency on deacidifying, and it has relatively strong operability, high reliable feasibility, and is economical. Furthermore, its chief limitation for industrialization, which is caused by the high cost of enzymes, can be solved by enzyme reusability. Hence, in times of a growing interest in low environmental factors and nutrient-rich requirements, this enzymatic deacidifying technology will contribute to the green production of nutritive RBO in the future.

5. ENZYMATIC INTERESTERIFICATION OF RICE BRAN OIL

5.1 Introduction

EIE is an enzymatic catalysis reaction, which can rearrange fatty acids on the glycerol backbone of TAG. It is one modification process applied in the edible oil industry to change the physicochemical properties of lipids for certain food products. Interesterification (IE) was developed initially to replace hydrogenation to produce trans-free margarine fat products.

The principle of the IE can be demonstrated in Fig. 5. IE is attributed to the ester-ester exchange, which resulted in the rearrangement of the position of fatty acids on the glycerol backbone. According to the catalyst difference,

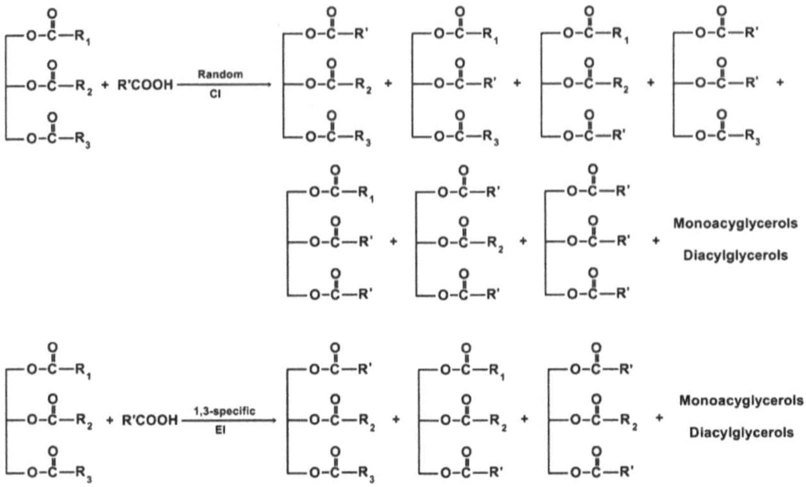

Fig. 5 Principle of IE reaction under chemical and enzymatic catalysts (Norizzah et al., 2018).

IE can be classified into chemical and enzymatic category. Chemical interesterification (CIE) exchanges the position of the fatty acids randomly (da Silva et al., 2010). In contrast, EIE can rearrange FFAs either randomly or region-selectively depending on the enzyme specificity. Some lipases exhibit stereo-specificity by working selectively to modify fatty acids in the position of TAGs (Speranza and MacEdo, 2012). Hence, EIE has certain advantages over CIE, which imply the potential nutritional benefit as maintaining of FFAs on *sn*-2 position (Farfán et al., 2013). As concern about environment and health issues increases, EIE is preferred in trans-free plastic fat production and has been greatly developed in recent years.

5.2 Industrialization in Food Industry

The aim of IE is to modify the melting properties of a fat or fat blend to produce a fat suitable for incorporation into margarine or other food fat product (Husum et al., 2004). It was developed as an alternative method of hydrogenation to produce trans-free plastic fat and used in industrial application since the 1940s (Xu et al., 2006).

Initially, the CIE was first industrialized due to its low price and short reaction time. And sodium methylate is mostly applied for CIE in factories

(Karabulut et al., 2004). However, the use of sodium methylate on an industrial scale has disadvantages. It is a strong basic chemical that needs careful operation. And the sodium ions have to be removed from the product by posttreatments, which adds to the cost and procedure time. On the other hand, the delay of EIE industrialization in food production can be caused by the high cost of immobilized lipase.

In recent years, with the developed commercially available immobilized lipase, there are great opportunities to produce cost-effective plastic fat products in a cost-effective process with an EIE approach. Compared with CIE, EIE reaction can be conducted at milder conditions (60–70°C), easier to control, and generating fewer byproducts. Because of the chemical catalysts and harsh reaction conditions, unwanted byproducts are produced after CIE process. Compared with EIE, CIE posttreatment is more complicated, including washing oil, drying, bleaching, and deodorization (Husum et al., 2004).

EIE as an edible oil and fat modification process has been studied for many years. And in the 1980s, Unilever has obtained a patent to produce plastic fat with lipase catalysis (Dayton, 2012). Afterward, ADM produced commercial margarine with the continuous EIE process, which launched in the U.S. market in 2002.

Nowadays, the commercialized immobilized enzymes (IMs) have been widely applied in the EIE. There are several types of immobilized lipases including Lipozyme TL IM, Lipozyme RM IM, Novozym 435, Lipase PS-C-I, and Lipase PS-D-I, which can be classified into randomly and specific clusters, are studied in EIE. Specific Lipozyme TL IM with quite high production efficiency is widely used in both academic and industrial EIE reaction. Lipozyme TL IM has been given particular concern in this study due to its activity and potential low cost in comparison with other lipases used, which was are discussed in detail in previous sections. This provides a potential possibility for industrial implementation. Meanwhile, Lipozyme RM and Novozym 435 are also used for some desired products in EIE research. The IMs information are listed in Table 3 (Rønne et al., 2005).

The existing industrialized EIE is operated in packed-bed continuous reactor. Immobilized lipases are packed into a vertical column. The blended oil is fed into the top of the reactor and continuously removed from the bottom though the filter screen. The degree of IE can be controlled by flow rate (Husum et al., 2004). The batch EIE reaction is only applied in pilot plant scale; further industrialization still need to be developed.

Table 3 The immobilized lipases used in EIE research and their characteristics

Name	Lipase species	Carrier	Specificity	Supplier
Lipozyme TL IM	*Thermomyces lanuginosus*	Silica granules	*sn*-1,3-specific	Novozymes A/S
Lipozyme RM IM	*Rhizomucor miehei*	Macroporous resin	*sn*-1,3-specific	Novozymes A/S
Novozym 435	*Candida antarctica lipase B*	Macroporous polymer based on methyl and butyl methacrylic esters	Nonspecific	Novozymes A/S
Lipase PS-C-I	*Burkholderia cepacia*	Ceramic particles	Nonspecific	/
Lipase PS-D-I	*Burkholderia cepacia*	Diatomaceous earth	Nonspecific	Amano Enzyme Inc.

5.3 EIE Application in RBO

RBO with its superior nutritional properties is a popular oil product in recent years. Crude RBO is extremely rich in bioactive phytochemicals such as γ-oryzanol, phytosterols, and tocols with proven health benefits. γ-Oryzanol is unique to RBO and is a potent hypocholesterolemic agent. The tocotrienols that comprise 70% of the tocols in crude RBO have been shown to be powerful antioxidants and antithrombotic agents (Arumughan et al., 2004).

The use of RBO in production of value-added products like margarines has been reported where EIE carried out with different kind oils such as palm oil, palm stearin (Neeharika et al., 2015). And in aspect of the process, there is no difference with general procedure compared with EIE applied in other vegetable oils.

Reshma et al. (2008) reported that the EIE products retained bioactive phytochemicals of tocols (839–1172 ppm), sterols (4318–9647 ppm), oryzanols (3000–6800 ppm), and carotene (121–180 ppm), which means there is no significant changes of micronutrients before and after EIE reaction. The same conclusion has also been reported in a study by Neeharika et al. (2015).

Meanwhile, RBO also provided a desirable fatty acid profile in terms of polyunsaturated fatty acids (PUFA), consequently reducing the saturated fatty acid in the blend, besides its contribution of bioactive phytochemicals to the blend. And there are no changes in FFA composition and no trans-fatty acids (TFAs) produced in EIE products, hence the functional FFAs can be introduced to get functional structured lipids (SLs) products. RBO thus could be an ingredient for nutritional shortenings (Reshma et al., 2008).

EIE is also used to achieve SLs, which are lipids that have been modified from their natural form for specific food and nutraceutical applications (Jennings and Akoh, 2009). Jianping et al. (Hou and Chang, 2016) also reported that the RBO is suitable for production of human milk fatty acid substitutes in infant formulas. Meanwhile, RBO provides the potential to improve the product's shelf life as its higher oxidation stability other than normal vegetable oils.

Therefore EIE as an ecofriendly approach applied in the utilization of RBO and can exploit its intrinsic nutritional properties for making margarine fat and SLs.

5.4 Section Summary

EIE is a new way to modify edible oil and has been greatly developed in industrialization in recent years. Compared with traditional modification processes, such as hydrogenation and CIE, EIE can not only produce trans-free fats products but also does it in an ecofriendly and cost efficient way. Meanwhile, the commercialization of immobilized lipases provides a favorable precondition for EIE development. Hence, it must be a promising process to produce margarine fat and PLs in future.

6. CHAPTER SUMMARY

Due to the development of modern biotechnology and the needs for milder processes and lower chemical and energy consumption, enzymes have gained more and more interest in the oil and fat industry. RBO is a popular nutritional edible oil that is rich in valuable nutrients such as γ-oryzanol, and has unusually high content of FFA, wax, unsaponifiable constituents, polar lipids, and pigments, making it an interesting oil for bioprocessing. The application of enzymes in RBO processing ranges from upstream raw material preservation to oil refining to downstream oil modification.

Although different enzymes are applied in each individual process, the main superiority of using enzymatic method is either reduced oil loss, better preservation of functional nutrients, or reduced byproduct, besides its reduced environmental impact. Amongst these bioprocesses, enzymatic degumming is the most mature process applied in oil processing, and its application in RBO refining is already state-of-the-art technology. Comparably, the industrial application of enzymatic deacidification and IE for RBO will largely benefit from the development of the enzyme immobilization technique. As development of enzyme techniques and the popularization of RBO progresses, bioprocessing will be applied more and more extensively in RBO processing in the future.

REFERENCES

Ajay, L., 1992. Lipase catalyzed deacidification of high free fatty acid rice bran oil. Biotechnol. Tech. 6 (2), 169–172.

Arumughan, C., Skhariya, R., Arora, R., 2004. Rice bran oil: an untapped health food. INFORM 15 (11), 706–707. Retrieved from:http://www.scopus.com/inward/record.url?eid=2-s2.0-10244263482&partnerID=40&md5=ef3230cedbe8cacfddfc383eba1917d1.

Bhattacharyya, S., Bhattacharyya, D.K., 1989. Biorefining of high acid rice bran oil. J. Am. Oil Chem. Soc. 66 (12), 1809–1811. https://doi.org/10.1007/BF02660752.

Bhosle, B.M., Subramanian, R., 2005. New approaches in deacidification of edible oils—a review. J. Food Eng. 69 (4), 481–494. https://doi.org/10.1016/j.jfoodeng.2004.09.003.

Craft, B.D., Nagy, K., Seefelder, W., Dubois, M., Destaillats, F., 2012. Glycidyl esters in refined palm (*Elaeis guineensis*) oil and related fractions. Part II. Practical recommendations for effective mitigation. Food Chem. 132 (1), 73–79. https://doi.org/10.1016/j.foodchem.2011.10.034.

da Silva, R.C., Soares, D.F., Lourenço, M.B., Soares, F.A.S.M., da Silva, K.G., Gonçalves, M.I.A., Gioielli, L.A., 2010. Structured lipids obtained by chemical interesterification of olive oil and palm stearin. LWT Food Sci. Technol. 43 (5), 752–758. https://doi.org/10.1016/j.lwt.2009.12.010.

Dayton, C.L.G., 2012. Enzymatic interesterification. In: Green Vegetable Oil Processing, pp. 217–237. https://doi.org/10.1201/9781420046649.ch30.

Dayton, C.L.G., Galhardo, F., 2012. Enzymatic Degumming. Green Vegetable Oil Processing. https://doi.org/10.1016/B978-0-9888565-3-0.50009-1.

Dayton, C.L.G., Galhardo, F., 2015. Enzymatic degumming utilizing a mixture of PLA and PLC phospholipases. US08956853.

De, B.K., 2006. Comparison of bio- and autocatalytic esterification of oils using mono- and diglycerides. JAOCS—J. Am. Oil Chem. Soc. 83 (5), 443–448. https://doi.org/10.1007/s11746-006-1224-9.

De Maria, L., Vind, J., Oxenbøll, K.M., Svendsen, A., Patkar, S., 2007. Phospholipases and their industrial applications. Appl. Microbiol. Biotechnol. 74 (2), 290–300. https://doi.org/10.1007/s00253-006-0775-x.

Dijkstra, A.J., 2010. Enzymatic degumming. Eur. J. Lipid Sci. Technol. 112, 1178–1189. https://doi.org/10.1016/B978-0-9888565-3-0.50009-1.

Dijkstra, A.J., 2011. Enzymatic degumming. Lipid Technol. 23 (2), 36–38. https://doi.org/10.1002/lite.201100085.

Farfán, M., Villalón, M.J., Ortíz, M.E., Nieto, S., Bouchon, P., 2013. The effect of interesterification on the bioavailability of fatty acids in structured lipids. Food Chem. 139 (1–4), 571–577. https://doi.org/10.1016/j.foodchem.2013.01.024.

Ghosh, M., 2007. Review on recent trends in rice bran oil processing. J. Am. Oil Chem. Soc. 84 (4), 315–324. https://doi.org/10.1007/s11746-007-1047-3.

Hammond, N.A., Cordova, T., 1994a. Stabilizing rice bran and rice bran products. US005292537A.

Hammond, N.A., Cordova, T., 1994b. Method for stabilizing rice bran and rice bran products. US005376390A.

Hammond, N.A., Cordova, T., 1996. Method for stabilizing rice bran and rice bran products. US005512307A.

Hammond, N.A., Cordova, T., 1998. Method for stabilizing rice bran and rice bran products. US005753283A.

Hou, J., Chang, G., 2016. Applications of rice bran oil in infant formulas. Cereal Food Ind. 23 (1), 33–40.

Huang, S., Liang, M., Xu, Y., Rasool, A., Li, C., 2014. Characteristics and vegetable oils degumming of recombinant phospholipase B. Chem. Eng. J. 237, 23–28. https://doi.org/10.1016/j.cej.2013.09.109.

Husum, T.L., Pedersen, L.S., Nielsen, P.M., Christensen, M.W., Kristensen, D., Holm, H.C., 2004. Enzymatic interesterification: process advantage and product benefits. Palm Oil Dev. 15 (3), 7–10.

Jahani, M., Alizadeh, M., Pirozifard, M., Qudsevali, A., 2008. Optimization of enzymatic degumming process for rice bran oil using response surface methodology. LWT Food Sci. Technol. 41 (10), 1892–1898. https://doi.org/10.1016/j.lwt.2007.12.007.

Jennings, B.H., Akoh, C.C., 2009. Effectiveness of natural versus synthetic antioxidants in a rice bran oil-based structured lipid. Food Chem. 114 (4), 1456–1461. https://doi.org/10.1016/j.foodchem.2008.11.031.

Jiang, F.Y., Wang, J.M., Ju, L.C., Kaleem, I., Dai, D.Z., Li, C., 2011. Optimization of degumming process for soybean oil by phospholipase B. J. Chem. Technol. Biotechnol. 86 (8), 1081–1087. https://doi.org/10.1002/jctb.2621.

Jiang, X., Chang, M., Jin, Q., Wang, X., 2015. Application of phospholipase A1and phospholipase C in the degumming process of different kinds of crude oils. Process Biochem. 50 (3), 432–437. https://doi.org/10.1016/j.procbio.2014.12.011.

Kaimal, T. N. B., Vali, S. R., Rao, B. V. S. K., Chakrabarti, P. P., Vijayalakshmi, P., Kale, V., et al. (2002). Origin of problems encountered in rice bran oil processing. Eur. J. Lipid Sci. Technol., 104(4), 203–211. https://doi.org/10.1002/1438-9312(200204)104:4<203::AID-EJLT203>3.0.CO;2-X.

Karabulut, I., Turan, S., Ergin, G., 2004. Effects of chemical interesterification on solid fat content and slip melting point of fat/oil blends. Eur. Food Res. Technol. 218 (3), 224–229. https://doi.org/10.1007/s00217-003-0847-4.

Kosugi, Y., Kunieda, T., Azuma, N., 1994. Continual conversion of free fatty acid in rice bran oil to triacylglycerol by immobilized lipase. J. Am. Oil Chem. Soc. 71 (4), 445–448. https://doi.org/10.1007/BF02540528.

Li, D., Faiza, M., Ali, S., Wang, W., Tan, C.P., Yang, B., Wang, Y., 2017a. Highly efficient deacidification of high-acid rice bran oil using methanol as a novel acyl acceptor. Appl. Biochem. Biotechnol., 1–12. https://doi.org/10.1007/s12010-017-2594-1.

Li, D., Liu, P., Wang, W., Wang, X., Yang, B., Wang, Y., 2017b. An innovative deacidification approach for producing partial glycerides-free rice bran oil. Food Bioprocess Technol. 10 (6), 1154–1161. https://doi.org/10.1007/s11947-017-1896-1.

Li, D., Wang, W., Durrani, R., Li, X., Yang, B., Wang, Y., 2016. Simplified enzymatic upgrading of high-acid rice bran oil using ethanol as a novel acyl acceptor. J. Agric. Food Chem. 64 (35), 6730–6737. https://doi.org/10.1021/acs.jafc.6b02518.

List, G.R., 2009. Bleaching and Purifying Fats and Oils: Theory and Practice. Bleaching and Purifying Fats and Oils: Theory and Practice. https://doi.org/10.1016/C2015-0-02456-X.

Lu, Y., Zou, X., Han, W., Jiang, Y., Jin, Q., Li, L., et al., 2016. Preparation of diacylglycerol-enriched Rice bran oil by lipase-catalyzed deacidification in packed-bed reactors by continuous dehydration. J. Oleo Sci. 65 (2), 151–159. https://doi.org/10.5650/jos.ess15238.

Manjula, S., Jose, A., Divakar, S., Subramanian, R., 2011. Degumming rice bran oil using phospholipase-A1. Eur. J. Lipid Sci. Technol. 113 (5), 658–664. https://doi.org/10.1002/ejlt.201000376.

Neeharika, T.S.V.R., Rallabandi, R., Ragini, Y., Kaki, S.S., Rani, K.N.P., Prasad, R.B.N., 2015. Lipase catalyzed interesterification of rice bran oil with hydrogenated cottonseed oil to produce trans free fat. J. Food Sci. Technol. 52 (8), 4905–4914. https://doi.org/10.1007/s13197-014-1563-8.

Norizzah, A.R., Nur Azimah, K., Zaliha, O., 2018. Influence of enzymatic and chemical interesterification on crystallisation properties of refined, bleached and deodourised (RBD) palm oil and RBD palm kernel oil blends. Food Res. Int. 106, 982–991. https://doi.org/10.1016/j.foodres.2018.02.001.

Panesar, P.S., Sharma, H.K., Sarkar, B.C., 2011. Enzymatic degumming of crude rice bran oil. Bio-Process. Foods 79 (8), 99.

Prabhakar, J.V., Venkatesh, K.V.L., 1986. A simple chemical method for stabilization of rice bran. J. Am. Oil Chem. Soc. 63 (5), 1983–1985.

Reshma, M.V., Saritha, S.S., Balachandran, C., Arumughan, C., 2008. Lipase catalyzed interesterification of palm stearin and rice bran oil blends for preparation of zero trans shortening with bioactive phytochemicals. Bioresour. Technol. 99 (11), 5011–5019. https://doi.org/10.1016/j.biortech.2007.09.009.

Rønne, T.H., Yang, T., Mu, H., Jacobsen, C., Xu, X., 2005. Enzymatic interesterification of butterfat with rapeseed oil in a continuous packed bed reactor. J. Agric. Food Chem. 53 (14), 5617–5624. https://doi.org/10.1021/jf050646g.

Sengar, G., Kaushal, P., Sharma, H.K., Kaur, M., 2014. Degumming of rice bran oil. Rev. Chem. Eng. 30 (2), 183–198. https://doi.org/10.1515/revce-2013-0030.

Sheelu, G., Kavitha, G., Fadnavis, N.W., 2008. Efficient immobilization of lecitase in gelatin hydrogel and degumming of rice bran oil using a spinning basket reactor. JAOCS—J. Am. Oil Chem. Soc. 85 (8), 739–748. https://doi.org/10.1007/s11746-008-1261-7.

Singhania, T., Sinha, H., Das, P., Mukherjee, A.K., 2015. Effi cient degumming of rice bran oil by immobilized PLA1 from *Thermomyces lanuginosus*. Food Technol. Biotechnol. 53 (1), 91–95. https://doi.org/10.17113/ftb.53.01.15.3740.

Song, Z., Liu, Y., Jin, Q., Li, L., Wang, X., Huang, J., Liu, R., 2012. Lipase-catalyzed preparation of diacylglycerol-enriched oil from high-acid rice bran oil in solvent-free system. Appl. Biochem. Biotechnol. 168 (2), 364–374. https://doi.org/10.1007/s12010-012-9780-y.

Speranza, P., MacEdo, G.A., 2012. Lipase-mediated production of specific lipids with improved biological and physicochemical properties. Process Biochem. 47 (12), 1699–1706. https://doi.org/10.1016/j.procbio.2012.07.006.

Su, L., Ji, D., Tao, X., Yu, L., Wu, J., Xia, Y., 2017. Recombinant expression, characterization, and application of a phospholipase B from *Fusarium oxysporum*. J. Biotechnol. 242, 92–100. https://doi.org/10.1016/j.jbiotec.2016.12.009.

Wang, X., Lu, J., Liu, H., Jin, Q., Wang, X., 2016. Improved deacidification of high-acid rice bran oil by enzymatic esterification with phytosterol. Process Biochem. 51 (10), 1496–1502. https://doi.org/10.1016/j.procbio.2016.08.013.

Wang, X., Wang, X., Wang, T., 2017. An effective method for reducing free fatty acid content of high-acid rice bran oil by enzymatic amidation. J. Ind. Eng. Chem. 48, 119–124. https://doi.org/10.1016/j.jiec.2016.12.028.

Wei, T., Xu, C., Yu, X., Jia, W., Yang, K., Jia, C., Mao, D., 2015. Characterization of a novel thermophilic phospholipase B from *Thermotoga lettingae* TMO: applicability in enzymatic degumming of vegetable oils. J. Ind. Microbiol. Biotechnol. 42 (4), 515–522. https://doi.org/10.1007/s10295-014-1580-7.

Xu, X., Guo, Z., Zhang, H., Vikbjerg, A.F., Damstrup, M.L., 2006. Chemical and Enzymatic Interesterification of Lipids for Use in Food. Modifying Lipids for Use in Food. https://doi.org/10.1533/9781845691684.2.234.

Yang, B., Zhou, R., Yang, J.G., Wang, Y.H., Wang, W.F., 2008. Insight into the enzymatic degumming process of soybean oil. JAOCS—J. Am. Oil Chem. Soc. 85 (5), 421–425. https://doi.org/10.1007/s11746-008-1225-y.
Yu, D., Ma, Y., Xue, S.J., Jiang, L., Shi, J., 2013. Characterization of immobilized phospholipase A 1 on magnetic nanoparticles for oil degumming application. LWT Food Sci. Technol. 50 (2), 519–525. https://doi.org/10.1016/j.lwt.2012.08.014.
Zhao, K., Cao, X., Di, Q., Wang, M., Cao, H., Deng, L., et al., 2017. Synthesis, characterization and optimization of a two-step immobilized lipase. Renew. Energy 103, 383–387. https://doi.org/10.1016/j.renene.2016.11.035.
Zheng, M., Zhu, J., Huang, F., Xiang, X., Shi, J., Deng, Q., et al., 2015. Enzymatic deacidification of the rice bran oil and simultaneous preparation of phytosterol esters-enriched functional oil catalyzed by immobilized lipase arrays. RSC Adv. 5 (86), 70073–70079. https://doi.org/10.1039/c5ra11533g.

FURTHER READING

Chakrabarti, P.P., Rao, B.V.S.K., Roy, S.K., Devi Bethala, L.A.P., 2004. Process for the pre-treatment of vegetable oils for physical refining. US20040005399.

CHAPTER 5

Micronutrients in Rice Bran Oil

Riantong Singanusong*,†, Umar Garba†,‡
*Centre of Excellence in Fats and Oils, Faculty of Agriculture, Natural Resources and Environment, Naresuan University, Phitsanulok, Thailand
†Department of Agro-Industry, Faculty of Agriculture, Natural Resources and Environment, Naresuan University, Phitsanulok, Thailand
‡Department of Food Science and Technology, Faculty of Agriculture, Bayero University, Kano, Nigeria

1. INTRODUCTION

Rice bran is a byproduct of rice milling industries, obtained after polishing of brown rice (Sengar et al., 2014). The bran represents 8–11 wt% of the rice grain and contains 16%–22% oil (Pestana-Bauer et al., 2012). Rice bran oil (RBO) is mainly produced through solvent extraction or hydraulic pressing of rice bran (Kordsmeier et al., 2015). A crude RBO contains between 90%–96% saponifiable lipid components (mono-, di-, and triglycerides; free fatty acids; and waxes) and 3%–5% unsaponifiable (sterols, tocotrienols, triterpene alcohols) (Oliveira et al., 2011). Refined RBO is a very good source of micronutrients (Table 1) and natural antioxidants such as γ-oryzanol, tocopherols, and tocotrienols (Jaiswal et al., 2015; Kordsmeier et al., 2015); phospholipids or lecithin (Liu et al., 2013; Jala and Prasad, 2015); and phytosterols, squalene, and ferulic acid (Huang et al., 2009; Sakunpak et al., 2014; Sawadikiat and Hongsprabhas, 2014). RBO has been widely used for food, pharmaceuticals, and cosmetics applications because of the health benefits of its abundant micronutrients (Kaur et al., 2012; Pali, 2013; Sengar et al., 2014; Sohail et al., 2017).

Micronutrients are needed in the body in small quantities for normal functioning and development of the body (Lohry, 2007). They are popularly known as bioactive compounds in RBO, mostly concentrated in the unsaponifiable fraction of the oil. This fraction consists of 43% phytosterols, 10% steryl esters, and 1% tocopherol (Dunford and King, 2000). Most of these micronutrients are often lost or reduced to a large degree during oil refining processes. They become concentrated in a number of byproducts or residues such as gum and soap stock from chemical refining steps, deodorizer distillate, and wax (Fig. 1) from both physical and chemical refining (Sawadikiat and Hongsprabhas, 2014; Sengar et al., 2014). For example, 90% of tocopherols and tocotrienols are lost during the deodorization process (Kordsmeier

Rice Bran and Rice Bran Oil
https://doi.org/10.1016/B978-0-12-812828-2.00005-6

Table 1 Typical profile of micronutrients in RBO and its by products

Micronu. trient (%)	Crude RBO	Refine RBO	Soap stock	Acid oil	Deodorizer distillate	Gum
Oryzanol	1.7–2.1 (1)	1.2–1.7 (2)	6.3–6.9 (1)	3.3–7.4 (1)	0.79 (1)	0.0029 (1)
Tocopherol	0.0264 (1)	0.02–0.08 (2)	0.0039 (1)	–	0.576 (1)	0.0045 (1)
Tocotrienols	0.007 (6)	0.025–0.17 (2)	–	–	0.05 (6)	–
Phytosterols	1.362–1.37 (4)	0.858–1.037 (4)	–	–	8.5 (3)	–
Squalene	0.756 (5)	0.3–0.4 (2)	–	–	3.5 (3)	–
Phospholipids	4.2 (5)	1.3 (2)	–	–	–	–

Data from: (1) Pestana-Bauer et al. (2012), (2) Pal and Pratap (2017), (3) Yukihiro and Hara (2012), (4) Sawadikat and Hongsprabhas (2014), (5) Rao (2000), and (6) Arumughan et al. (2005).

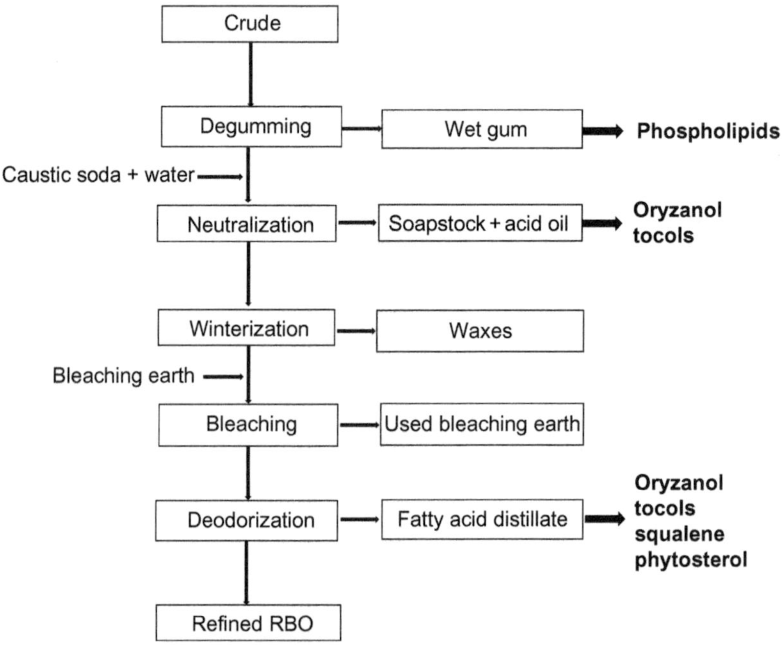

Fig. 1 Flowchart for RBO refining displacing the various sources of micronutrients.

et al., 2015; Howard, 2016), and 83%–89% of γ-oryzanol (i.e., from 1.1%–1.74% oryzanol content in RBO) is lost in chemical refining (Krishna et al., 2001). Patel and Naik (2004) stated that during alkali treatment of crude RBO, up to about 93%–94.6% of oryzanol can be removed while only 1.1% and 5.9% oryzanol was lost during degumming and dewaxing of crude

RBO, respectively. The deodorization distillate typically contains squalene (8.5%), phytosterols (3.5%, containing sitosterol 2.05%, stigmasterol 0.69%, and campesterol 0.76%), tocopherol (1.6%), triacylglycerol (58.0%), and diacylglycerol + fatty acids (26.5%) (Yukihiro and Hara, 2012).

Several methods such as solvent extraction and crystallization, supercritical fluid extraction (SFE), saponification, esterification, and molecular distillation are used to recover these bioactive compounds (Sawadikiat and Hongsprabhas, 2014; Sengar et al., 2014). In this chapter, the most important of these micronutrients, such as oryzanol, tocopherols, tocotrienols, phytosterols, and phospholipids, as well as their benefits, will be discussed.

2. GAMMA-ORYZANOL

Gamma-oryzanol is one of the important micronutrients of RBO with the potential to be used in pharmaceuticals, cosmeceutical, and nutraceutical industries (Joshi et al., 2016). It was mostly extracted from the soap stock byproduct of the RBO refining process (Pal and Pratap, 2017). Other important sources are rice bran acid oil and biodiesel residues (Zullaikah et al., 2009). The unsaponifiable fraction of RBO (which is 4.2%) contains 1.6% γ-oryzanol (Pali, 2013; Al-Okbi et al., 2014). Ghatak and Panchal (2012) reported that the γ-oryzanol content in crude RBO can range from 1.5% to 2.9%. A chemically refined RBO contains 0.19%–0.20% γ-oryzanol, or 248–887 mg/100 g of oil as reported by Sawadikiat and Hongsprabhas (2014). On the other hand, a physically refined RBO has a considerably higher γ-oryzanol of 1.1%–1.74% (Zullaikah et al., 2009). The complete oryzanol is known to be unique to RBO, although its exact composition may differ depending on the rice cultivar (Al-Okbi et al., 2014); long and medium grain rice was reported to contain 6.4 and 5.17 mg/g, respectively (Orthoefer, 2005).

Most of the γ-oryzanol lost during the RBO refining process finds its way into different byproducts in high concentration. Among the various refining steps, degumming, soap precipitation, bleaching earth filtering, dewaxing, and deodorization distillation produced byproduct with γ-oryzanol (Pestana-Bauer et al., 2012). Degumming process removed about 1.1% of oryzanol, whereas during dewaxing, 5.9% oryzanol is removed from crude RBO. During alkali treatment, about 93%–94.6% oryzanol is removed from original crude oil (Zullaikah et al., 2009). The soap precipitation was found to be the best source of γ-oryzanol with 43.1 mg/g representing 11.5% of total γ-oryzanol in crude RBO (Pestana-Bauer et al.,

2012). Depending on the refining process, the oryzanol content of physically refined RBO remained 1.1%–1.74% as original, and reduced to 0.19%–0.20% for chemically refined RBO (Nagendra et al., 2015).

2.1 Chemical Structure

The γ-oryzanol was originally considered a single component until it was later found to be a mixture of steryl ferulates (Orthoefer, 2005). The molecular structure of oryzanol contains a ferulic acid aromatic phenolic nucleus, which is esterified to cyclopentanoperhydrophenanthrene (Ghatak and Panchal, 2012). The γ-oryzanol consisted of a mixture of ferulic acid ester of sterol and triterpene alcohols such as 24-methyelene cycloartenol (494 mg%) and cycloartenol (106 mg%) (Al-Okbi et al., 2014; Patel and Naik, 2004). The four major components in γ-oryzanol (Fig. 2) are cycloartenylferulate, 24-methylenecycloartenyl ferulate, campesterylferulate, and β-sitosterylferulate (Patel and Naik, 2004; Sakunpak et al., 2014).

2.2 Properties

Pure γ-oryzanol is in the form of a crystalline powder, white, odorless, with a molecular weight of 602.89 (Kangcare, 2012). The alcoholic group (-OH) in the ferulic acid portion of γ-oryzanol (C40-H58-O4) rendered its relatively high polarity and possible solubility in polar and nonpolar solvents (Xu and Godber, 2000). It is insoluble in water, but it becomes soluble following treatment with urea, nicotinamide, and/or thioctamide dissolved in acetone or a mixture of ethanol-acetone. It can also be dissolved in medicinal drinks by using sucrose fatty acid ester and ethoxylated hydrogenated castor oil (Patel and Naik, 2004).

2.3 Analysis, Extraction, and Purification

Isolation of γ-oryzanol from RBO has continued to receive much attention particularly in the major rice-producing countries such as India, China, Japan, Thailand, and the United States. Numerous patents on isolation of oryzanol are being filed globally (Narayan et al., 2006). Various methods were reported to be used for the extraction of γ-oryzanol from the byproducts including direct solvent extraction, liquid-liquid phase extraction, solid phase extraction, and supercritical fluid extraction (Xu and Godber, 2000). Methods of purification ranging from simple crystallization to

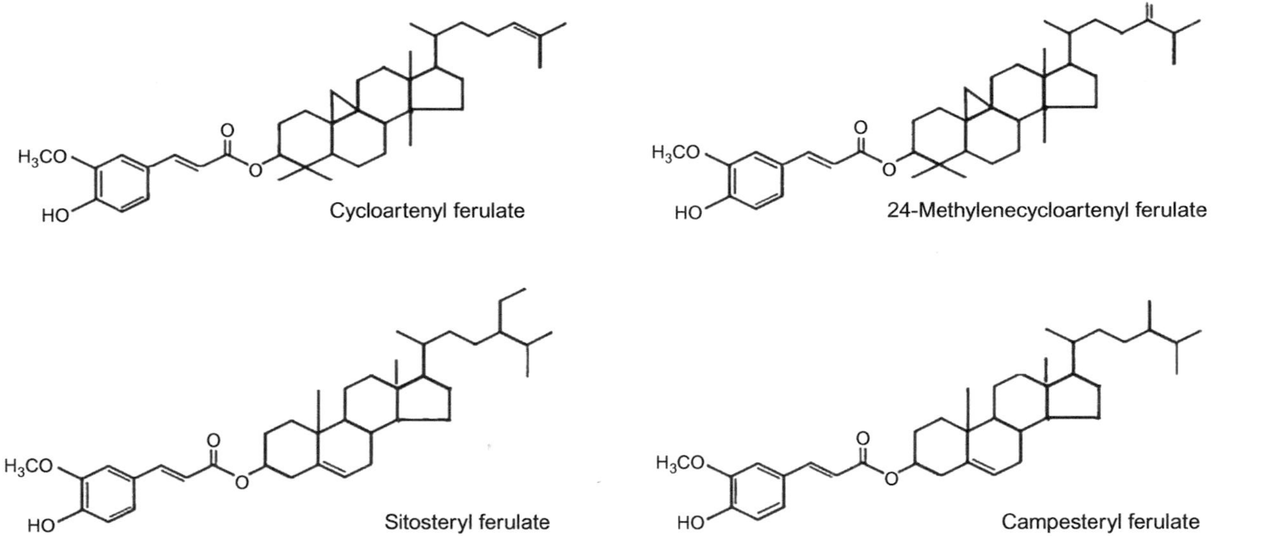

Fig. 2 Chemical structures of four major components of γ-oryzanol.

nonporous polymeric membranes have been employed to concentrate oryzanol by three-fold (Angelis et al., 2011). The critical parameters considered for extraction-based processes are solid to solvent ratio, temperature, and time. For crystallization or precipitation-based processes, the temperature of solvent or mixture of solvents with their proportion, rate of nucleation, and rate of crystal growth are important parameters to be considered (Zullaikah et al., 2009). High performance liquid chromatography (HPLC) and ultraviolet spectrophotometry (UV) are the most common techniques for quantification of γ-oryzanol content. The HPLC method has greater accuracy and efficiency, however the high cost involved is a disadvantage when compared to the later method, which is simple, inexpensive, and practical. Nevertheless, HPLC analysis of γ-oryzanol content is not always accurate because absorbance readings may be influenced by the matrix of sample components (Heidtmann-bemvenuti et al., 2012). Thongchai and Liawruangrath (2016) have reported the use of HPLC with molecular imprinted solid-phase extraction as another alternative for a rapid and easy method for separation and determination of γ-oryzanol.

Ruen-ngam et al. (2016) studied the extraction of γ-oryzanol from Khao Dawk Mali 105 rice bran (origin of Thailand) using different methods of pretreatments. Highest γ-oryzanol yield (9.8 mg/g dried rice bran) was obtained by parboiling at 70°C using hexane. Extraction of the γ-oryzanol from soap stock of RBO and further crystallization (at 4°C) and drying (at 80–140°C) was studied. A yield of 33.6% pure γ-oryzanol in a crystallized form was obtained (Joshi et al., 2016). A γ-oryzanol content of 13.98 mg/g was obtained by using 75 mL of hexane:isopropanol ratio (1:3) heated at 40°C for 40 min (Heidtmann-bemvenuti et al., 2012). Jesus et al. (2010) investigated the recovery of γ-oryzanol from fatty acid distillate of RBO soap stock, using supercritical fluid extraction (SFE) at pressure of 30 MPa and temperature of 303 K. A recovery rate and yield of 31.3% (w/w) and 3.2% (w/w), respectively, was obtained. A typical process for extraction of γ-oryzanol from soap stock is presented in Fig. 3.

Two-step crystallization process was also adopted to isolate oryzanol to a purity of 93%–95%, and recovery of 59% was achieved. This method involved first crystallization of 10 g RBO mixed with 400 mL methanol/acetone (ratio of 7:3 v/v) using magnetic stirring (at 700 rpm) for 1 h at room temperature (20.5°C) and stored at −60°C. The solid phase was separated by rotary evaporation (at 47.5°C and 550 mmHg) and used for second phase crystallization. In the second phase, the solid crystals from the first phase were kept at room temperature for 72 h and then washed with 50 mL hexane

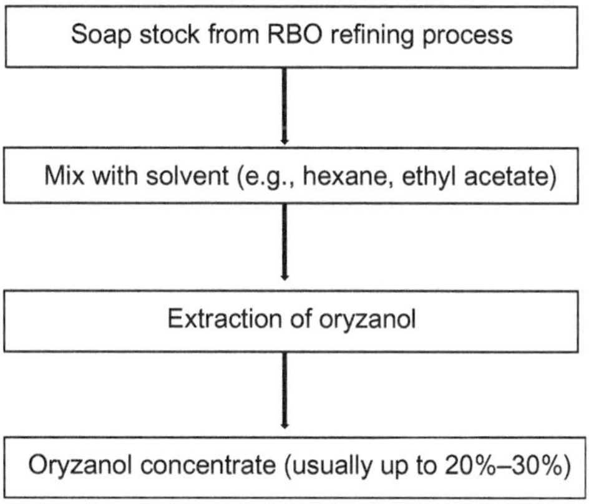

Fig. 3 A typical process for extraction of γ-oryzanol from soap stock.

and finally separated using filter paper (Zullaikah et al., 2009). In a similar method for isolation and purification of oryzanol using two-step crystallization, white oryzanol crystals with purity and recovery of 95.03% and 55.61%, respectively, were obtained (Ghatak and Panchal, 2012). Kaewboonnum et al. (2010) also studied γ-oryzanol recovery from soap stock. In the process, the soap stock was dehydrated and extracted using ethyl acetate. The extract was purified by double crystallization. The optimum conditions are 20% (v/v) ethyl acetate in methanol and temperature of 30°C for 1 h during first crystallization step, and 5°C and 24 h for the second step. The yield and purity obtained are 55.17 and 74.6 wt%, respectively. Imsanguan et al. (2008) have demonstrated the use of supercritical carbon dioxide for extraction of γ-oryzanol from rice bran. The extraction yielded 11,371.79 mg/kg (dry basis) at a pressure of 48 MPa and temperature of 65° C in a continuous mode.

The major impurities removed after isolation and recovery of oryzanol, mainly from soap stock or acid oil, are free fatty acids, glycerides, soap, phospholipids, sterols, waxes, glycolipids, resinous matter, tocopherols, and pigments. Understanding the impurities associated with the process are essential to designing large-scale γ-oryzanol isolation and recovery processes. Details on these impurities are discussed by Narayan et al. (2006).

2.4 Uses and Applications

Gamma-oryzanol has many applications in nutraceuticals, pharmaceutical, and cosmeceutical preparations (Patel and Naik, 2004; Nagendra et al., 2015; Sohail et al., 2017), due to its numerous health benefits such as hypolipidemic effect, promotion of growth, stimulation of hypothalamus, and increased bile excretion (Dunford and King, 2000; Nayik et al., 2015). It possesses great potential for lowering blood cholesterol (Pal and Pratap, 2017) and has an antioxidant activity four times higher than tocopherols. Gamma-oryzanol can prevent cancer causing free radicals (Nagendra et al., 2015; Nayik et al., 2015). Oryzanol has demonstrated the ability to modulate pituitary secretion, inhibit gastric acid secretion and platelet aggregation, and has in vitro antioxidant properties (Ghatak and Panchal, 2012). Ferulic acid esters of triterpene alcohol in RBO are shown to exhibit a strong stabilization effect during frying (Wang, 2002). It is also worth mentioning that oryzanol remains stable even when RBO was heated at 180°C for 8 h (Sanghi and Tiwle, 2015). Additionally, oryzanol is known to possess anti-dandruff, anti-itching and anti-aging properties, and is effective in treating many gastrointestinal disorders such as stress-induced gastric and duodenal ulcers (Al-Okbi et al., 2014).

Ferulic acid of oryzanol can be used for production of vanillin (3-methoxy4-hydroxybenzaldehyde). Ferulic acid can be purified from the waste residue of RBO and further used as a substrate for vanillin production through microbial hydrolysis (Zheng et al., 2007; Mazhar et al., 2017). Globally, vanillin is considered one of the most important flavors and fragrances in food and cosmetics (Furuya et al., 2015). It is ranked the second largest aroma of the world, with an annual demand of about 12,000 tons (Mazhar et al., 2017). Vanillin is known as a good flavoring agent and has been widely used in food and beverages, pharmaceutical, fragrance industries, and other industries. Chocolate and ice cream industries accounted for about 75% of the total market for vanillin (Kumar and Pruthi, 2014).

3. TOCOPHEROLS

Tocopherol is a member of the vitamin E family (Ahsan et al., 2015). It was named as "tocopherol" in 1924 and first synthesized in 1938 (Aggarwal et al., 2010). There are four naturally occurring tocopherols, namely α-, β-, γ-, and δ-tocopherol. They are widely distributed in animals, cereal bran and germ oils, vegetable oils, fruits and vegetables, and nuts (Fiume, 2014; Ahsan et al., 2015). Crude RBO contains about 2%–4% tocopherols, with nutritional and antacid effects (Sanghi and Tiwle, 2015). A 100 g tocopherol consist of 19–46 mg α-tocopherol, 1–3 mg β-tocopherol, 1–10 mg γ-tocopherol, and 0.4–0.9 mg δ-tocopherol (Orthoefer, 2005). Refined RBO was reported to contain 48–70 mg% tocopherols (Krishna et al.,

2006). Among all the vitamin E compounds, α-tocopherol is the only one that meets human dietary requirements. Other tocols have recently become subject of many research interests because of their potential health benefits (Kordsmeier, 2013).

3.1 Chemical Structure

Tocopherols (Fig. 4) and tocotrienols are subgroups that are collectively known as vitamin E (Al-Okbi et al., 2014), also referred to as tocols (Kordsmeier et al., 2015). There are four naturally occurring structural analogues of tocopherol: namely α-, β-, γ-, and δ-analogues. Tocopherols exist with a six chromanol ring structure that is methylated at different degrees at positions five, seven, and eight. It has C16 saturated side chain at position two (Ahsan et al., 2015). They differ from tocotrienols in the presence of additional methyl groups around the aromatic ring of the chromanol core (Ko et al., 2003). Tocopherols are saturated and possess an isoprenoid side chain (Ahsan et al., 2015). It is an analogue to tocotrienol and shares certain similar structures with common chromanol head and side chain at C-2 position (Ko et al., 2003).

3.2 Properties

Tocopherols are one of the most abundant natural antioxidants with excellent biological antioxidants that protect cellular membranes and increase stability in fat and oil (Bruscatto et al., 2009). Pure form of tocopherols and tocotrienols are yellow viscous liquids, easily decomposed upon exposure to oxygen, light, alkaline pH, and traces of transition metals. They are

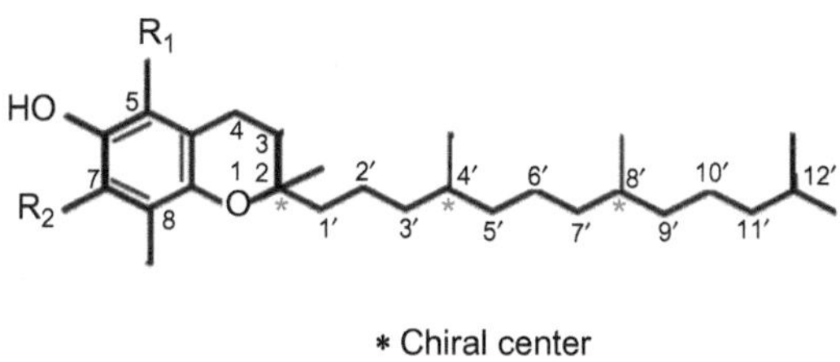

$R_1 = R_2 = CH_3$: α-Tocopherol

$R_1 = CH_3$, $R_2 = H$: β-Tocopherol

$R_1 = H$, $R_2 = CH_3$: γ-Tocopherol

$R_1 = R_2 = H$: δ-Tocopherol

Fig. 4 Chemical structures of tocopherols.

soluble in organic solvents but insoluble in water (Sarmento et al., 2006). The pure forms are also known to be high-boiling liquids, moderately soluble in lower aliphatic alcohols and ketones, and highly soluble in hydrocarbons and nonpolar organic solvents like ether (Dobbins et al., 2008). The hydrophobic properties of tocopherols decrease in the order of α->β->γ->δ-tocopherols. They are oxidized in the presence of heat, light, and alkali, readily oxidized, and darken upon exposure to light. Tocopherol esters are considered more stable than tocotrienols (Lampi et al., 2002). Alpha-tocopherol was historically considered the homolog of vitamin E with greatest value, because of its important level of physiological activity until recent times (Chen and Bergman, 2005). When α-tocopherol is exposed to light, its stability decreases by half with every increase in 10°C and above in lipid material (Lampi et al., 2002).

3.3 Analysis, Extraction, and Purification

The principal source of tocopherol is the deodorizer distillate of the edible oil refining process (Dobbins et al., 2008). Extraction of tocopherol from this byproduct requires three major steps: (1) removal of free fatty acids and their esters, (2) separation of tocopherol, and (3) purification. The removal of fatty acids can be achieved by vacuum distillation after chemical/enzymatic methylation by saponification, supercritical fluid extraction, or fractional solvent extraction. Purification can be achieved by distillation or chromatography to remove sterols that are also present in the distillate, to obtain pure tocopherols and/or tocotrienols (Lampi et al., 2002). Another method such as solvent fractionation in combination with cold crystallization was used for isolation and purification of tocopherols and/or tocotrienols or gamma and delta isomers from RBO deodorizer distillate of oil (Howard, 2016). Several methods of tocopherols and tocotrienols quantification such as gas chromatography (GC), normal or reverse phase-HPLC, capillary electrochromatography, nanoliquid chromatography, capillary liquid chromatography, and thin layer chromatography (TLC) have been extensively reviewed and discussed by Saini and Keum (2016).

Kordsmeier (2013) studied extraction of tocopherols by using different solvents (acetonitrile, methanol, and ethanol) from RBO deodorizer distillate (RBODD). The ratio of solvent:RBODD ratios (5:1, 10:1, 15:1, and 20:1) and two temperatures (4°C and −20°C) were used. At optimum conditions of acetonitrile and temperature (10:1, and 4°C), an extract was obtained with 57.4 mg tocopherol/g of concentrate, which was 1.55 and

1.34 times higher than that extracted using ethanol and methanol, respectively. The tocopherol [Ts] composition was 34.0% α-T, 3.4% β-T, 44.9% γ-T, and 17.4% δ-T. Ko et al. (2008) concentrated tocopherols from RBODD with 14.9% tocopherols. An optimum concentration was achieved using acetonitrile as a solvent. The final concentration of 21.4mg/g was obtained, with 16.1mg/g α-T, 0.4mg/g β-T, 4.6mg/g γ-T, and 0.3mg/g δ-T. In another work, the RBODD was extracted to concentrate tocols (tocopherols and tocotrienols). Tocols concentration have increased from 4% in a RBODD sample to 34% tocols after extraction and concentration process (Gomes et al., 2015). The use of supercritical carbon dioxide (55°C, 48MPa, and 0.45mL/min CO_2 flow rate) yielded high α-tocopherols (31.83mg/kg/h) from rice bran, and about 7.18mg/kg/h was obtained by using soxhlet extraction with hexane (Imsanguan et al., 2008). A supercritical fluid extraction (SFE) of tocopherol concentrate from olive oil using optimum condition of 25–45MPa pressure, 313–333K temperature, 0.5–1.5SL/min particle solvent flow, and 0.25–1.5mm particle size was reported. An extract with 74.5% and 97.1% (w/w) tocopherol concentration was obtained by these conditions (de Lucas et al., 2002).

Dobbins et al. (2008) patented a process for isolating tocopherols from deodorizer distillate. The process involved saponification of the deodorizer distillate using potassium hydroxide in a solvent medium of methanol and water to obtain a solvent medium of methanol, water, and potassium soap of fatty acids. Tocopherols and other unsaponifiable matter remains dissolved in the solvent medium. Upon distillation, a pure form of tocopherol is recovered. An example of method for purification of tocopherol from deodorizer distillate is presented in Fig. 5.

3.4 Uses and Applications

Tocopherols are known to suppress tumor-cell proliferation, inhibit cholesterol synthesis, and lower serum cholesterol levels in many animal models (Chen and Bergman, 2005). They have gained much recognition in the cosmetics industry for their strong antioxidant properties. Tocopherols are also reported to exhibit soothing and moisturizing effects, improve skin elasticity, and have anti-aging properties (Aggarwal et al., 2010). The α-T is the isomer with highest vitamin-E activity. It has many applications in all kinds of food products (de Lucas et al., 2002). Tocopherols are used in products such as baby lotions, oils, and cream formulation (at 0.1%). They have been used in several products such as those that can be inhaled. For example, up to

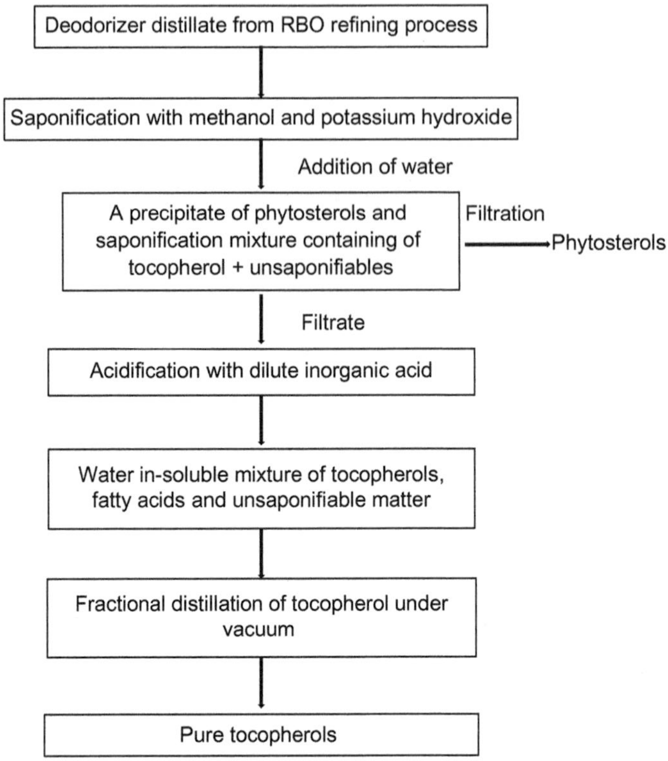

Fig. 5 Extraction process for tocopherols from deodorizer distillate.

5% of tocopheryl-acetate is used in foot powders and sprays, and in aerosol hair spray formulations, up to 0.2% is used. Tocopherol (up to 1%) is also used in formulations for pump hair sprays (Bergfeld et al., 2014).

4. TOCOTRIENOLS

Tocotrienols (T3s) are members of the vitamin E family. They attracted less or no attention until around the 1980s and 1990s when their potential for lowering cholesterol was recognized (Ahsan et al., 2015). Tocotrienols together with tocopherols are known as tocols, consisting of four isomers (Fig. 6) of different saturated analogues (α, β, γ, and δ) (Aggarwal et al., 2010). Tocotrienols consist of 14–33 mg α-T3 and 9–69 mg γ-T3 per 100 g (Orthoefer, 2005). However, according to the definition by *Cosmetic Ingredient Review*, tocotrienols do not include the β-analogues (Fiume, 2014; Zielinska and Nowak, 2014). They differ from tocopherols in the number and location of methyl (—CH_3[subscript]) groups on the chromanol rings (Aggarwal et al., 2010).

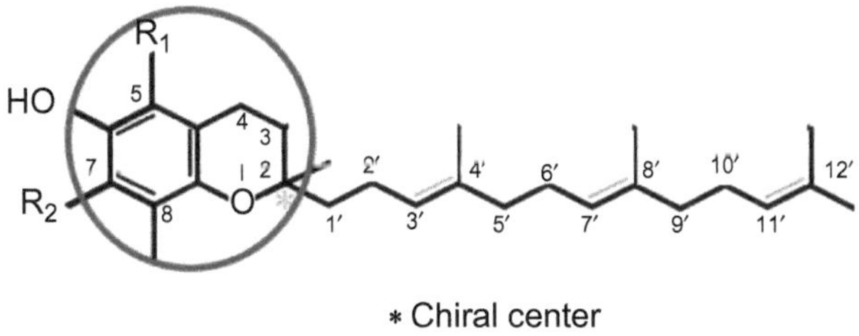

$R_1=R_2=CH_3$:	α-Tocotrienol
$R_1=CH_3$, $R_2=H$:	β-Tocotrienol
$R_1=H$, $R_2=CH_3$:	γ-Tocotrienol
$R_1=R_2=H$:	δ-Tocotrienol

Fig. 6 Chemical structures of tocotrienols.

4.1 Chemical Structure

Tocotrienols have a single stereocenter and three double bonds, that is, unsaturated at positions 3, 7, and 11 (Ahsan et al., 2015), which are in the *R*-configuration and all *trans*-geometries (Fiume, 2014). They differ from tocopherols for having farnesyl rather than saturated isoprenoid C16 side chain (Colombo, 2010). The unsaturation in tocotrienol tails gives them only one stereoisomeric carbon with one possible isomer per structural formula. There are four specific tocotrienols that differ in number and position of methyl group on the six-chromanol ring. Thus α-T3 is 5,7,8-trimethyl; β-T3 is 5,8-dimethyl; γ-T3 is 7,8-dimethyl, and δ-T3 is 8-monomethyl (Ahsan et al., 2015).

4.2 Properties

Tocotrienols are pale yellow viscous liquids at room temperature, soluble in lipids and lipophilic solvents, but insoluble in water. They are easily oxidized in the presence of light, heat, and alkali. Tocotrienols are slowly oxidized by atmospheric oxygen in the dark. In the absence of oxygen, they are resistant to alkali and temperatures up to 200°C (Lampi et al., 2002). Generally, the L-isomers of tocotrienols do not possess vitamin activity, and their naturally occurring forms are D-forms (Ahsan et al., 2015). Tocotrienols possess a lower vitamin E activity than tocopherols (Lampi et al., 2002). Among the tocols family, γ- and δ-analogues of tocotrienol are the most effective and strongest natural preventive agents against cancer and harmful effects of exposure to radiation (Kordsmeier et al., 2015). Despite the highest absolute antioxidant activity of α-T3, it is the most key factor involved in degradation of tocotrienol-rich foods. This is due to the hydrogen-donating power of α-T3 toward free radicals. However, in fat and oil, it exhibits the fastest degradation and lowest relative antioxidant activity against lipid peroxidation as compared to remaining tocotrienol isomers (Ahsan et al., 2015).

4.3 Analysis, Extraction, and Purification

Many researchers have employed different methods for extraction and purification of tocotrienol. These methods include solvent extraction (soxhlet, direct, supercritical, and pressurized liquid), chemical method (saponification and esterification), enzymatic method, adsorption, molecular distillation, membrane technology, and microwave-assisted extraction (Ahsan et al., 2015). A patented isolation and purification method of individual γ- and δ-T3s by flash chromatography (packed with a short column of 40–60 μm) was found to be a promising technique for the individual isomer's purification (Howard, 2016). The methods of extraction, purification, and quantification of tocotrienols are similar to that of tocopherols as discussed in Section 3.3.

Kordsmeier (2013), in similar work as described previously in Section 3.3, studied the extraction of tocotrienols (T3s) using different solvents (acetonitrile, methanol, and ethanol) from RBODD, a solvent: RBODD ratio (5:1, 10:1, 15:1, and 20:1) and two temperatures (4 and −20°C). At these conditions, an extract with acetonitrile was obtained with 2.23 and 1.82 times higher tocotrienols than the extracts obtained with ethanol and methanol. The concentration of 61.1 mg/g tocotrienol with composition of 25.5% α-T3, 72.8% γ-T3, and 1.7% δ-T3 was achieved. In another work, Ko et al. (2008) concentrated tocotrienol (16.6 mg/g) from RBODD using acetonitrile as the optimum solvent. The final concentration of 63.6 mg/g was obtained, with 22.3 mg/g α-T3, 39.2 mg/g γ-T3, and 2.1 mg/g δ-T3.

A U.S.-based patent invented a process of producing purified γ- and δ-T3s from tocols-rich oil/distillates (95% total tocols) of RBO deodorizer distillate or palm oil. The process yields a fraction containing high γ- and δ-T3, yet minimizing the presence of alpha isomers from the distillate/oil. A fraction containing approximately 10% γ-T3 and 3% δ-T3, each with purity of about 95%, was obtained. The process employed the use of flash or other low-pressure chromatography for rapid isolation of the γ-T3 and δ-T3. Fractionation of the mixtures uses a flash chromatography with short columns, packed with intermediate size particles of 40–60 μm, with accelerated solvent flow and moderate pressure. In the process, a binary mobile phase gradient consisted of 0.8% ethyl acetate-acetic acid (99.1:0.9 v/v) to hexane-acetic acid (99.1:0.9 v/v) for about 5 min was used (Howard, 2016).

4.4 Uses and Applications

Tocotrienols are used in sunscreens and body lotions to protect from harmful UV radiations and to help in skin softening and repair (Nayik et al., 2015). Ko et al. (2003) have reported that tocotrienols are substantially different to tocopherols in their capacity to suppress proliferation of a tumor cell. When tocotrienols are applied to the skin, it rapidly penetrates and is absorbed, thereby acting as a first line of defense, because of their antioxidant properties. Tocotrienols have great potential to stabilize free radicals generated in the skin during exposure to oxidative rays and to help repair skin by protecting against UV radiation-induced skin damage and aging (Zielinska and Nowak, 2014; Nagendra et al., 2015). In an in vitro study, α-T3 was reported to have more than three times the free radical scavenging activity than that of α-T3 (Chen and Bergman, 2005). The δ-T3 has been suggested to provide some important radioactive potential. The radioactive effects of γ-T3 and δ-T3 may be of great benefit in treatment of radiation in cancer patients. However, α-T3 was proposed to interfere with the radioactive and anticarcinogenic effect of γ-T3 and δ-T3 (Kordsmeier, 2013).

5. PHYTOSTEROL

Phytosterols are also known as plant sterols found mainly in plant cell walls and membranes (Özdestan et al., 2014). There are >200 distinct types of phytosterols in plant species (Dunford and King, 2000; Özdestan et al., 2014). The phytosterols and stanols occur naturally having similar structure to cholesterol with a cellular function like human cholesterol, but the phytosterols have a higher degree of absorption than plant stanols (Harland, 2016). Phytosterols are synthesized from squalene. One of the first intermediate products is cycloartenol (a 4,4′-dimethylsterol) (Al-Okbi et al., 2014). An unsaponifiable fraction of RBO contains 1.8% phytosterols (Pali, 2013). Phytosterols in crude and commercially refined RBO ranges from 1362 to 1376 mg/100 g and 858–1034 mg/100 g, respectively (Sawadikiat and Hongsprabhas, 2014). Niawanti and Zullaikah (2017) stated that RBO contains as much phytosterol as 30 g/kg higher than other vegetable oils, which ranged between 1 and 5 g/kg. The main phytosterols in RBO are β-sitosterol (c.50%), campesterol (c.20%), stigmasterol (c.15%), and isofucosterol (c.5%) (Yukihiro and Hara, 2012). In an RBO extracted from an Egyptian rice bran using n-hexane and supercritical CO_2, total sterols of 12.0% and 7.6% were reported (Al-Okbi et al., 2014). RBO also contains a mixture of ferulic acid esters of 4-desmethylsterol like campesterol and β-sitosterol, an end product of plant sterol synthesis from squalene (Vissers et al., 2000). Phytosterols are known for cholesterol-reduction properties by inhibiting cholesterol absorption in the small intestine (Yukihiro and Hara, 2012).

5.1 Chemical Structure

Phytosterols occur in free and conjugate forms. There are four types of conjugates in which the 3β —OH group is esterified to moieties like fatty acid, hydroxycinnamic acid, glycosylated with hexose, 6-fatty-acyl hexose, or ferulic acid. When the moiety is ferulic acid ester, the phytosterol is known as oryzanol (Dunford and King, 2000; Özdestan et al., 2014). The major components of phytosterols (Fig. 7) in lipid extract are sitosterol, stigmasterol, and campesterol. Phytosterols are classified as either sterols or stanols, depending on the presence of a double bond within the $\triangle 5$ position and are found in a free or esterified form (Özdestan et al., 2014).

5.2 Properties

The free form of phytosterol or stanols tends to be poorly soluble in water or oil. However, the esterified form becomes more soluble and easy to use in food product formulations (e.g., margarine, salad dressings, and spreads). They are mostly esterified with polyunsaturated fatty acids. This makes them more soluble in vegetable oil. Phytosterols are usually unsaturated and can be converted to phytostanols (a subgroup of phytosterol) through chemical hydrogenation. The phytostanols are a fully saturated form of phytosterols and occur in a small amount in different plant species and in high concentrations in tissues of a few species of cereal crops (Osborn and Morley, 2016).

5.3 Analysis, Extraction, and Purification

Methods of extraction of phytosterols from deodorizer distillates involve three major steps, namely: (1) saponification, to convert phytosterol fatty acid ester to free phytosterol; (2) esterification of free fatty acid; and (3) distillation to recover phytosterol or its concentrate. The final phytosterol or its concentrate can be extracted by three different methods of extraction. These

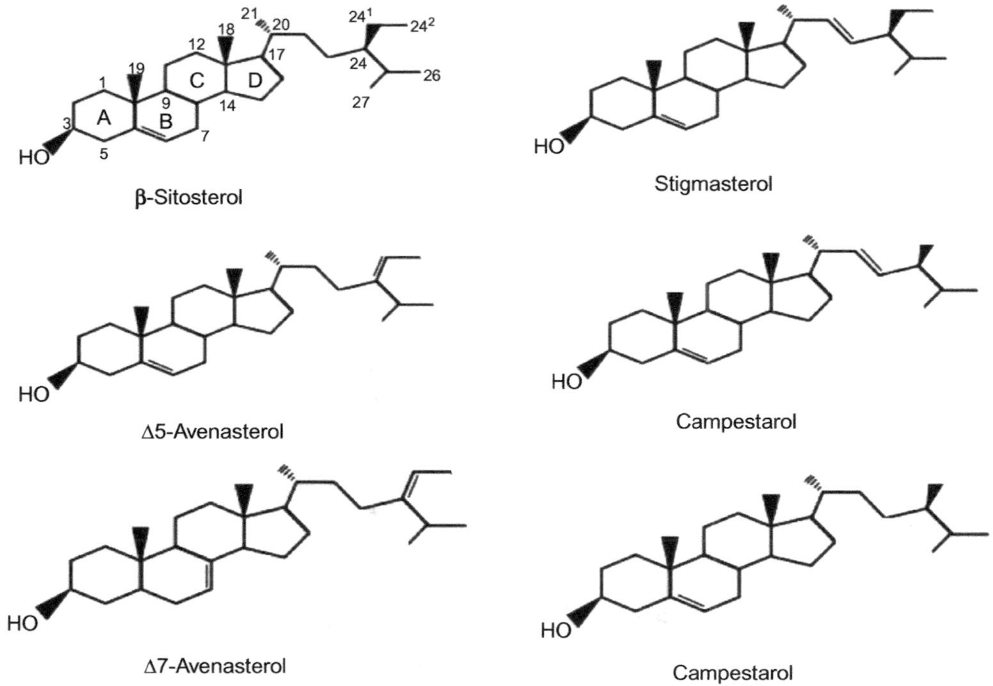

Fig. 7 Chemical structures of some phytosterols.

include physical extraction (plant crystal separation), chemical extraction (using solvent), and physical and chemical extraction (phytosterol crystal separation from additive through an addition compound) (Wang, 2016). Methods such as HPLC and GC-MS are used for quantification and analysis of a phytosterol mixture (Rajan and Krishna, 2014). A typical example of extraction steps for phytosterol from deodorization distillate is presented in Fig. 8.

Arumughan et al. (2007) patented an invention for isolation and purification of sterols from soybean oil deodorized distillate (SODD) by using simple acid catalyzed esterification and separation by distillation and water washing, followed by crystallization and purification of phytosterols. The process is applicable to RBO or other vegetable oil distillate containing >1% phytosterols in free form. Resulting in a highly purified phytosterol (95%–99%) with 80%–90% yield, containing β-sitosterol, stigmasterol, and campesterol, the process involves mixing the SODD with methanolic solution of sulfuric acid in the ratio of 1:1 to 1:1.5 v/v, at 50–85°C with stirring for a period of 2–4 h. The excess methanol is distilled under a vacuum, and a brine solution was added to the reaction mixture to separate the sulfuric acid and the lower layer containing the ester layer, which was removed. The ester layer was washed with hot water until all acid and salts were removed and then crystallized at a temperature of 0–25°C for 1–10 h,

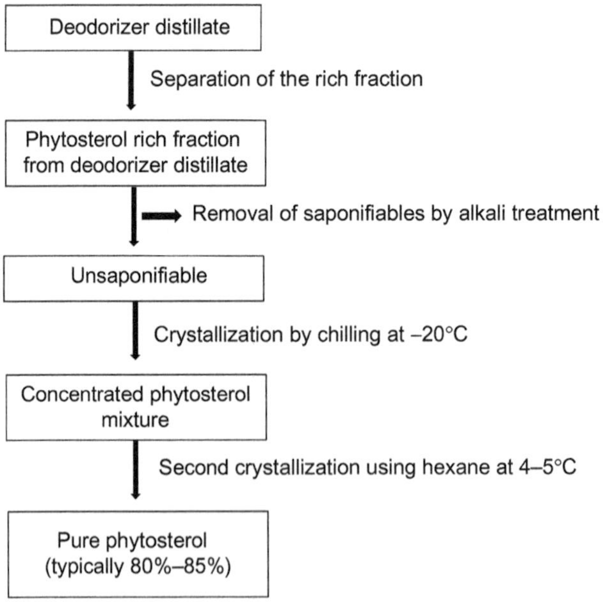

Fig. 8 Extraction of phytosterol from deodorization byproduct.

filtered, and the resultant crystals was washed with chilled *n*-hexane (at 0–10°C). This process yielded high purity phytosterols.

A highly purified phytosterol (97% purity) was achieved by using SFE, supercritical fluid chromatography (SFC), followed by solvent fractionation of deodorized distillate of RBO with hexane (Sugihara et al., 2010). The SFC was also used to separate the plant sterol in the soybean and corn fiber oil with original concentration of 2.2 mg/g, and 13.2 mg/g, respectively. After fractionation by SFC, the sterol concentration increased to 64.4 mg/g (Snyder et al., 1999). Rajan and Krishna (2014) concentrated the phytosterol in RBO deodorizer distillate from 8.71% and 4.22% to 12.4% and 5.9% concentrations. A simple method of percolation process using silica gel (100–200 mesh size) was developed to separate the phytosterol's concentrate fraction and brown color/bad odor fraction. The silica gel was heated at 105°C for 6 h to remove moisture and then later hydrated with 5% water addition. Column of 60 cm length and 4 cm internal diameter was packed with the silica gel (100 g) using petroleum ether. A deodorizer distillate mixed with hexane (1:5 w/w) was introduced into the column and allowed to percolate at 2–3 mL/min flow rate. HPLC analysis of phytosterol mixture revealed the composition contained 38.2% β-sitosterol, 34.9% stigmasterol, 9.5% campesterol, and 17.4% other sterols.

5.4 Uses and Applications

Phytosterols are important in food and pharmaceutical industries (Özdestan et al., 2014) due to their important bioactive properties for human health. Since the 1950s, several human and animals studies have demonstrated that the consumption of phytosterols can lead to significant reduction in serum cholesterol (Osborn and Morley, 2016). They can reduce low density lipoprotein (LDL) and total cholesterol level in the blood by preventing cholesterol absorption. They also have antibacterial, antifungal, and antiulcer effects (Özdestan et al., 2014). The phytosterols or stanols in esterified forms are extensively used in formulation of different food products such as dressings, margarines and spreads, and yogurt (Osborn and Morley, 2016). Consumption of plant sterol ester (2–2.5 g/day) incorporated in food products was found to lower plasma LDL cholesterol levels by 10%–14% (Plat and Mensink, 2005). Nguyen (1999) has studied the cholesterol-lowering activity of plant stanol esters. It was observed that consumption of 2–3 g/day of plant stanol esters incorporated in margarine demonstrated the ability to reduce LDL cholesterol by 10%–15% without side effects. Other research has revealed that administration of 2.1 g of sterol from RBO by humans lowered serum total cholesterol by 5% and LDL cholesterol by 9%; this has been related to its inhibition of cholesterol-absorption properties (Vissers et al., 2000). Özdestan et al. (2014) stated that the effect of RBO sterol is probably due to the 4-desmethysterol β-sitosterol and not due to 4,4′-dimethylsterols, for example, cycloartenol and 24-methylene cycloartenol. The cycloartenol and 2.4-methylene cycloartanol that formed >40% of the total phytosterols are present in very high concentration in RBO (Özdestan et al., 2014). Phytosterols also have a strong antioxidant activity. In RBO, the avenasterol in phytosterol acts as antioxidant at elevated frying temperature because of its ethylidene group on the side chain of the molecules (Wang, 2002). This is because RBO contains mainly 4,4′-dimethylsterols and less 4-desmethylsterol (Vissers et al., 2000).

6. SQUALENE

Squalene was first discovered by Mitsumaru Tsujimoto, a Japanese researcher and expert in fats and oils at Tokyo Industrial Testing Station in 1906. Later in 1930, Dr. Keijiro Kogami studied its health benefits. In 1963, Nobel Prize winner Paul Kerrer discovered its presence in the human body (Gunes, 2013). Squalene (Fig. 9) is a triterpene (Huang et al., 2009; Al-Okbi et al., 2014) that is known primarily for its role as an intermediate metabolite during synthesis of cholesterol (Gunes, 2013). The name "squalene" was derived from shark fish (*Squalus* spp.) liver oil, which is the most abundant source of squalene in large quantity (Charan et al., 2015). It is a principal component of human surface lipids (21%) and acts as a biochemical precursor of sterols and terpenoids, with a central role in

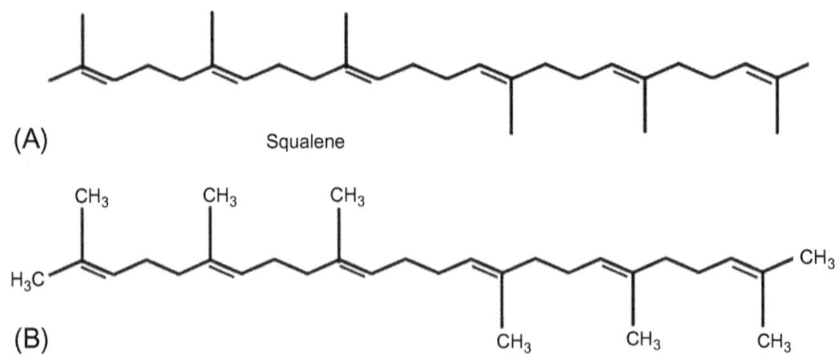

Fig. 9 Chemical structures of squalene (A) and squalene (B).

plant, animal, and human functions (Naziri et al., 2011). Squalene is widely found in RBO and other sources such as olive oil, wheat-germ oil, palm oil, and amaranth oil (Huang et al., 2009). Squalene attracts much attention of scientists because of its beneficial effects in human health (Nandi and Bhattacharyya, 2017). Commercially extracted squalene from olive oil is popularly marketed as "vegetable squalene" with about 97.5%, and squalene from shark liver oil has up to 99.9% purity (Perma Health Care, 2018). The RBO a has higher content of squalene (320 mg/100 g) (Pali, 2013). Other cheap and rich sources of squalene include deodorizer distillate, a byproduct of deodorization step of RBO refining (Popa et al., 2015). The deodorization distillate contains squalene (c.8%) as unsaponifiable components (Sugihara et al., 2010).

6.1 Chemical Structure

The chemical structure (Fig. 9) of squalene (C30-H50) contains anhydrocarbonated chain and a triterpene with six unsaturated bonds (Rosales-garcía et al., 2017). The strong antioxidant effect of squalene is due to the double bond structure of the six CH_3. It has similar structure to other isoprenoids like lycopene, vitamin A, β-carotene, vitamin E, and coenzyme Q10 (Ronco et al., 2013).

6.2 Properties

Squalene is a colorless, tasteless, and odorless substance. It has a molecular weight of 410.7 g/mol and density of 0.858 g/mol (Popa et al., 2016). The boiling point, melting point, surface tension, and iodine number are 285°C, −75°C, −32 mN/m, and 381 g/100 g, respectively (Naziri et al., 2011). Squalene has hydrophobic properties (Rosales-garcía et al., 2017) with solubility in fat. Its high hydrophobicity and surface tension make it very important for preparation of oil-in-water stable emulsions for drug delivery and vaccines, as well as formulations for skin disorders and cancer treatments (Naziri et al., 2011).

6.3 Analysis, Extraction, and Purification

Extraction of squalene is mainly carried out from plant or animal sources. Squalene extraction from shark liver oil has been challenged due to environmental concern, and thus plants and microorganisms are now considered as the important alternative sources (Spanova and Daum, 2011). The major source of squalene is deodorized fatty acid distillate (FAD) of the oil refining process (Nandi and Bhattacharyya, 2017). Pietsch and Jaeger (2007) stated that a deodorizer distillate may contain up to 28% of squalene and, after concentration, up to 70% can be achieved. A higher concentration up to 90% has been obtained using SFE (Pietsch and Jaeger, 2007). Various methods have been reported for extraction of squalene from different sources, for example, supercritical carbon dioxide was reported for extraction of squalene from shark liver oil (Catchpole et al., 1997) and amaranth seed (Kraujalis and Venskutonis, 2013). Soxhlet extraction and fractional crystallization was reported for Lebanese olive oil (El Zakhem et al., 2016). Fractionation of unsaponifiable amaranths grain by silica gel column chromatography to remove saponifiable gave a highly purified squalene fraction containing as high as 94% concentrate (He et al., 2002). However, SFE is the most widely used due to its high recovery rate, economy, safety, and easy removal (El Zakhem et al., 2016; Rosales-garcía et al., 2017). A short path distillation as an alternative to SFE was reported to give squalene with a purity over 97% from shark liver oil (Pietsch and Jaeger, 2007). For quantification of squalene, methods used include gas chromatography (GC), GC-coupled with mass spectroscopy (GC-MS), HPLC, or HPLC coupled with mass spectroscopy (HPLC-MS) (Grigoriadou et al., 2007; Popa et al., 2015).

Nandi and Bhattacharyya (2017) have extracted squalene from RBO fatty acid distillate using a combination of biohydrolysis, bioesterification, and molecular distillation. In the study, the neutral glyceride in RBOFAD was hydrolyzed in the presence of water using lipase Amano 30 (*Candida rugosa*). The hydrolyzed FAD was followed by esterification with butanol using nonspecific enzyme NS40013 (*Candida antarctica*) for 8 h. The hydrolyzed FAD was then distilled by two-stage molecular distillation at 175 ± 2°C and 185 ± 2°C in the first and second stages, respectively, at 2–3 Pa. The final distillate contained 88.2% squalene, which is further concentrated to 96.5% by boiling with alcoholic potassium hydroxide and extraction with hexane. To fractionate squalene from deodorization distillate, supercritical carbon dioxide can be used (Yukihiro and Hara, 2012). A fractionation method for concentration of squalene from deodorization distillate of RBO was studied by SFE with carbon dioxide at 30°C, pressure at 100 kg/cm^2, and flow rate of carbon dioxide 7 mL/min. At these conditions, squalene was concentrated to about 25%. A higher purity was obtained by further concentrating using SFC with silica gel packed into extraction vessel to 50% purity. A more pure squalene was also achieved by double-SFC concentration to 68% and further to 81% and 100% by using combined SFC with the solvent fractionation technique (Sugihara et al., 2010).

Squalene extraction and purification from deodorizer distillate of soybean oil was studied by Gunawan et al. (2008). The starting material contained 1.83%, and a modified soxhlet extraction method using hexane was employed. A yield of 9.83% was obtained with purity and recovery of 95.90% and 93.09%, respectively. Bondioli et al. (1993) also studied the extraction of squalene from olive oil deodorizer distillate by supercritical carbon dioxide extraction. The initial concentration of squalene in deodorizer distillate ranged between 10 and 30 wt%. The process involved converting the free fatty acids and methyl and ethyl esters into triglycerides, and the latter is extracted using supercritical carbon dioxide. Squalene of higher purity and yield of about 90% was obtained. A solvent (hexane) extraction method and isolation of squalene using silica gel-chromatography is presented in Fig. 10.

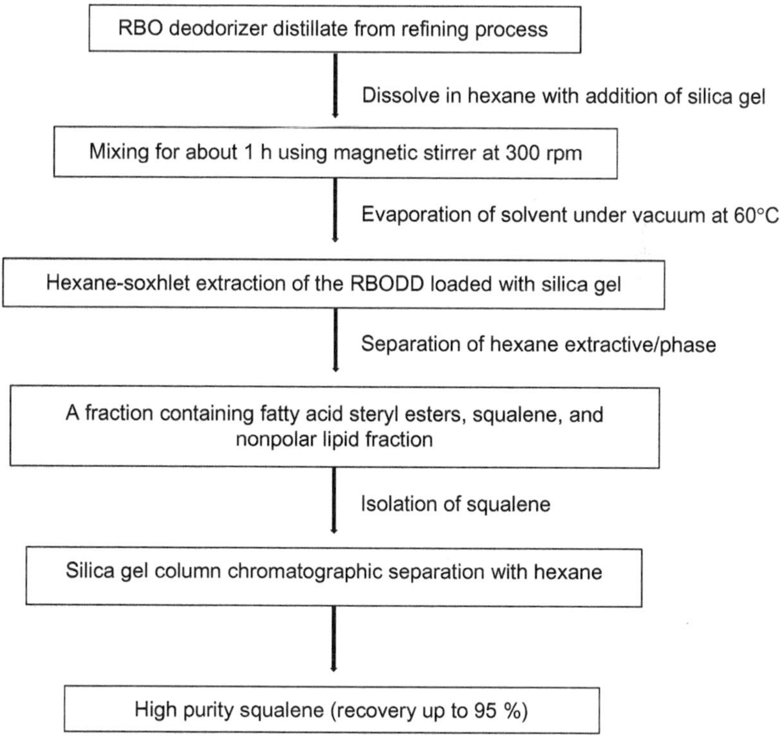

Fig. 10 Extraction and purification of squalene from deodorizer distillate.

6.4 Uses and Applications

Squalene market is considered high in cosmetics (69.2%), food (22.8%), and pharmaceuticals (8%), with demand amounted to 267,000 tons, representing $102.4 billion (Rosales-garcía et al., 2017). Squalene has been reported to possess an inhibitory effect for cancer and antitumor activity (Gunes, 2013). It is a main component of polyunsaturated lipid of skin surface, and helps to lubricate and soften skin (Huang et al., 2009). Squalene, just like tocotrienols, is used in sunscreen to help protect from harmful UV radiation, improve skin softening, and repair (Nayik et al., 2015). Intake of dietary squalene helps to reduce plasma cholesterol. It is also used as a major component in many cosmetic and pharmaceutical products (El Zakhem et al., 2016; Rosales-garcía et al., 2017). Squalene helps in the formation of steroid hormones, steroids, bile acids, and sterols synthesis through mevalonic acid pathways (Rosales-garcía et al., 2017). It has several other beneficial effects in medicine, fine chemicals, and functional food applications, as bactericidal and fungicidal agents, moisturizers or emollients, and an antistatic agent (Nandi and Bhattacharyya, 2017).

7. PHOSPHOLIPIDS

Phospholipids (PLs) are found in all living cells of both plant and animal origins (Patil et al., 2010). They are one of the major classes of lipid in RBO (Liu et al., 2013), constituting up to 10% of the total rice grain lipid content (Sumit, 2017). The term "lecithin" is used synonymously as pure phosphatidylcholine, a main component of PLs (Machado et al., 2015).

Crude RBO contains about 1.6% PLs (Nayik and Muzaffar, 2015). Water degumming is the simplest method of removing the hydratable PL from the crude oil. The nonhydratable are often removed by pretreating the oil with phosphoric or citric acid (Tyagi et al., 2012). Phospholipase enzymes are sometimes used to hydrolyze PLs into calcium and magnesium salt of phosphatidic acid (Hammond et al., 2005). PLs (Table 2) consist predominantly of hydratable phosphatidylcholine (PC), phosphatidylinositol (PI), phosphatidylethanolamine (PE), and phosphatidic acids (PA) of calcium and magnesium salt (Sengar et al., 2014). The PLs in oil are mainly in hydratable and nonhydratable forms. The nonhydratable form is obtained when the PLs are combined with metal cations. When this form is reacted with acid, a hydratable gum is obtained. Hydratable PLs can also be converted to hydratable gum by adding water (Liu et al., 2012).

Table 2 Composition of RBO phospholipids

Components	Crude PLs (%)	De-oiled PLs (%)
PC	20.4	46.2 ± 0.256
PA	–	12.2 ± 0.104
PE	17.8	23.5 ± 0.188
PI	5.8	9.8 ± 0.023
PS	–	0.9
Other	9.4	7.4

Data from Adhikari and Adhikari (1986) and Sumit (2017).

7.1 Chemical Structure

The PLs in RBO contain different components (Fig. 11), namely glycerol, fatty acids, phosphate, and polyhydroxy groups, and the phosphate is always linked to the sn-3 position of glycerol molecule. PLs have two fatty-acyl groups and one polar head group (Jangle et al., 2013). The PLs differ in the functional groups, namely choline, ethanolamine, inositol, or serine (Liu et al., 2013).

7.2 Properties

Pure PLs are white solid (at room temperature), colorless, and odorless. Depending on the refining method, product categories, and storage conditions, PLs may range from yellow to brown. They are unstable when exposed to air or sunlight, which leads to oxidative rancidity. They are also unstable at high temperature and decompose at temperatures over 150°C (Liu and Ma, 2011). PLs are amphiphilic in nature, with different affinity

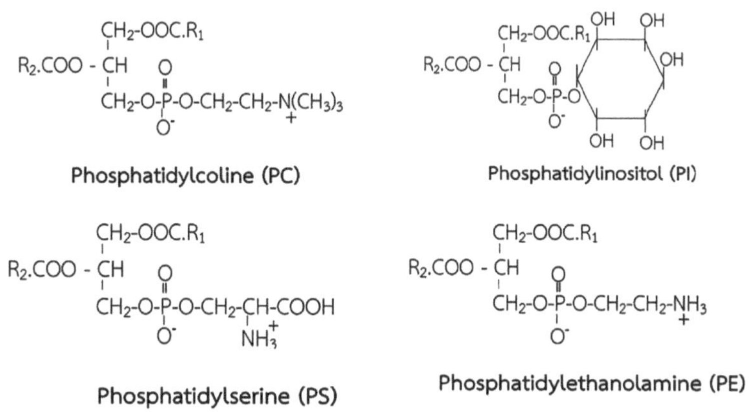

Fig. 11 Chemical structures of major phospholipids.

to both oil and water due to their simultaneous hydrophilic (water-loving) and hydrophobic (water-repelling) properties. Due to their characteristics, they act as emulsifiers, wetting and instantizing agents, viscosity modifiers, release agents (promote separation of food from contact surfaces in dip tanks and spray applications), separating agents (in processed cheese slices to form a stable film barrier that prevent sticking), extrusion aids, antidusting agents (to enhance wettability by reducing static interface), shelf life aids, and nutritional supplements (American Lecithin company, 2009).

7.3 Analysis, Extraction, and Purification

Phospholipids are purified mainly to obtain value-added nutritional and pharmaceutical compounds with better surface-active properties. The interest to purify PLs has mainly focused on the two main PL fractions, namely PC and PE, due to their various applications. High purity of PC can be obtained via fractionation of PLs (Jangle et al., 2013). The fractionation process makes use of the solubility differences between the phospholipids in ethanol. PC is readily soluble than PI and PA that are virtually insoluble, whereas the PE is found in both fractions (Bueschelberger, 2004). Methods for PL fractionation include extraction, precipitation, and column chromatographic purification (Patil et al., 2010). Other methods of PC purification are solvent extraction and SFE. The SFE and column chromatography are considered to have high efficiency and low solvent residue (Siriluck et al., 2012). Crude lecithin has to firstly undergo a de-oiling process (Fig. 12) using acetone extraction to separate neutral and polar lipids such as PLs and glycolipids (Jangle et al., 2013). Liu et al. (2013) have separated PLs from jatropha oil using two-step methods. The first step involves the use of a conventional method to remove the hydratable (by water degumming) and nonhydratable PLs (acid degumming), followed by the second step of membrane separation process for the degummed oil to remove the additional PLs. After these two treatments, PLs are reduced from 1200 ppm to about 60 ppm. PLs composition are usually analyzed using methods such as HPLC (Wu and Wang, 2003), normal phase-HPLC (Siriluck et al., 2012), and nuclear magnetic resonance (NMR) (Cabezas et al., 2009).

Fractionation of crude soy lecithin (CSL) (Fig. 12) using absolute ethanol:CSL ratio of 1:1 at 40°C gave a high purity PC-enriched fraction (64.68% PC; 6.61% PE; 1.21% PI; and 21.72% neutral oil). A further purification of this fraction with 0.2 mL of water to 1 mL of the PC-enriched fraction resulted in 84.11% purity with 19.77% yield of PC

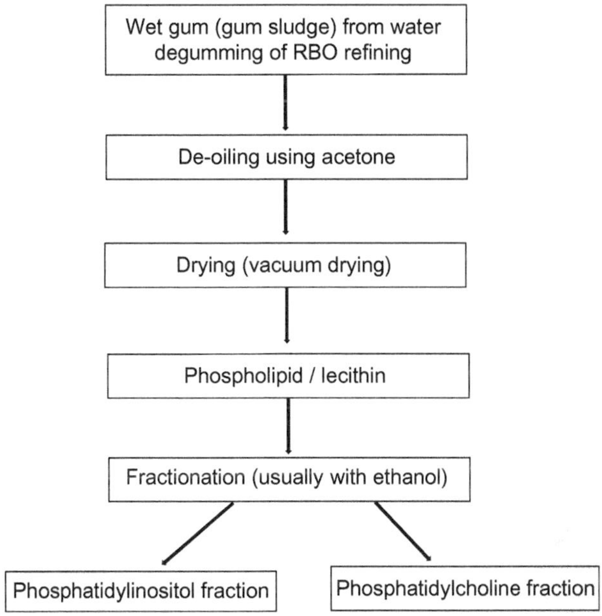

Fig. 12 Flowchart for extraction and fractionation of phospholipids from wet gum.

(Siriluck et al., 2012). Another fractionation of soy lecithin using ethanol 90% concentration yielded a PC-enriched fraction containing 73% PC, 24% PE, and 3% PI based on the total PL content, and PI fraction containing 26% PC, 35% PE, and 39% PI (Wu and Wang, 2003). Jangle et al. (2013) have also studied separation and purification of PC from de-oiled lecithin by solid-liquid extraction and liquid-liquid partitioning. At an optimum lecithin-to-ethanol ratio of 1:7 in solid-liquid extraction, a purity and yield of 62% and 98%, respectively, were obtained. The liquid-liquid extraction using hexane and acetonitrile (5% isopropyl alcohol) at ratio of 1:15 and 40°C for 8h stirring time gave the PC purity of 85%.

7.4 Uses and Applications

Dietary PLs have a number of health benefits for a range of human diseases such as cancer, coronary heart diseases, and inflammation (Liu et al., 2013; Sumit, 2017). Pure form of PLs is used as value-added nutritional, pharmaceutical, and cosmetic ingredients. They also have good surface-active properties for improving emulsion stability and shelf life (Jangle et al., 2013). Highly purified PC (≥85%) is used for pharmaceutical applications (Siriluck et al., 2012). PLs are also used as emulsifying agents in foods, agrochemicals, cosmetics, and pharmaceuticals (Sumit, 2017). They are fundamental components of cell membrane and therefore vital for growth, maturing, and good function of cells (Jangle et al., 2013). When PLs are used in pharmaceutical formulation, they help to improve bioavailability of drugs that have low aqueous solubility or membrane penetration potential. They also improve/alter drug uptake and release, protect sensitive active agents from degradation in the gastrointestinal tract, and improve the bitter taste of drugs (orally administered) (Ugandar et al., 2015).

8. CONCLUSION

RBO is a highly value-added product from byproduct (rice bran) of the rice milling process that is getting global popularity and wide application mainly due to its health benefits associated with the various micronutrients, namely oryzanol, tocols, phytosterols, squalene, and phospholipids. Many researchers have confirmed their health benefits. These micronutrients are mostly concentrated in different byproducts of the oil refining process such as deodorization, degumming, deacidification, and dewaxing. Thus their potentials are not properly utilized most of the time especially when these byproducts are considered as waste or animal products. Therefore extraction of these micronutrients from the byproducts will generate more income and add value to the RBO extraction sectors.

REFERENCES

Adhikari, S., Adhikari, J., 1986. Indian rice bran lecithin. J. Am. Oil Chem. Soc. 63 (10), 1367–1369.

Aggarwal, B.B., Sundaram, C., Prasad, S., Kannappan, R., 2010. Tocotrienols, the vitamin E of the 21st century: its potential against cancer and other chronic diseases. Biochem. Pharmacol. 80 (11), 1613–1631. https://doi.org/10.1016/j.bcp.2010.07.043.

Ahsan, H., Ahad, A., Siddiqui, W.A., 2015. A review of characterization of tocotrienols from plant oils and foods. J. Chem. Biol. 8 (2), 45–59. https://doi.org/10.1007/s12154-014-0127-8.

Al-Okbi, S., Ammar, N., Mohamed, D., Hamed, I., Desoky, A., El Bakry, H., Helal, A., 2014. Egyptian rice bran oil: chemical analysis of the main phytochemicals. Rivista Italiana Delle Sostanze Grasse 91 (1), 47–58.

American Lecithin company, 2009. Lecithins and phospholipids: a simple guide to use and selection. Retrieved from http://www.americanlecithin.com/lecithin_2009.pdf.

Angelis, A., Urbain, A., Halabalaki, M., Aligiannis, N., Skaltsounis, A.L., 2011. One-step isolation of γ-oryzanol from rice bran oil by non-aqueous hydrostatic countercurrent chromatography. J. Sep. Sci. 34 (18), 2528–2537. https://doi.org/10.1002/jssc.201100192.

Arumughan, C., Sobankumar, D.R., Sundaresan, A., Nair, S.S., Yohesh, K., Rajam, L., 2007. A process for the preparation of high purity phytosterols from deodourizer distillate from vegetable oils. In: Google Patents. Retrieved from http://www.google.com/patents/WO2007054759A1?cl=en.
Bergfeld, W.F., Donald, V., Hill, R.A., Klaassen, C.D., Liebler, D.C., Marks, J.G., Shank, C., Slaga, T.J., Snyder, P.W., 2014. Safety Assessment of Boron Nitride as Used in Cosmetics. Vol. 20036, 1–3. Cosmetic Ingredient Review, Washington.
Bondioli, P., Mariani, C., Lanzani, A., Fedeli, E., Muller, A., 1993. Squalene recovery from olive oil deodorizer distillates. J. Am. Oil Chem. Soc. 70 (8), 763–766. https://doi.org/10.1007/BF02542597.
Bruscatto, M.H., Zambiazi, R.C., Sganzerla, M., Pestana, V.R., Otero, D., Lima, R., Paiva, F., 2009. Degradation of tocopherols in Rice bran oil submitted to heating at different temperatures. J. Chromatogr. Sci. 47, 762–765.
Bueschelberger, H., 2004. Lecithins. In: Whitehurst, R.J. (Ed.), Emulsifiers in Food Technology. Blackwell Publishing Ltd., Oxford, UK, pp. 1–39.
Cabezas, D.M., Diehl, B.W.K., Tomás, M.C., 2009. Sunflower lecithin: application of a fractionation process with absolute ethanol. J. Am. Oil Chem. Soc. 86 (2), 189–196. https://doi.org/10.1007/s11746-008-1336-5.
Catchpole, O.J., von Kamp, J.-C., Grey, J.B., 1997. Extraction of squalene from shark liver oil in a packed column using supercritical carbon dioxide. Ind. Eng. Chem. Res. 36 (10), 4318–4324. https://doi.org/10.1021/ie9702237.
Charan, S., Bavisetty, B., Narayan, B., 2015. An improved RP-HPLC method for simultaneous analyses of squalene and cholesterol especially in aquatic foods. J. Food Sci. Technol. 52 (9), 6083–6089. https://doi.org/10.1007/s13197-015-1793-4.
Chen, M.H., Bergman, C.J., 2005. A rapid procedure for analysing rice bran tocopherol, tocotrienol and γ-oryzanol contents. J. Food Compos. Anal. 18 (2–3), 139–151. https://doi.org/10.1016/j.jfca.2003.09.004.
Colombo, M.L., 2010. An update on vitamin E, tocopherol and tocotrienol-perspectives. Molecules 15 (4), 2103–2113. https://doi.org/10.3390/molecules15042103.
de Lucas, A., Ossa, E.M., Rinco, J., Blanco, M.., Gracia, I., 2002. Supercritical fluid extraction of tocopherol concentrates from olive tree leaves. J. Supercrit. Fluids 22, 221–228.
Dobbins, T.A., Wiley, D.B., Dobbins, D.C., 2008. Process for Isolating Phytosterols and Tocopherols from Deodorizer Distillate, United States. Retrieved from https://patentimages.storage.googleapis.com/02/4f/53/e1e2226ad76c31/US20080015367A1.pdf.
Dunford, N.T., King, J.W., 2000. Phytosterol enrichment of rice bran oil by a supercritical carbon dioxide fractionation technique. J. Food Sci. 65 (8), 1395–1399. https://doi.org/10.1111/j.1365-2621.2000.tb10619.x.
El Zakhem, H., Romanos, C., Bakhos, C., Chahal, H., Koura, J., 2016. Extraction of squalene from Lebanese olive oil. Int. J. Chem. Mol. Eng. 10 (3), 308–311.
Fiume, M.M., 2014. Safety assessment of tocopherols and tocotrienols as used in cosmetics status. Cosmet. Ingred. Rev. 20036, 1–38. https://doi.org/10.1177/1091581814533354.
Furuya, T., Miura, M., Kuroiwa, M., Kino, K., 2015. High-yield production of vanillin from ferulic acid by a coenzyme-independent decarboxylase/oxygenase two-stage process. New Biotechnol. 32 (3), 335–339. https://doi.org/10.1016/j.nbt.2015.03.002.
Ghatak, S.B., Panchal, S.J., 2012. Investigation of the immunomodulatory potential of oryzanol isolated from crude rice bran oil in experimental animal models. Phytother. Res. 26 (11), 1701–1708. https://doi.org/10.1002/ptr.4627.
Gomes, T.A., Gu, X., Brownmiller, C.R., Howard, L., Compadre, C.M., Lee, S.-O., 2015. Effects of tocols rich fraction isolated from rice bran oil deodorizer distillate on plasma and hepatic lipid concentrations in rats. J. Biosci. Med. 3 (9), 1. https://doi.org/10.4236/jbm.2015.39001.

Grigoriadou, D., Androulaki, A., Psomiadou, E., Tsimidou, M.Z., 2007. Solid phase extraction in the analysis of squalene and tocopherols in olive oil. Food Chem. 105 (2), 675–680. https://doi.org/10.1016/j.foodchem.2006.12.065.

Gunawan, S., Kasim, N.S., Ju, Y.H., 2008. Separation and purification of squalene from soybean oil deodorizer distillate. Sep. Purif. Technol. 60 (2), 128–135. https://doi.org/10.1016/j.seppur.2007.08.001.

Gunes, F., 2013. Medical use of squalene as a natural antioxidant. J. Marmara Univ. Inst. Health Sci. 3 (4), 1. https://doi.org/10.5455/musbed.20131213100404.

Hammond, E.G., Johnson, L.A., Su, C., Wang, T., White, P.J., 2005. Soybean oil. In: Shahidi, F. (Ed.), Bailey's Industrial Oil and Fat Products. sixth ed. John Wiley and Sons, Inc., Ames, IA, pp. 577–650. Retrieved from https://www.pharosproject.net/uploads/files/cml/1360080729.pdf.

Harland, J.I., 2016. Functional dietary lipids. In: Woodhead Publishing Series in Food Science, Technology and Nutrition. Woodhead Publishing, Cambridge, UK, pp. 249–285. https://doi.org/10.1016/B978-1-78242-247-1.00010-7.

He, H.P., Cai, Y., Sun, M., Corke, H., 2002. Extraction and purification of squalene from Amaranthus grain. J. Agric. Food Chem. 50 (2), 368–372. https://doi.org/10.1021/jf010918p.

Heidtmann-bemvenuti, R., Nora, N.S., Badiale-furlong, E., 2012. Extraction of γ-Oryzanol from rice bran. Ciênc. Agrotec., Lavras 36 (6), 665–673.

Howard, L., 2016. Process of Producing Purified Gamma- and Deltab- Tocotrienols from Tocol-Rich Oils or Distillates. U.S. Provisional Patent, United State. Retrieved from https://encrypted.google.com/patents/US9512098.

Huang, Z.R., Lin, Y.K., Fang, J.Y., 2009. Biological and pharmacological activities of squalene and related compounds: potential uses in cosmetic dermatology. Molecules 14 (1), 540–554. https://doi.org/10.3390/molecules14010540.

Imsanguan, P., Roaysubtawee, A., Borirak, R., Pongamphai, S., Douglas, S., Douglas, P.L., 2008. Extraction of alpha-tocopherol and gamma-oryzanol from rice bran. LWT Food Sci. Technol. 41 (8), 1417–1424. https://doi.org/10.1016/j.lwt.2007.08.028.

Jaiswal, S.G., Pradhan, S., Patel, M., Naik, M., Naik, S., 2015. Rice bran oil distillate, a choice for γ-oryzanol: separation and oxidative stability study. J. Food Res. 4 (2), 36–43. https://doi.org/10.5539/jfr.v4n2p36.

Jala, R.C.R., Prasad, R.B.N., 2015. Rice bran lecithin: composition, nutritional and functional characteristics. In: Ahmad, M.U., Xu, X. (Eds.), Polar Lipids Biology, Chemistry, and Technology. AOAC Press, Urbana, IL, pp. 35–55.

Jangle, R.D., Magar, V.P., Thorat, B.N., 2013. Phosphatidylcholine and its purification from raw de-oiled soya lecithin. Sep. Purif. Technol. 102 (2013), 187–195. https://doi.org/10.1016/j.seppur.2012.10.002.

Jesus, S.P., Grimaldi, R., Hense, H., 2010. Recovery of γ-oryzanol from rice bran oil by product using supercritical fluid extraction. J. Supercrit. Fluids 55 (1), 149–155. https://doi.org/10.1016/j.supflu.2010.08.004.

Joshi, M., Kaur, R., Kanwar, P., Dhiman, G., Lata, S., Tilak, K., Gupta, N., Mishra, T., 2016. To evaluate antioxidant activity of gamma-oryzanol extracted from rice bran oil. Int. J. Pharm. Res. 6 (3), 17–25.

Kaewboonnum, P., Vechpanich, J., Santiwattana, P., Shotipruk, A., 2010. γ-Oryzanol recovery from rice bran oil soapstock. J. Sep. Sci. Technol. 45 (9), 1186–1195. https://doi.org/10.1080/01496391003775790.

Kangcare, 2012. Oroyal TM Gamma Oryzanol 99% (Vol. August 16). Retrieved from http://www.kangcare.com/editor/attached/file/20131030/20131030191112_51946.pdf.

Kaur, A., Jassal, V., Thind, S.S., Aggarwal, P., 2012. Rice bran oil an alternate bakery shortening. J. Food Sci. Technol. 49 (1), 110–114. https://doi.org/10.1007/s13197-011-0259-6.

Ko, S., Kim, C., Kim, C., Kim, H., Chung, S., Lee, S., Yoon, H., Diego, S., 2003. Changes of vitamin E content in rice bran with different heat treatment. Eur. J. Lipid Sci. Technol. 105, 225–228. https://doi.org/10.1002/ejlt.200390045.

Ko, S.N., Lee, S.M., Kim, I.H., 2008. The concentration of tocols from rice bran oil deodorizer distillate using solvent. Eur. J. Lipid Sci. Technol. 110 (10), 914–919. https://doi.org/10.1002/ejlt.200700310.

Kordsmeier, M., 2013. Extraction and Purification of Gamma-Tocotrienol From Rice Bran Oil Deodorizer Distillate. University of Arkansas. Retrieved from http://scholarworks.uark.edu/cgi/viewcontent.cgi?article=1704andcontext=etd.

Kordsmeier, M., Howard, L.R., Brownmiller, C., Proctor, A., Hauer-Jensen, M., 2015. Isolation of gmma and delta tocotrienols from rice bran oil deodorizer distillate using flash chromatography. J. Am. Oil Chem. Soc. 92 (9), 1243–1252. https://doi.org/10.1007/s11746-015-2696-2.

Kraujalis, P., Venskutonis, P.R., 2013. Supercritical carbon dioxide extraction of squalene and tocopherols from amaranth and assessment of extracts antioxidant activity. J. Supercrit. Fluids 80, 78–85. https://doi.org/10.1016/j.supflu.2013.04.005.

Krishna, A.G.G., Khatoon, S., Shiela, P.M., Sarmandal, C.V., Indira, T.N., Mishra, A., 2001. Effect of refining of crude Rice bran oil on the retention of oryzanol in the refined oil. J. Am. Oil Chem. Soc. 78 (2), 127–131.

Krishna, A.G.G., Hemakumar, K.H., Khatoon, S., 2006. Study on the composition of rice bran oil and its higher free fatty acids value. J. Am. Oil Chem. Soc. 83 (2), 117–120.

Kumar, N., Pruthi, V., 2014. Potential applications of ferulic acid from natural sources. Biotechnol. Rep. 4 (1), 86–93. https://doi.org/10.1016/j.btre.2014.09.002.

Lampi, A.-M., Kmal-Eldin, Piironen, V., 2002. Functional foods and nutraceutical series. In: Functional Foods Biochemical and Processing Aspects. second ed. CRC Press, London, pp. 1–29.

Liu, D., Ma, F., 2011. Krezhova, D. (Ed.), Soybean Phospholipids, Recent Trends for Enhancing the Diversity and Quality of Soybean Products. InTech, Rijeka, Croatia. ISBN: 978-953-307-533-4. Available from:http://www.intechopen.com/books/recent-trends-for-enhancing-the-diversity-and-quality-of-soybean-products/soybean-phospholipids.

Liu, K.-T., Gao, S., Chung, T.-W., Huang, C., Lin, Y.-S., 2012. Effect of process conditions on the removal of phospholipids from Jatropha curcas oil during the degumming process. Chem. Eng. Res. Des. 90 (9), 1381–1386. https://doi.org/10.1016/j.cherd.2012.01.002.

Liu, L., Waters, D.L.E., Rose, T.J., Bao, J., King, G.J., 2013. Phospholipids in rice: significance in grain quality and health benefits: a review. Food Chem. 139 (1–4), 1133–1145. https://doi.org/10.1016/j.foodchem.2012.12.046.

Lohry, R., 2007. Micronutrients: functions, sources and application methods. Indiana CCA Conference Proceedings (Cl)p. 15.

Machado, A.R., De Assis, L.M., Machado, M.I.R., Souza-soares, L.A., 2015. Importance of lecithin for encapsulation processes. African J. Food Sci. 8, 176–183. https://doi.org/10.5897/AJFS2013.1092.

Mazhar, B., Jahan, N., Ali, N.M., Andleeb, S., Ali, S., 2017. Production of vanillin by a novel bacterium from waste residues of rice bran oil. Punjab Univ. J. Zool. 32 (1), 137–142. https://doi.org/10.3724/SP.J.1258.2011.00337.

Nagendra, P.M., Sanjay, K.R., Shravya, K.M., Vismaya, M., Nanjunda, S.S., 2015. Health benefits of rice bran—a review. Nutr. Food Sci. 1 (3), 1–8. https://doi.org/10.4172/2155-9600.1000108.

Nandi, S., Bhattacharyya, R., 2017. Isolation of squalene from rice bran oil fatty acid distillate using bioprocess technology. Int. J. Res. Appl. Sci. Eng. Technol. 5 (IX), 509–512.

Narayan, A.V., Barhate, R.S., Raghavarao, K.S.M.S., 2006. Extraction and purification of oryzanol from rice bran oil and rice bran oil soapstock. J. Am. Oil Chem. Soc. 83 (8), 663–670. https://doi.org/10.1007/s11746-006-5021-2.

Nayik, G.A., Muzaffar, K., 2015. Future edible oil of India: Rice bran oil – "The wonder oil" SM J. Food Nutri. Disord. 1 (2), 1007.
Nayik, G.A., Majid, I., Gull, A., Muzaffar, K., 2015. Rice bran oil, the future edible oil of India: a mini review. Rice Res.: Open Access 3 (4), 4–6. https://doi.org/10.4172/2375-4338.1000151.
Naziri, E., Mantzouridou, F., Tsimidou, M.Z., 2011. Squalene resources and uses point to the potential of biotechnology. Lipid Technol. 23 (12), 270–273. https://doi.org/10.1002/lite.201100157.
Nguyen, 1999. Recent Advances in Nutritional Sciences. The Cholesterol-Lowering Action of Plant Stanol Esters. American Society for Nutritional Sciences, .pp. 2109–2112. https://academic.oup.com/jn/article-abstract/129/12/2109/4722154.
Niawanti, H., Zullaikah, S., 2017. Removal of bioactive compound (γ-oryzanol) from rice bran oil-based biodiesel using deep eutectic solvent. Chem. Eng. Trans. 56, 1513–1518. https://doi.org/10.3303/CET1756253.
Oliveira, M.S., Feddern, V., Kupski, L., Cipolatti, E.P., Badiale-Furlong, E., Souza-Soares, L.A., 2011. Changes in lipid, fatty acids and phospholipids composition of whole rice bran after solid-state fungal fermentation. Bioresour. Technol. 102 (17), 8335–8338. https://doi.org/10.1016/j.biortech.2011.06.025.
Orthoefer, F.T., 2005. Rice bran oil. Bailey's Indust. Oil Fat Prod. 2 (7), 465–489. https://doi.org/10.1002/047167849X.bio015.
Osborn, S., Morley, W., 2016. Developing food products for consumers with specific dietary needs. In: Woodhead Publishing Series in Food Science, Technology and Nutrition. . ISBN: 978-0-08-100340-4pp. 1–278. https://doi.org/10.1016/C2014-0-03773-2.
Özdestan, Ö., Erol, T., Acar, B., 2014. Phytosterols in rice bran and usage of rice bran in food industry. 9th Baltic Conference on Food Science and Technology" Food for Consumer Well-Being" FoodBalt 2014, Jelgava, Latvia, 8–9 May, 2014vol. 1. , pp. 24–27.
Pal, Y.P., Pratap, A.P., 2017. Rice bran oil: a versatile source for edible and industrial applications. J. Oleo Sci. 556 (6), 551–556.
Pali, V., 2013. Rice bran oil- unique gift of nature: a review. Agric. Rev. 34 (4), 288–294.
Patel, M., Naik, S.N., 2004. Gamma-oryzanol from rice bran oil—a review. J. Sci. Ind. Res. 63 (7), 569–578.
Patil, V.V., Galge, R.V., Thorat, B.N., 2010. Extraction and purification of phosphatidylcholine from soyabean lecithin. Sep. Purif. Technol. 75 (2010), 138–144. https://doi.org/10.1016/j.seppur.2010.08.006.
Perma Health Care, 2018. Squalene—Sources and Properties. Retrieved February 10, 2018, from https://www.permahealthcare.com/squalene-sources.html.
Pestana-Bauer, V.R., Zambiazi, R.C., Mendonca, C.R.B., Beneito-Cambra, M., Ramis-Ramos, G., 2012. Gamma-oryzanol and tocopherol contents in residues of rice bran oil refining. Food Chem. 134 (3), 1479–1483. https://doi.org/10.1016/j.foodchem.2012.03.059.
Pietsch, A., Jaeger, P., 2007. Concentration of squalene from shark liver oil by short-path distillation. Eur. J. Lipid Sci. Technol. 109 (11), 1077–1082. https://doi.org/10.1002/ejlt.200700039.
Plat, J., Mensink, R.P., 2005. Plant stanol and sterol esters in the control of blood cholesterol levels: mechanism and safety aspects. Am. J. Cardiol. 96(1 Suppl)https://doi.org/10.1016/j.amjcard.2005.03.015.
Popa, O., Bəbeanu, N.E., Popa, I., Niţə, S., Dinu-Pârvu, C.E., 2015. Methods for obtaining and determination of squalene from natural sources. Biomed. Res. Int. 2015, 1–16. https://doi.org/10.1155/2015/367202.
Popa, I., Babeanu, E.N., Nita, S., Popa, O., 2016. Squalene-natural resources and applications. Farmacia 62 (5), 840–862.

Rajan, R.R.G., Krishna, A.G.G., 2014. A simple method for purification of deodorizer distillate from Indian rice (*Oryza sativa*) bran oil and preparation of phytosterols. Grasas Aceites 65 (4), e050. https://doi.org/10.3989/gya.0229141.

Rao, B.N., 2000. Nutritive value of rice bran. Nutr. Found. India Bull. 21 (4), 5–7. Retrieved from http://www.nutritionfoundationofindia.org/pdfs/BulletinArticle/Pages_from_nfi_10_00_2.pdf.

Ronco, A.L., De Stéfani, E., Ronco, A., 2013. Squalene: a multi-task link in the crossroads of cancer and aging. Funct. Foods Health Dis. 3 (12), 462–476.

Rosales-garcía, T., Jimenez-martinez, C., Dávila-ortiz, G., 2017. Squalene extraction: biological sources and extraction methods. Int. J. Environ. Agric. Biotechnol. (4), 1662–1670.

Ruen-ngam, D., Thawai, C., Sukonthamut, S., 2016. Pretreatment to increase yield and antioxidant activity of γ-oryzanol in rice bran oil. Sci. Asia 42, 75–82. https://doi.org/10.2306/scienceasia1513-1874.2016.42.075.

Saini, R.K., Keum, Y.-S., 2016. Tocopherols and tocotrienols in plants and their products: a review on methods of extraction, chromatographic separation, and detection. Food Res. Int. 82 (February), 59–70. https://doi.org/10.1016/j.foodres.2016.01.025.

Sakunpak, A., Suksaeree, J., Pathompak, P., Sermkaew, N., 2014. Antioxidant individual γ-oryzanol screening in cold pressed rice bran oil of different thai rice varieties by HPLC-DPPH method. Int J Pharm Pharm Sci 6 (6), 2–7.

Sanghi, D., Tiwle, R., 2015. A rievew of comparative study of rice bran oil and rice bran wax. Int. J. Pharm. Rev. Res. 5 (4), 403–410.

Sarmento, C.M.P., Ferreira, S.R.S., Hense, H., 2006. Supercritical fluid extraction (SFE) of rice bran oil to obtain fractions enriched with tocopherols and tocotrienols. Braz. J. Chem. Eng. 23 (2), 243–249. https://doi.org/10.1590/S0104-66322006000200012.

Sawadikiat, P., Hongsprabhas, P., 2014. Phytosterols and γ-oryzanol in rice bran oils and distillates from physical refining process. Int. J. Food Sci. Technol. 49 (9), 2030–2036. https://doi.org/10.1111/ijfs.12506.

Sengar, G., Kaushal, P., Sharma, H.K., Kaur, M., 2014. Degumming of rice bran oil. Rev. Chem. Eng. 30 (2), 183–198. https://doi.org/10.1515/revce-2013-0030.

Siriluck, P., Aryusuk, K., Krisnangkura, K., 2012. Purification of phosphatidylcholine from crude soybean lecithin. RMUTP Res. J. Spec. Issue 111–119.

Snyder, J.M., King, J.W., Taylor, S.L., Neese, A.L., 1999. Concentration of phytosterols for analysis by supercritical fluid extraction. J. Am. Oil Chem. Soc. 76 (6), 717–721. https://doi.org/10.1007/s11746-999-0165-5.

Sohail, M., Rakha, A., Butt, M.S., Iqbal, M.J., Rashid, S., 2017. Rice bran nutraceutics: a comprehensive review. Crit. Rev. Food Sci. Nutr. 57 (17), 3771–3780. https://doi.org/10.1080/10408398.2016.1164120.

Spanova, M., Daum, G., 2011. Squalene—biochemistry, molecular biology, process biotechnology, and applications. Eur. J. Lipid Sci. Technol. 113 (11), 1299–1320. https://doi.org/10.1002/ejlt.201100203.

Sugihara, N., Kanda, A., Nakano, T., Nakamura, T., Igusa, H., Hara, S., 2010. Novel fractionation method for squalene and phytosterols contained in the deodorization distillate of rice bran oil. J. Oleo Sci. 59 (2), 65–70. https://doi.org/10.5650/jos.59.65.

Sumit, N., 2017. Enzymatic synthesis of Lysophospholipids from crude rice bran phospholipids. Indian J. Sci. 24 (89), 1–6.

Thongchai, W., Liawruangrath, B., 2016. Determination of gamma oryzanol in rice bran oil by HPLC with molecularly imprinted solid-phase extraction. Int. Food Res. J. 23 (4), 1389–1395.

Tyagi, K., Ansari, M.A., Tyagi, S., Tyagi, A., Tech, S.D.I., College, B.B., Meerut, M.C., 2012. A novel process for physically refining rice bran oil through degumming. Adv. Appl. Sci. Res. 3 (3), 1435–1439.

Ugandar, R., Nilugal, C.K., Lyn, D.L., 2015. In-vitro and in-vivo evaluation of dirctly compressed tablets of simvastatin with soy lecithin. World J. Pharm. Res. 4 (8), 2693–2718.
Vissers, M.N., Zock, P.L., Meijer, G.W., Katan, M.B., 2000. Effect of plant sterols from rice bran oil and triterpene alcohols from sheanut oil on serum lipoprotein concentrations in humans. Am. J. Clin. Nutr. 72 (March), 1510–1515. Retrieved from http://www.scopus.com/inward/record.url?eid=2-s2.0-84905252059andpartnerID=40andmd5=98cf0506cae284ca62dd9c34a7898c89.
Wang, T., 2002. Antioxidant activity of phytosterols, oryzanol, and other phytosterols. J. Am. Oil Chem. Soc. 79 (12), 1201–1206. https://doi.org/10.1007/s11746-002-0628-x.Rights.
Wang, Q., 2016. Peanut byproducts utilization technology. In: Peanuts: Processing Technology and Product Development. Academic Press, Beijing, China, pp. 211–325. https://doi.org/10.1016/B978-0-12-809595-9.00005-3.
Wu, Y., Wang, T., 2003. Soybean lecithin fractionation and functionality. J. Am. Oil Chem. Soc. 80 (4), 319–326. https://doi.org/10.1007/s11746-003-0697-x.
Xu, Z., Godber, J.S., 2000. Comparison of supercritical fluid and solvent extraction methods in extracting γ-oryzanol from rice bran. J. Am. Oil Chem. Soc. 77 (5), 547–551. https://doi.org/10.1007/s11746-000-0087-4.
Yukihiro, Y., Hara, S., 2012. Novel fractionation method for squalene and phytosterols contained in the deodorization distillate of rice bran oil. In: Valdez, B. (Ed.), Scientific, Heath and Social Aspect of the Food Industry. IN TECH Open Science/Open Minds, pp. 72–82. Retrieved from www.intechopen.com.
Zheng, L., Zheng, P., Sun, Z., Bai, Y., Wang, J., Guo, X., 2007. Production of vanillin from waste residue of rice bran oil by *Aspergillus niger* and *Pycnoporus cinnabarinus*. Bioresour. Technol. 98 (5), 1115–1119. https://doi.org/10.1016/j.biortech.2006.03.028.
Zielinska, A., Nowak, I., 2014. Tocopherols and tocotrienols as vitamin E. Chem. Aust. 68 (7), 585–591.
Zullaikah, S., Melwita, E., Ju, Y.H., 2009. Isolation of oryzanol from crude rice bran oil. Bioresour. Technol. 100 (1), 299–302. https://doi.org/10.1016/j.biortech.2008.06.008.

FURTHER READING

McCaskill, D.R., Zhang, F., 1999. Use of rice bran oil in foods: developing nutraceuticals for the new millenium. Food Technol. 53 (2), 50–53.
Sugano, M., Tsuji, E., 1996. Rice bran oil and human health. Biomed. Environ. Sci. 9 (2–3), 242–246.
Xu, Z., Godber, J.S., 1999. Purification and identification of components of γ-oryzanol in rice bran oil. J. Agric. Food Chem. 47 (7), 2724–2728. https://doi.org/10.1021/jf981175j.

CHAPTER 6

Applications of Rice Bran Oil

Yong Wang
Wilmar (Shanghai) Biotechnology Research & Development Center Co., Ltd, Shanghai, China

1. INTRODUCTION

Rice bran oil (RBO) has a balanced fatty acid composition and high levels of functional ingredients such as oryzanol, phytosterols, tocopherols, tocotrienols, and other nutrients. It has been demonstrated to have positive effects on modulation of cholesterol, hypertension, hyperglycemia, insomnia, and other chronic diseases. The unique characteristics of RBO has rendered it useful in a broad application of areas including food, pharmaceutics, cosmetics, industries, etc. Currently, there are many global interests to diversify the applications of RBO (Liang et al., 2014).

2. FOOD APPLICATIONS

2.1 Cooking

Unlike soybean oil and rapeseed oil, which have strong aggressive flavors that might cover the natural flavor of food, RBO is a mild-flavor oil that can be used in various cuisines all across the world. RBO has a very high smoke point (>200°C) and ignition point (~350°C). Hence, it is very stable with a low level of degradation and polymerization during cooking (Sharma and Das, 2013). Another unique characteristic of RBO is the viscosity. Fig. 1 demonstrates the changes in viscosity with temperature of various oils. RBO, which has an oryzanol content of >10,000 mg/kg (oryzanol content may influence viscosity), has the highest viscosity among the various oils, even higher than that of extra virgin olive oil. This indicates a very good dressing performance for RBO, especially in some Chinese cuisine. Due to high viscosity, RBO can be easily retained on the surface of food, making the food shiny and appetizing.

Rice Bran and Rice Bran Oil
https://doi.org/10.1016/B978-0-12-812828-2.00006-8

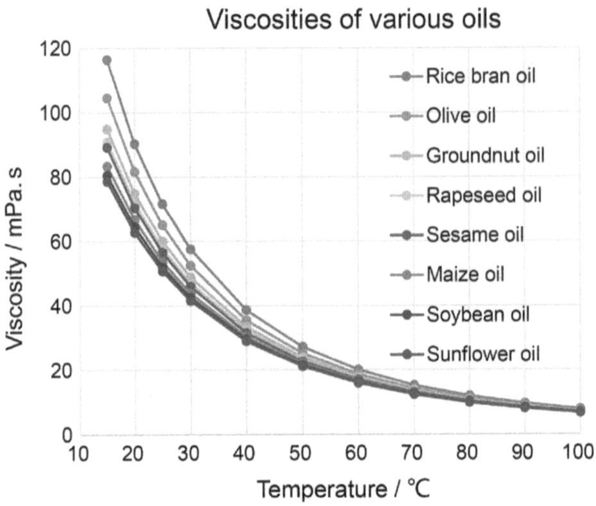

Fig. 1 Changes of RBO viscosities with temperature of various oils.

2.2 Deep Frying

2.2.1 Stability

Oxidative stability of RBO is equivalent to that of peanut and cottonseed oils in deep-frying applications (Shahidi, 2005a). Presence of natural antioxidants such as tocopherols and γ-oryzanol in RBO can decrease the oxidation rate of oil during frying periods (Fan et al., 2013). In comparison to other vegetable oils, quality parameters of RBO including acid value (AV), peroxide value (PV), total polar compounds (TPC), and degree of polymerization (DP) demonstrated by a slower increment upon deep frying. Low degree of polymerization makes it easier to clean cooking appliances. Studies have shown that snacks prepared using RBO absorbed 12%–25% lesser oil as compared to that prepared using groundnut oil. Another advantage of RBO is that food fries faster and absorbs lesser oil, yet it also has an excellent quality and oxidative stability (Choudhary and Grover, 2013). Oxidative stability index (OSI) measurement (120°C, 20 L/h) shows RBO has a higher stability than other vegetable oils (Fig. 2).

2.2.2 Flavor

RBO has a very pleasant sweet flavor upon heating, which might be attributed to presence of vanillin. Vanillin is one of the most widely used flavoring agents. Flavor compound present in unheated and heated RBO (180°C) are evaluated using GC-MS. Fig. 3 shows no vanillin was detected in

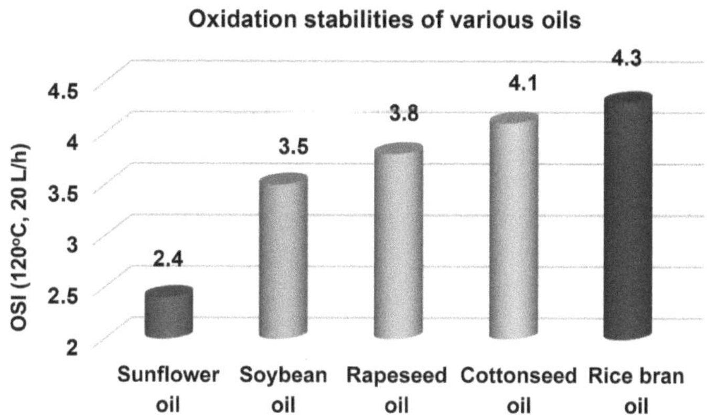

Fig. 2 Oxidation stabilities evaluation of various oils.

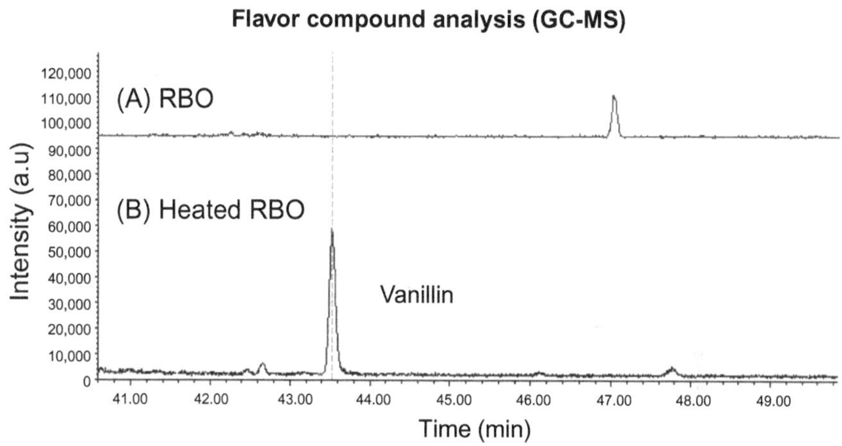

Fig. 3 Flavor compound analysis in RBO.

unheated RBO (19 flavor substances in total, threshold: 0.1%); meanwhile, vanillin can be detected in the heated RBO (55 flavor substances in total, threshold: 0.1%).

Presence of vanillin upon heating might be due to hydrolysis of a unique element in RBO-oryzanol into sterols/triterpene alcohols and ferulic acid. Ferulic acid is the precursor of vanillin (fermentation) (Muheim and Lerch, 1999). Although the actual reaction route or mechanism remains unreported, we believe that vanillin derived from oryzanol resulted in the pleasant flavor in heated RBO. At the same time, RBO produces less bad flavor substance (acrolein) that causes an unpleasant flavor when heated than other vegetable oils. Due to the pleasant aroma upon heating, the price of RBO in

Japan is three times more expensive than other edible oils. Nevertheless, RBO remained a common frying oil used in mid- to high-end restaurants in Japan.

2.2.3 Cuisine Case: Tempura

Tempura is a classical Portuguese dish brought to and popularized by Japan, consisting of seafood or vegetables that have been battered and deep-fried at 170–180°C (Wikipedia, 2018). Tempura has a strict demand on the frying oil, which needs gentle flavor, but not aggressive, with less oil absorption. Thus RBO is an excellent choice for tempura preparation, especially in Kyota cuisine where nonsticky nor smeary tempura is sought after. Japanese chefs including Kikunoi (three Michelin stars) and Nakahigashi (two Michelin stars) have been reported to favor RBO as cooking oil. One of Nakahigashi's specialties is crispy fish bone deep-fried with RBO. Although RBO has not been very popular in Japanese homemade cooking yet, high-end chefs of Kyoto cuisine are very interested in using RBO for cooking (ICRBO, 2016). Kondo Fumio, Japanese Michelin's two-star Tempura chef for 6 consecutive years who devoted 52 years frying tempura, also speaks highly of RBO as a very healthy oil with good flavor.

2.3 Salad Dressing and Baking

Winterized RBO is an acceptable oil for salad dressing and mayonnaise. The hard fraction of RBO may be used to replace plastic fats in margarines and shortening. Hydrogenated RBO is adaptable to specialty shortenings and margarines (Shahidi, 2005b). RBO is good for grilling/baking food and also for salad dressings as it is easily emulsified (Sharma and Das, 2013). RBO without winterizing is also used as a demolding oil (release agent).

3. FUNCTIONAL FOOD APPLICATIONS

RBO comes from rice bran and rice germ. The weight of rice bran is only 7%–10% (by weight) of the rice grain, but 64% of its nutrient is enriched in them. Considering the oil yield (12%–18%) and practical refining yield (50%–70%), RBO is a very precious oil. Almost 150 kg paddy can only produce 1 L of end product. RBO has balanced fatty acid composition (closest to WHO recommendation: SFA/MUFA/PUFA = 24%/42%/34%) and is rich in unsaponifiable matter (~4%) including phytosterols, tocopherols, tocotrienols, squalene, and most importantly, oryzanol (Rukmini, 1988).

History has recorded the use of RBO as a functional food. RBO originated from Japan and has over a 100-year-old history dating back to the "Edo period". Records in Chinese masterworks including "Compendium of Materia Medica," "Chinese Medicated Diet Dictionary" states that paddy essence is called "poor man's ginseng soup". In Japan, about 40% of kindergartens, nurseries, elementary, and middle schools use RBO for cooking during lunch and dinner meals. Multiple associations across the world have stated that RBO is a healthy oil, including World Health Organization (WHO), American Heart Association (AHA), National Institute Of Nutrition (NIN, India), Indian Council Of Medical Research (ICMR, India), the Japan Oil Chemists' Society (JOCS, Japan), and Chinese Cereals And Oils Association (CCOA, China) (Ahmad Nayik et al., 2015; Japan Oilseeds Processors Association, 2018). As well, Japanese nutrition expert Dr. Teruo Miyazawa and the deputy prime minister of Thailand (Dr. Prajin Juntong) have advocated the consumption of nutritious RBO publicly in their respective countries (ICRBO, 2017).

RBO is popular in many Asian countries such as Japan, India, China, Thailand, Vietnam, Pakistan, and Bangladesh as a cooking oil. New RBO products are released every year. In recent years, research interest has been growing in RBO processing to obtain good quality oil with low refining loss and high nutrient reservation (Pali, 2013). In 2017, Wilmar launched a new product called "Double 10,000 RBO", which emphasized high concentrations of both oryzanol and phytosterol contents in their product ($\geq$10,000 mg/kg). This is a balanced combination and represents the future trend that nutrients should be reserved as much as possible. However, too high nutrient concentration approaching crude RBO are not recommended; that oil quality is hard to assure.

4. PHARMACEUTICAL APPLICATIONS

Clinical trials or animal experiments have demonstrated that RBO with high oryzanol content has many positive effects on modulating cholesterol, hypertension, hyperglycemia, and sleeping. This is an excellent food to reduce risk of chronic diseases caused by modern stressful life. Dietary therapy is the best way as it works silently and consistently. Also it has higher consumer acceptance as every medicine has its side effects. More importantly, dietary therapy can help primary prevention and treatment.

4.1 Cholesterol-Lowering Effect

Multiple cases have proven that RBO has a cholesterol-lowering effect (Lichtenstein et al., 1994; Ausman et al., 2005). The hypercholesterolemia-inhibition mechanism of RBO can inhibit HMG-CoA to reduce synthesis of cholesterol. In addition, it increased CYP7A1 activity to promote cholesterol synthesis of bile acids, and excreted bile acid can result in a decrease of plasma cholesterol (Liang et al., 2014).

4.2 Antihypertension

In 2016, Devarajan conducted a large-scale clinical trial, which demonstrated that a blend of RBO and sesame oil can lower blood pressure and improve blood lipid profile in mild-to-moderate hypertensive patients. The oil blend consisted of 80% physically refined γ-oryzanol-rich (10,000 mg/kg) RBO and 20% unrefined, cold-pressed, lignans-rich sesame oil. Three-hundred hypertensive patients and 100 normotensives were divided into four groups as: (1) normotensives treated with blend oil, (2) hypertensives treated with blend oil, (3) hypertensives treated with medicine, and (4) hypertensives receiving a combination of blend oil and medicine. The result demonstrated that using a blend of RBO and sesame oil as cooking oil showed a significant antihypertensive and lipid-lowering action, and had a noteworthy additive effect with antihypertensive medication (Devarajan et al., 2016a).

4.3 Hyperglycemia-Lowering Effect

Another large-scale clinical trial published in 2016 demonstrated that a blend of RBO and sesame oil lowers hyperglycemia and improves blood lipids profile. The oil blend consists of 80% physically refined γ-oryzanol-rich (10,000 mg/kg) RBO and 20% unrefined, cold-pressed, lignans-rich sesame oil. Three-hundred type 2 diabetes mellitus patients and 100 normoglycemic subjects were grouped as (1) normoglycemic subjects treated with blend oil, (2) type 2 diabetes mellitus patients treated with blend oil, (3) type 2 diabetes mellitus patients treated with medicine, and (4) type 2 diabetes mellitus patients treated with a combination of medicine and blend oil. The result demonstrated that a blend of RBO and sesame oil as cooking oil lowered hyperglycemia and improved the lipid profile in type 2 diabetes mellitus patients and had noteworthy additive effect with medication (Devarajan et al., 2016b).

4.4 Insomnia Alleviation

Yang et al. (2014) conducted an animal study to evaluate the effects of RBO rich in oryzanol on pentobarbital-induced sleeping behaviors in partial sleep-deprived mice through modulation of monoamines. Sixty mice were randomly divided into five groups (12 for each). Control and partial sleep deprivation (PSD) model groups were fed with basic rodent chow with 8% soybean oil (oryzanol-free). PSD-low oryzanol (PSD-lOZ), PSD-medium oryzanol (PSD-mOZ), and PSD-high oryzanol (PSD-hOZ) groups were fed basic rodent chow with 8% RBO, containing 3000, 7000, and 15,000 mg/kg of oryzanol, respectively for 25 d. The result demonstrated that RBO rich in oryzanol may alleviate fatigue and improve sleep in mice with PSD through modulation of monoamines (Yang et al., 2014).

4.5 Other Functions

Besides reducing cholesterol, hypertension, hyperglycemia, and alleviating insomnia, RBO has been proven to have multiple functions such as antioxidation (Hsieh et al., 2005), anticancer (Shih et al., 2011), immune modulatory (Sierra et al., 2005), and anti-inflammatory (Joshi et al., 2015) effects. It is also believed that oryzanol within RBO can help eliminate pica in children by adjusting brain and autonomic function. In addition, oryzanol also has a growth-promoting effect on teenagers and has an effect on curing sweating syndrome of children (https://e.uuuwell.com). However, more systematic experiments need to be done to scientifically evaluate the functions of RBO.

5. COSMETIC APPLICATIONS

Use of RBO grows as a specialty ingredient in the cosmetic/personal care market. The demand is for natural, value-added healthy ingredients. On one hand, nanoemulsions of RBO could improve physical stability and moisturizing activity on skin, so that RBO could be used in cosmetics (Bernardi et al., 2011). RBO is beneficial for baby care as a moisturizer, especially for those with sensitive skin. On the other hand, as RBO is rich in antioxidants such as vitamin E and gamma-oryzanols, it is used in skin creams and soaps, which claim to slow down aging and the appearance of facial wrinkles. Japanese women, who apply RBO on their faces to keep their skin smooth and shiny, are called Nuka-Bijin (bran beauties). Massage oils containing oils of rice bran, sesame, jojoba, apricot, and almond are already available in the market. RBO is also used in cosmetics like lipsticks, sunscreen products (as it intercepts ultraviolet rays and thus impedes the melanin pigmentation), and hair conditioners (Sharma and Das, 2013). The probable mechanism is that RBO contains approximately 500 ppm of tocotrienols. When applied to the skin, tocotrienols will penetrate and is absorbed rapidly by skin. Tocotrienols accumulates in the skin and acts as the first defense layer with antioxidant properties. As a consequence, tocotrienols stabilize the free radicals generated in the skin when exposed to oxidative rays. Tocotrienols will thus protect against skin damage induced by ultraviolet rays (Rohman, 2014).

6. INDUSTRIAL APPLICATIONS

6.1 Byproducts

During the processing of RBO, quite a lot of byproducts can be acquired such as oryzanol, lecithin, rice bran wax, rice bran stearin, rice bran meal, fatty acid, and sterols. Among these byproducts, oryzanol is a unique product derived from RBO soap stock (Indira et al., 2005), which is almost the only source of oryzanol in industry (Patel and Naik, 2004).

6.2 Biodiesels

Biodiesel received increased attention as a nontoxic, biodegradable, and renewable diesel fuel. The main concern with biodiesel fuel is its high price. One of the future aims in biodiesel research is on the selection of inexpensive feedstock with high value-added byproducts. RBO is a relatively inexpensive raw material for the production of biodiesel. The utilization of byproduct such as defatted rice bran for the production of proteins, carbohydrates, phytochemical, and the isolation and purification of value-added nutraceuticals generated during biodiesel production from RBO are attractive options to lower the cost of biodiesel. Production of biodiesel from RBO can be carried out either via in situ esterification, lipase-catalyzed esterification, and acid-catalyzed or base-catalyzed reactions (Ju and Vali, 2005).

7. CONCLUSION

RBO is still an underutillized oil. Due to its novel stability and pleasant flavor, RBO is ideal for frying and cooking. RBO contains multiple nutrients such as γ-oryzanols, phytosterols, tocopherols, and tocotrienols, and it has been reported to have several beneficial modulating effects on cholesterol, hypertension, hyperglycemia, insomnia, etc. RBO can also be used in cosmetics and industries. Researchers all across the world should join hands and conduct more studies on this nutritious oil, and we should advocate dietary therapy and the consumption of RBO to prevent chronic disease and to have a healthier life.

REFERENCES

Ahmad Nayik, G., Majid, I., Gull, A., Muzaffar, K., 2015. Rice bran oil, the future edible oil of India: a mini review. Rice Res.: Open Access 03(04).

Ausman, L.M., Rong, N., Nicolosi, R.J., 2005. Hypocholesterolemic effect of physically refined rice bran oil: studies of cholesterol metabolism and early atherosclerosis in hypercholesterolemic hamsters. J. Nutr. Biochem. 16 (9), 521–529.

Bernardi, D.S., Pereira, T.A., Maciel, N.R., Bortoloto, J., Viera, G.S., Oliveira, G.C., Rocha-Filho, P.A., 2011. Formation and stability of oil-in-water nanoemulsions containing rice bran oil: in vitro and in vivo assessments. J. Nanobiotechnol. 9, 44.

Choudhary, M., Grover, K., 2013. Blended rice bran and olive oil—moving towards a new cooking media. Int. J. Life Sci. Educ. Res. 1 (1), 14–20.

Devarajan, S., Singh, R., Chatterjee, B., Zhang, B., Ali, A., 2016a. A blend of sesame oil and rice bran oil lowers blood pressure and improves the lipid profile in mild-to-moderate hypertensive patients. J. Clin. Lipidol. 10 (2), 339–349.

Devarajan, S., Chatterjee, B., Urata, H., Zhang, B., Ali, A., Singh, R., Ganapathy, S., 2016b. A blend of sesame and rice bran oils lowers hyperglycemia and improves the lipids. Am. J. Med. 129 (7), 731–739.

Fan, H.Y., Sharifudin, M.S., Hasmadi, M., Chew, H.M., 2013. Frying stability of rice bran oil and palm olein. Int. Food Res. J. 20 (1), 403.

Hsieh, R.H., Lien, L.M., Lin, S.H., Chen, C.W., Cheng, H.J., Cheng, H.H., 2005. Alleviation of oxidative damage in multiple tissues in rats with streptozotocin-induced diabetes by rice bran oil supplementation. Ann. N. Y. Acad. Sci. 1042 (1), 365–371.

ICRBO, 2016. From International Conference on Rice Bran Oil, ICRBO2016, Japan.

Indira, T.N., Narayan a, V., Barhate, R.S., et al., 2005. Process for the production of oryzanol enriched fraction from rice bran oil soapstock: US. US6896911.

Japan Oilseeds Processors Association, 2018. http://www.oil.or.jp/.

Joshi, S., Devaraju, C.J., Upadya, H., 2015. Anti-inflammatory properties of blended edible oil with synergistic antioxidants. Indian J. Endocrinol. Metab. 19 (4), 511.

Ju, Y., Vali, S.R., 2005. Rice bran oil as a potential resource for biodiesel: a review. J. Sci. Ind. Res. 64, 866–882.

Liang, Y., Gao, Y., Lin, Q., Luo, F., Wu, W., Lu, Q., Liu, Y., 2014. A review of the research progress on the bioactive ingredients and physiological activities of rice bran oil. Eur. Food Res. Technol. 238 (2), 169–176.

Lichtenstein, A.H., Ausman, L.M., Carrasco, W., Gualtieri, L.J., Jenner, J.L., Ordovas, J.M., Nicolosi, R.J., Goldin, B.R., Schaefer, E.J., 1994. Rice bran oil consumption and plasma lipid levels in moderately hypercholesterolemic humans. Arterioscler. Thromb. Vasc. Biol. 14 (4), 549–556.

ICRBO, 2017. 4th International Conference on Rice Bran Oil, Bangkok, Thailand.

Muheim, A., Lerch, K., 1999. Towards a high-yield bioconversion of ferulic acid to vanillin. Appl. Microbiol. Biotechnol. 51 (4), 456–461.

Pali, V., 2013. Rice bran oil-unique gift of nature: a review. Agric. Rev. 34 (4), 288–294.

Patel, M., Naik, S.N., 2004. Gamma-oryzanol from rice bran oil—a review. J. Sci. Ind. Res. 63, 569–578.

Rohman, A., 2014. Rice bran oil's role in health and cooking. In: Wheat and Rice in Disease Prevention and Health. Academic Press, Cambridge, MA, pp. 481–490 (Chapter 37).

Rukmini, C., 1988. Chemical, nutritional and toxicological studies of rice bran oil. Food Chem. 30 (4), 257–268.

Shahidi, F., 2005a. Edible oil and fat products: products and applications. In: sixth ed Bailey's Industrial Oil and Fat Products, vol. 4. John Wiley & Sons, Inc., Hoboken, NJ.
Shahidi, F., 2005b. Edible oil and fat products: edible oils. In: sixth ed Bailey's Industrial Oil and Fat Products, vol. 2. John Wiley & Sons, Inc., Hoboken, NJ.
Sharma, S.G., Das, A., 2013. Rice bran oil—A cooking medium with health benefits. In: CRRI Technology Bulletin-61.
Shih, C., Ho, C., Li, S., Yang, S., Hou, W., Cheng, H., 2011. Preventive effects of rice bran oil on 1,2-dimethylhydrazine/dextran sodium sulphate-induced colon carcinogenesis in rats. Food Chem. 126 (2), 562–567.
Sierra, S., Lara Villoslada, F., Olivares, M., Jiménez, J., Boza, J., Xaus, J., 2005. Increased immune response in mice consuming rice bran oil. Eur. J. Nutr. 44 (8), 509–516.
Wikipedia, 2018. Tempura. Wikipedia website, https://en.m.wikipedia.org/wiki/Tempura.
Yang, Y., Chen, G., Zhu, C., Luo, X., Huang, B., Zhu, H., 2014. Effects of rice bran oil rich in oryzanol on pentobarbital-induced sleeping behaviors in partial sleep-deprived mice through modulation of monoamines. Acta Nutrimenta Sin. 36 (6), 577–583.

CHAPTER 7

Analytical Aspects of Rice Bran Oil

Dongping He, Lingyi Liu
Wuhan Polytechnic University, Wuhan, China

Among the vegetable oils, rice bran oil (RBO) is an excellent source of nutritionally beneficial compounds, such as sterols, tocopherols, and tocotrienols (Yang et al., 2018). RBO has a balanced fatty acid composition, with 47% of monounsaturated fatty acids (MUFAs), 33% of polyunsaturated fatty acids (PUFAs), and 20% of saturated fatty acid (SFA). As an essential part in understanding the composition and quality of RBO, the related methods will be summarized in this chapter.

1. COMPOSITIONAL ANALYSIS

1.1 Fatty Acid Composition

Palmitic (17.0%), oleic (47%), and linoleic (32.65%) acids make up more than 90% of the fatty acid in RBO (Yang et al., 2018). With respect to its structure, fatty acid can be categorized into several groups with different functions, including saturated fatty acid (SFA), monounsaturated fatty acid (MUFA), and polyunsaturated fatty acid (PUFA). RBO is characterized by a balanced MUFA/PUFA (1:1) (Friedman, 2013). The most common analysis for fatty acid profiles could be determined according to the Association of Official Agricultural Chemists' (AOAC) official method 996.06 (AOAC, 1995), or the American Oil Chemist's Society (AOCS) official method Ce 1–62 (AOCS, 2011) by gas chromatography coupled with flame ionization detector (GC-FID). Both of these two official methods was carried out by preparing fatty acid methyl esters (FAMEs), which would be further separated and quantified by GC-FID. Fatty acid profiles can also be obtained using gas chromatography-mass spectrometry (GC-MS) technology, which is the most sensitive and accurate method for identifying fatty acid compositions.

Rice Bran and Rice Bran Oil
https://doi.org/10.1016/B978-0-12-812828-2.00007-X

1.2 γ-Oryzanol

Antioxidants plays an important role in providing stability to vegetable oils. RBO contains a considerable amount of various phytochemicals with antioxidant potential in addition to other health benefits. Among these, γ-oryzanol and vitamin E (four tocopherols and four tocotrienols) factions have gained special attention of researchers owing to their antioxidant potential (Akihisa et al., 2000; Foo Wong et al., 2014; Sakunpak et al., 2014a,b). A minor faction of phenolic compounds were also reported in RBO (Aleksander et al., 2008; Janu et al., 2014).

γ-Oryzanol contains a mixture of steryl ferulates found in RBO. The reported values for γ-oryzanols in RBO range from 0.2% to 2.72%, depending on the method of extraction, rice variety, weather, and area of cultivation (Butsat and Siriamornpun, 2010). γ-Oryzanol is readily soluble in conventional organic solvents such as hexane, which has been classically employed in its extraction. However, the components of γ-oryzanol have an alcoholic group in the ferulate part (Patel and Naik, 2004) (Fig. 1), which makes the molecule highly polar. The extraction rate of γ-oryzanol can be significantly affected by the polarity of solvent.

Nowadays, several methods for quantitative determination of γ-oryzanol in RBO and dietary supplements have been used. High-performance liquid chromatography (HPLC) is usually the most commonly used method (White et al., 2012; Sakunpak et al., 2014a,b), whereas an ultraperformance liquid chromatography (UPLC) method, similar to HPLC in principle, is faster with greater sample throughput and uses less solvent as compared to the HPLC technique (Xu and Howard, 2012). Lu et al. (2014) proposed a method to quantify and detect steryl ferulates in RBO using HPLC and UPLC, respectively. However, there are no studies using the UPLC technique alone in the simultaneous quantification and identification of steryl ferulates directly in RBO without further sample preparation or extraction. Thus Cuevas et al. (2017) developed a method based on UPLC coupled with electrospray ionization mass spectrometry for the simultaneous quantification of γ-oryzanol and identification of five major steryl ferulates directly in refined RBO samples.

Although methods based on HPLC or UPLC are highly sensitive and specific, the analytical instruments are quite costly, and expertise is usually required. Sakunpak et al. (2014a,b) provided a validated thin layer chromatography (TLC)-image analysis method for quantitative analysis of

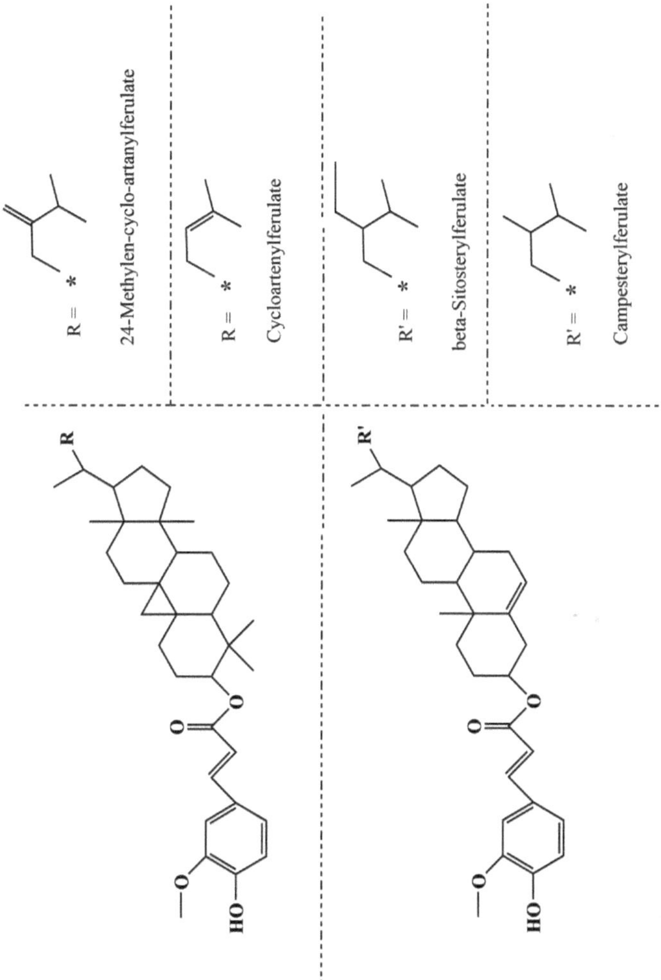

Fig. 1 Chemical structures of the main components of oryzanol (Patel and Naik, 2004).

γ-oryzanol in cold-pressed RBO. This method costs less, thus it can also be useful for small RBO manufacturers due to its simplicity and low operating costs. Nevertheless, TLC-image analysis system is not applicable for quantitative assays requiring high accuracy and high confidence (Tie-Xin and Hong, 2008).

1.3 Vitamin E Homologues and Derivatives

Vitamin E is a term used to designate a group of essential fat-soluble compounds that share a common structure, that is, a chromanol ring and either a phytyl tail (tocopherols) or isoprenoid chain (tocotrienols). RBO contains a particularly high content of tocotrienols, at an amount of 0.025%–0.17% mg/kg, which varies with the different rice cultivars, planting area, extracting, etc. (Ahsan et al., 2015; Pal and Pratap, 2017). Tocopherols in RBO varies from 0.02% to 0.08% (Pal and Pratap, 2017). Both tocopherols and tocotrienols exist naturally in four of their corresponding homologues (α-, β-, γ-, and δ-).

The analysis of vitamin E is normally carried out either using high performance liquid chromatography (HPLC) or gas chromatography (GC). Generally, GC methods require derivatization procedure to increase the volatility. These derivatization reactions often produce incomplete conversion of the compounds and undesirable interfering side products (Cunha et al., 2006). Therefore HPLC is the most commonly used technique for the analysis of vitamin E and derivatives. Several reversed-phase (RP) and normal-phase (NP) HPLC methods are available. The NP-HPLC method enabled the separation of all homologues of tocopherols and tocotrienols, whereas separations using RP-HPLC, which usually employed a C_{18} column, were less successful. However, NP-HPLC methods are plagued by the lengthy analysis time, are less compatible with aqueous biological sample matrices, and consume more hazardous solvents (Grebenstein and Frank, 2012). Various HPLC methods have been reported for the determination of tocopherols and tocotrienols, however, most of the described methods were incapable of separating either tocopherols or tocotrienols for β- and γ-type. Grebenstein and Frank reported a breakthrough in the separation of vitamin E where all eight homologues were separated using RP-HPLC (Grebenstein and Frank, 2012). Foo Wong et al. (2014) also provided a novel UPLC method using PFP column for the separation of 10 vitamin E components. Compared with RP-HPLC, the UPLC method exhibited considerable advantages as it is not only simpler, faster (≈9.5 min), and sensitive (especially with FL detection), but it also managed to overcome the pressure buildups inherent of the conventional RP-HPLC systems. Using this method, α-, β-, γ- and δ-isomers of tocopherols in RBO were found.

1.4 Phenolic Compounds

Phenolic compounds have been reported to be present in some vegetable oils, which is very important for the oxidative stability of the polyunsaturated fatty acids (Kochhar, 2002). The general analysis for the phenolic compounds was to determine the total phenolic content or use HPLC to identify the specific compound. Total phenolic content is generally measured using the Folin-Ciocalteu colorimetric method. Aleksander et al. (2008) and Janu et al. (2014) tried to extract the phenolic compounds from refined RBO and found that total phenolic content in RBO did not exceed 2 mg/100 g. Then HPLC was followed to identify phenolic compounds in RBO, with only ferulic acid being detected by Aleksander et al., whereas sinapic acid and cinnamic acid was detected by Janu et al. Nevertheless, crude RBO was reported to contain desmethylsterols (3225 mg/100 g, reduced to 1055 mg/100 g in refined oil), monomethylsterols (420 mg/100 g), and dimethylsterols or triterpene alcohols (176 mg/100 g), making a total of 4.8 g/100 g.

1.5 Phospholipids

Phospholipids have a polar phosphate group attached to a lipid group via electrostatic forces and hydrogen bonds, which may be transferred to the oil when more polar solvents, able to break these bonds and release them, are used (Brum et al., 2009). Extraction of phospholipids in crude RBOs using different solvents have been reported in several studies. For industrial RBO extraction with hexane, the phospholipid content ranges from 1% to 4% (Sengar et al., 2014; Sharif et al., 2014), which could be measured by the standard molybdenum blue method as the AOCS method Ca 12–55 (AOCS, 2011). Usually, the absorbance of the color developed due to phosphomolybdic acid complex was measured at 460 nm. Phosphorus concentration in the sample was measured from the standard curve determined by known concentration of KH_2PO_4 (1–20 mmol/L).

2. QUALITY ANALYSIS

RBO is recommended as a good source for cooking and frying, as it exhibits good thermal stability and gives a pleasant flavor to fried food. RBO readily undergo structural decomposition or transformation with the generation of compounds with documented toxicity when subjected to processes that require high temperature and/or oxygen. Oxidized oil can produce a rancid flavor that can be detected by consumers. In addition to a rancid flavor, the unsaturated fatty acids (UFAs) can be converted into trans-fatty acids (TFAs) and other hazardous components by thermal oxidative deterioration during cooking (Ganguly et al., 2016). Apart from structural alterations of the major components, that is, the triacylglycerols, endogenous minor components in the oil such as phytosterols, tocopherols, carotenoids, and phenolic acids can also be sufficiently modified during processing resulting in hazard products. The methods most commonly used for the evaluation of RBO's quality, composition, and properties include sensory analysis, techniques in which titration is utilized, spectrophotometry, and gas chromatography. The brief comparisons of methods in quality evaluation are listed in Table 1.

2.1 Sensory Analysis

Sensory evaluation of oils is considered a critical test of quality. Generally, sensory analysis is based on the evaluation of the quality of a product by a team of trained panelists. It is among the oldest and commonly used

Table 1 Comparison of methods for quality evaluation of rice bran oil

Method	Advantages	Disadvantages
Sensory analysis	• Direct evaluation of the oil's quality • Well-documented procedures and official standards • Ease of use	• Risk of sensory fatigue • Need to employ highly qualified personnel • Limited sensitivity
Chemical analysis	• Straightforward procedure • Well-documented procedures and official standards • Relatively low cost of a single analysis	• Use of organic solvents • Hard to fully automatize • Limited sensitivity • Lack of qualitative information • Relatively time-consuming
Chromatography method	• Qualitative and quantitative analysis • High sensitivity • Well-researched methodology • Guidelines present in official standards	• Relatively time-consuming • Often entails a sample preparation stage • Complex equipment • No possibility to employ in situ • Relatively high cost of equipment and use

techniques for food quality control. Sensory odor evaluation of edible oil is governed by factors such as the medium in which the oil is dispersed, temperature of oil during testing, and panel training for unambiguous under standing of descriptors (Raj et al., 2006). This method is easy to use, which is usually applied with some statistic method, such as principal component analysis (PCA) (Ravi et al., 2005). However, it is important to note that the occurrence of olfactory sensation depends on concentration and individual detection threshold of these substances. Also, panelists are professionals whose employment entails high expenses and a long period of training.

2.2 Chemical Analysis

Chemical methods constitute a separate group of oil quality assessment methods. These include the peroxide value (PV), *p*-anisidine value (*p*-AV), iodine value (IV), spectrophotometric measurement of color, or determination of free fatty acids (FFA) and polar compounds by chromatography.

2.2.1 Color

Color is considered to be an important quality parameter for edible oils. Crude RBO is greenish yellow, dark brown, and yellowish green in color with a hazy appearance, due to some pigments, such as carotenoids, xanthophylls, and gums. The color value of the oils was measured by some specific instrument, such as Lovibond Tintometer (Latha and Nasirullah, 2014). After the refining process, the color of RBO becomes lighter, and unlike the other vegetable oils, the color of refined RBO cannot be completely removed.

2.2.2 Primary Oxidation Products

Like most vegetable oils, RBO is prone to oxidation, which resulted in formation of some nonvolatile toxic compounds during storage or processing (Alkahtani, 1991). Peroxides are the primary oxidized products produced, which on further oxidation would degrade to aldehydes, ketones, esters, etc., which are the secondary oxidized products. Peroxide value (PV) is a measure of oxidation during storage and the freshness of lipid matrix (Malheiro et al., 2013). In addition, it gives important information about lipid autoxidation. PV measures only the earlier stage of oxidation and primary oxidation products (Atinafu and Bedemo, 2011; Srivastava and Singh, 2015). PV was estimated by using sodium thiosulfate solution as a titrating agent against the evolved iodine in the sample; after reacting, the peroxides present in the sample with salt of iodine (KI). In addition, conjugated dienes

(CDs) and conjugated trienes (CTs) could also be used to measure the primary oxidation products of hydroperoxide,

2.2.3 Secondary Oxidation Products

Unsaturated fatty acids are the primary targets of thermal oxidation as well as autoxidations, leading to formation of secondary oxidation products (Latha and Nasirullah, 2014). *p*-Anisidine value (*p*-AV) is a reliable indicator of oxidative rancidity in fats and oils. *p*-AV measures the secondary oxidation product (aldehyde and ketone, etc.) produced during oxidative degradation of oils, which is usually estimated by using isooctane as the solvent under 350 nm of wavelength (AOCS, 2011).

2.2.4 Free Fatty Acid (FFA)

Free fatty acid (FFA) content is considered to be an indicator of oil quality in the food industry as it leads to development of off-flavor in oils and fried products. FFAs were generally determined by titrating the FFA with alkali in presence of ethyl alcohol as a solvent. FFA was found to increase with an increase with the oxidation of RBO (Debnath et al., 2012).

2.2.5 Iodine Value (IV)

Iodine value (IV) is an index of the unsaturation, which is the most important analytical characteristic of oil (Otunola et al., 2009). It has been found that there is a relative loss of the C18:2 fatty acid and a decrease in the iodine value of oil after heating due to more intensive thermo-oxidative transformations that occur as compared to heated oil containing food (Tynek et al., 2001). IV was determined by treating the sample with an excess of solutions of iodine monochloride (ICl) in glacial acetic acid. Unreacted iodine monochloride reacted with potassium iodide, converting it to iodine, whose concentration was determined by titration with sodium thiosulfate. The decrease in IV can be attributed to the destruction of double bonds by oxidation, scission, and polymerization.

2.2.6 Polar Compounds (PC)

Polar compounds (PC), which are generated due to thermal oxidation and auto-oxidation, provide a reliable measure of the extent of oxidative degradation, especially during deep-fat frying. During frying, as peroxides/hydroperoxides break down, short chain acids, aldehydes, ketones, alcohols, and nonvolatile end products are formed in the frying oil. These cause some molecules in the oils to become polar. The polar material is an indicator for the extent of deterioration of oils used for frying. As a measurement of quality control, a polar material concentration of 25%–27% can be used as indicator for discarding frying oil (Debnath et al., 2012).

2.2.7 Oxidative Stability Index (OSI)

Oxidative stability index (OSI) can be determined according to AOCS Official Method Cd 12b-92, or using a Metrohm Rancimat apparatus (Rudzinska et al., 2016). The oxidative stability is usually evaluated by the induction period (IP) of the oils under certain temperature.

2.3 Analysis of Hazardous Materials

2.3.1 Trans-Fatty Acid (TFA)

Vegetable oils and fats have fatty acids in *cis* configuration, which are nutritionally important, while during partial hydrogenation of fats and oil, some *cis* form of fatty acids are converted into a trans form, which is reported to be harmful for human health, such as increasing the risk of coronary heart disease (Aro et al., 2006). Trans-fatty acid, namely elaidic acid, is formed during hydrogenation of fat, and microbial action on fat molecule. Also trans-fatty acids are formed when oil is exposed to heat for a longer period of time. Due to the increasing concern on human health, formation of trans-isomers during partial hydrogenation and high-temperature physical refining has become an important quality parameter.

Analysis of TFA involves the use of capillary gas chromatography (GC), infrared spectroscopy, and both techniques coupled together. Fourier Transform IR analysis is recommended for TFA analysis wherever the fats have a TFA content of 18:1n9t, more than 5%. FTIR analysis allows quantification of isolated trans-isomers, whereas GC provides quantification of all fatty acids except the 12t and 16t isomers of 18:1 (Ratnayake et al., 1998). Gas chromatograph analysis of the TFA content can be optimized to measure the TFA isomers found in hydrogenated fats as well as the TFA isomers found in refined oils. Gas chromatographic analysis is suitable for the reliable determination of a wide range of TFA levels in refined and crude RBO, whereas fatty acid profiles are obtained using the gas chromatography-mass spectrometry (GC-MS) technology, which is the most sensitive and accurate method for determining the fatty acid compositions and thus is particularly suitable for the purpose of detecting and quantifying TFA (Ecker et al., 2012).

2.3.2 Polycyclic Aromatic Hydrocarbons (PAH)

Polycyclic aromatic hydrocarbons (PAH) include more than 100 different compounds that have two or more fused aromatic rings (Guillén, 1994). Due to their lipophilic nature, oil and fats are a principle contamination source of PAH. PAHs are mostly formed by incomplete combustion of organic matter as a consequence of a series of natural and anthropogenic processes (Ciemniak et al., 2013). Sixteen priority PAH were defined by the European Food Safety Authority (EFSA). More than 16 PAH, which are potentially mutagenic and carcinogenic to humans, are formed at parts per billion (ppb) levels during cooking under conditions of high temperature (Wenzl et al., 2006). Several countries have established their own limits on concentrations of PAHs in edible oil. For example, China has set the upper residue level of $10\,\mu g\,kg^{-1}$ for benzo[*a*]pyrene (BaP) in edible oils (Wang and Cui, 2010).

GC-MS is often used to determine the types and content of PAHs. Kang et al., used liquid-liquid extraction and gas chromatography-mass spectroscopy to measure eight PAH (Kang et al., 2014). Whereas when GC-MS was used coupled with SPE column, 16 PAHs in RBO could be identified (An et al., 2017). In most of the analysis with GC-MS, the identification of PAH was shown by comparison of the standards' retention time and their retention time in the same conditions.

3. CHALLENGES CONCERNING THE ANALYSIS OF RICE BRAN OIL

The aforementioned methods provide an indication on the quality of RBO. Methods in which titration or measurement of absorbance are utilized are usually more popular due to their relative simplicity of measurement as compared to chemical methods. For this reason, they are used in analytical laboratories of sanitary agencies and manufacturing plants. However, most of these methods lack sensitivity, which limits the applications in routine measurements, especially for the growing awareness of proper nutrition and hazard production during storage and production. Nowadays, some new techniques for quality analysis involve usage of chromatography and spectroscopic methods like high-performance liquid chromatography (HPLC), Fourier transform IR (FTIR), nuclear magnetic resonance (NMR), Raman spectra, etc. have been developed (Yildiz et al., 2002; Rohman and Man, 2013). Nevertheless, these methods at present are unable to realistically replace the currently used methods because of the relatively low sensitivity of detection and high cost of equipment and operation.

REFERENCES

Ahsan, H., Ahad, A., Siddiqui, W.A., 2015. A review of characterization of tocotrienols from plant oils and foods. J. Chem. Biol. 8 (2), 45–59.

Akıhısa, T., Yasukawa, K., Yamaura, M., Ukiya, M., Kimura, Y., Shimizu, N., Arai, K., 2000. Triterpene alcohol and sterol ferulates from rice bran and their anti-inflammatory effects. J. Agric. Food Chem. 48 (6), 2313–2319.

Aleksander, S., Malgorzata, N., Eleonora, L., 2008. The content and antioxidant activity of phenolic compounds in cold-pressed plant oils. J. Food Lipids 15 (2), 137–149.

Alkahtani, H.A., 1991. Survey of quality of used frying oils from restaurants. J. Am. Oil Chem. Soc. 68 (11), 857–862.

An, K.J., Liu, Y.L., Liu, H.L., 2017. Relationship between total polar components and polycyclic aromatic hydrocarbons in fried edible oil. Food Addit. Contam. Part A Chem. Anal. Control Expo. Risk Assess. 34 (9), 1596–1605.

AOAC, 1995. Method 996.06, Fat (Total, Saturated and Unsaturated) in Foods. AOAC International, Gaithersburg, MD.

AOCS, 2011. Official Methods and Recommended Practices of the American Oil Chemist's Society. AOCS Press, Champaign, Il.

Aro, A., Becker, W., Pedersen, J.I., 2006. Trans fatty acids in the Nordic countries. Scand. J. Food Nutr. 50 (4), 151–154.

Atinafu, D.G., Bedemo, B., 2011. Estimation of total free fatty acid and cholesterol content in some commercial edible oils in Ethiopia, Bahir Dar. J. Cereals Oilseeds 5 (5), 71–76.

Brum, A.a.S., Arruda, L.F.D., Regitanod'arce, M.a.B., 2009. Extraction methods and quality of the lipid fraction of vegetable and animal samples. Quím. Nova 32 (4), 849–854.

Butsat, S., Siriamornpun, S., 2010. Antioxidant capacities and phenolic compounds of the husk, bran and endosperm of Thai rice. Food Chem. 119 (2), 606–613.

Ciemniak, A., Witczak, A., Mocek, K., 2013. Assessment of honey contamination with polycyclic aromatic hydrocarbons. J. Environ. Sci. Health, Part B 48 (11), 993–998.

Cuevas, M., De Souza, P.T., Rodrigues, C.E.D., Meirelles, A.J.A., 2017. Quantification and determination of composition of steryl ferulates in refined rice bran oils using an UPLC-MS method. J. Am. Oil Chem. Soc. 94 (3), 375–385.

Cunha, S.C., Amaral, J.S., Fernandes, J.O., Oliveira, M.B., 2006. Quantification of tocopherols and tocotrienols in Portuguese olive oils using HPLC with three different detection systems. J. Agric. Food Chem. 54 (9), 3351–3356.

Debnath, S., Rastogi, N.K., Krishna, A.G.G., Lokesh, B.R., 2012. Effect of frying cycles on physical, chemical and heat transfer quality of rice bran oil during deep-fat frying of poori: an Indian traditional fried food. Food Bioprod. Process. 90 (C2), 249–256.

Ecker, J., Scherer, M., Schmitz, G., Liebisch, G., 2012. A rapid GC-MS method for quantification of positional and geometric isomers of fatty acid methyl esters. J. Chromatogr. B Analyt. Technol. Biomed. Life Sci. 897 (4), 98–104.

Foo Wong, Y., Makahleh, A., Saad, B., Ibrahim, M.N.M., Abdul Rahim, A., Brosse, N., 2014. UPLC method for the determination of vitamin E homologues and derivatives in vegetable oils, margarines and supplement capsules using pentafluorophenyl column. Talanta 130 (Suppl. C), 299–306.

Friedman, M., 2013. Rice brans, rice bran oils, and rice hulls: composition, food and industrial uses, and bioactivities in humans, animals, and cells. J. Agric. Food Chem. 61 (45), 10626–10641.

Ganguly, R., Lavallee, R., Maddaford, T.G., Devaney, B., Bassett, C.M., Edel, A.L., Pierce, G.N., 2016. Ruminant and industrial trans-fatty acid uptake in the heart. J. Nutr. Biochem. 31, 60–66.

Grebenstein, N., Frank, J., 2012. Rapid baseline-separation of all eight tocopherols and tocotrienols by reversed-phase liquid-chromatography with a solid-core pentafluorophenyl column and their sensitive quantification in plasma and liver. J. Chromatogr. A 1243 (12), 39.

Guillén, M.D., 1994. Polycyclic aromatic compounds: extraction and determination in food. Food Addit. Contam. 11 (6), 669–684.
Janu, C., Kumar, D.R.S., Reshma, M.V., Jayamurthy, P., Sundaresan, A., Nisha, P., 2014. Comparative study on the total phenolic content and radical scavenging activity of common edible vegetable oils. J. Food Biochem. 38 (1), 38–49.
Kang, B., Lee, B.M., Shin, H.S., 2014. Determination of polycyclic aromatic hydrocarbon (PAH) content and risk assessment from edible oils in Korea. J. Toxicol. Environ. Health A 77 (22–24), 1359–1371.
Kochhar, S.P., 2002. Sesame, Rice-Bran and Flaxseed Oils. In: Vegetable Oils in Food Technology, Composition, Properties and Uses. Wiley-Blackwell, Oxford, UK, pp. 297–326.
Latha, R.B., Nasirullah, D.R., 2014. Physico-chemical changes in rice bran oil during heating at frying temperature. J. Food Sci. Technol. Mysore 51 (2), 335–340.
Lu, W., Niu, Y., Yang, H., Sheng, Y., Shi, H., Yu, L.L., 2014. Simultaneous HPLC quantification of five major triterpene alcohol and sterol ferulates in rice bran oil using a single reference standard. Food Chem. 148 (4), 329.
Malheiro, R., Rodrigues, N., Manzke, G., Bento, A., Pereira, J.A., Casal, S., 2013. The use of olive leaves and tea extracts as effective antioxidants against the oxidation of soybean oil under microwave heating. Ind. Crop. Prod. 44, 37–43.
Otunola, G.A., Adebayo, G.B., Olufemi, O.G., 2009. Evaluation of some physicochemical parameters of selected brands of vegetable oils sold in Ilorin metropolis. Int. J. Phys. Sci. 4 (5), 327–329.
Pal, Y.P., Pratap, A.P., 2017. Rice bran oil: a versatile source for edible and industrial applications. J. Oleo Sci. 66 (6), 551–556.
Patel, M., Naik, S.N., 2004. Gamma-oryzanol from rice bran oil—a review. J. Sci. Ind. Res. 63 (7), 569–578.
Raj, P.N., Prakash, M., Bhat, K.K., 2006. Quality assessment of oil blends by electronic nose technique and sensory methods. J. Sens. Stud. 21 (3), 322–332.
Ratnayake, W.M.N., Pelletier, G., Hollywood, R., Bacler, S., Leyte, D., 1998. Trans fatty acids in Canadian margarines: recent trends. J. Am. Oil Chem. Soc. 75 (11), 1587–1594.
Ravi, R., Prakash, M., Bhat, K.K., 2005. Sensory odour profiling and physical characteristics of edible oil blends during frying. Food Res. Int. 38 (1), 59–68.
Rohman, A., Man, Y.B.C., 2013. Application of FTIR spectroscopy for monitoring the stabilities of selected vegetable oils during thermal oxidation. Int. J. Food Prop. 16 (7), 1594–1603.
Rudzinska, M., Hassanein, M.M., Abdel-Razek, A.G., Ratusz, K., Siger, A., 2016. Blends of rapeseed oil with black cumin and rice bran oils for increasing the oxidative stability. J. Food Sci. Technol. 53 (2), 1055–1062.
Sakunpak, A., Suksaeree, J., Charoenchai, L., Monton, C., Pathompak, P., Charoonratana, T., Kraisintu, K., 2014a. Development and validation of RP-HPLC method for quantitative analysis of individual γ-oryzanol in cold pressed rice bran oil. Thai J. Pharm. Sci. 38, 124–128.
Sakunpak, A., Suksaeree, J., Monton, C., Pathompak, P., Kraisintu, K., 2014b. Quantitative analysis of gamma-oryzanol content in cold pressed rice bran oil by TLC-image analysis method. Asian Pac. J. Trop. Biomed. 4 (2), 119–123.
Sengar, G., Kaushal, P., Sharma, H.K., Kaur, M., 2014. Degumming of rice bran oil. Rev. Chem. Eng. 30 (2), 183–198.
Sharif, M.K., Butt, M.S., Anjum, F.M., Khan, S.H., 2014. Rice bran: a novel functional ingredient. Crit. Rev. Food Sci. Nutr. 54 (6), 807–816.
Srivastava, P., Singh, R.P., 2015. Frying stability evaluation of rice bran oil blended with soybean, mustard and palm olein oils. Orient. J. Chem. 31 (3), 1687–1694.

Tie-Xin, T., Hong, W., 2008. An image analysis system for thin-layer chromatography quantification and its validation. J. Chromatogr. Sci. 46 (6), 560–564.
Tynek, M., Hazuka, Z., Pawlowicz, R., Dudek, M., 2001. Changes in the frying medium during deep-frying of food rich in proteins and carbohydrates. J. Food Lipids 8 (4), 251–261.
Wang, J.H., Cui, G., 2010. Ultrasonication extraction and gel permeation chromatography clean-up for the determination of polycyclic aromatic hydrocarbons in edible oil by an isotope dilution gas chromatography-mass spectrometry. J. Chromatogr. A 1217 (28), 4732–4737.
Wenzl, T., Simon, R., Anklam, E., Kleiner, J., 2006. Analytical methods for polycyclic aromatic hydrocarbons (PAHS) in food and the environment needed for new food legislation in the European Union. Trends Anal. Chem. 25 (7), 716–725.
White, B., Rice, L., Howard, L.R., 2012. The procedure, principle, and instrumentation of antioxidant phytochemical analysis. In: Analysis of Antioxidant-Rich Phytochemicals. Wiley-Blackwell, Chichester, UK, pp. 25–68.
Xu, Z., Howard, L.R., 2012. Analysis of Antioxidant-Rich Phytochemicals. Wiley-Blackwell, Oxford, UK, pp. 353–386.
Yang, R., Zhang, L., Li, P., Yu, L., Mao, J., Wang, X., Zhang, Q., 2018. A review of chemical composition and nutritional properties of minor vegetable oils in China. Trends Food Sci. Technol. 74, 26–32.
Yildiz, G., Wehling, R.L., Cuppett, S.L., 2002. Monitoring PV in corn and soybean oils by NIR spectroscopy. J. Am. Oil Chem. Soc. 79 (11), 1085–1089.

CHAPTER 8

Development of Rice Bran Functional Food and Evaluation of Its Healthful Properties

Md. Alauddin*, Sadia Rahman†, Jahidul Islam‡, Hitoshi Shirakawa‡, Michio Komai‡, Md Zakir Hossen Howlader†
*Department of Nutrition and Food Technology, Jessore University of Science and Technology, Jessore, Bangladesh
†Department of Biochemistry and Molecular Biology, University of Dhaka, Dhaka, Bangladesh
‡Laboratory of Nutrition, Department of Science of Food Function and Health, Graduate School of Agricultural Science, Tohoku University, Sendai, Japan

1. INTRODUCTION

Multifactorial metabolic disorder (MMD) is characterized by its main components: hypertension, dyslipidemia, glucose impairment, inflammation, and cancer. MMD is associated with a two-fold increase in disease outcomes (Mottillo et al., 2010). It is noted that various functional foods, beverages, fruits, vegetables, grains, legumes, herbs, and spices are considered to prevent or moderate MMD (Mohamed, 2014). Rice bran, a byproduct of the rice milling process that contains various bioactive compounds, is one of the important food candidates (Abdul-Hamid and Luan, 2000). Rice bran ingredients have been widely used to increase functionality of some foods and as a functional component to enhance properties of foods against chronic disease (Quirós-Sauceda et al., 2014). Rice bran and its derivatives have been effective against dyslipidemia in different animal models (Kahlon et al., 1992; Nicolosi et al., 1991). An array of health-promoting value-added products have been derived from processed rice bran due to its identified active components such as oryzanols, tocopherols, tocotrienols, phytosterols, nucleotides, dietary fiber content, and phenolic compounds (Palou et al., 2015; Wang et al., 2015; Ardiansyah et al., 2009).

Biotechnological interventions such as the enzymatic treatment of rice bran are effective against MMD by attenuating hypertension, dyslipidemia, and inflammation, as well as functioning as a potent functional food component to prevent oxidative stress. A recent study of rice bran protein hydrolysate showed that it improved insulin resistance and metabolic disorder in a

Rice Bran and Rice Bran Oil
https://doi.org/10.1016/B978-0-12-812828-2.00008-1

mice model. This has lead to development of rice bran for human use as a functional food and dietary supplement (Boonloh et al., 2015; Ryan et al., 2011). Colored or pigmented rice (purple, black, and red rice) is one of the major food items in an Asian-based diet (Hu, et al., 2003). The constituents of colored rice are flavonoids, phenolics, tannin, sterols, tocols, γ-oryzanols, amino acids, fatty acids, phyto-antioxidant compounds, vitamins, and dietary fibers (Nakornriab et al. 2008; Min et al., 2009; Saenjum et al., 2012). Although rice bran usage is limited due to its rapid rancidity and unfavorable aroma, rice bran fermented with different types of microorganisms makes it an effective agent to retain its potential therapeutic efficacy.

Fermented rice bran (FRB) can be prepared from rice bran. It contains a variety of bioactive components (i.e., polyphenols, fatty acids, and peptides) that have been shown to have promising protective effects against several diseases such as cancer, metabolic syndrome, obesity and diabetes, and immune modulatory effects. There are limited studies regarding FRB against inflammation-related disease in animal models, for example, inflammatory bowel disease (IBD) and MMD. IBD is a chronic and relapsing inflammation in the gastrointestinal tract closely linked to an increased risk of colon cancer. Besides, the risk of development of colorectal cancer among IBD patients is increased about 10-fold (Seril et al., 2003). The most common IBD is ulcerative colitis (UC), a complex and debilitating disorder identified by the presence of sporadic lesions in the rectal and colonic mucosa (Da Silva et al., 2014; Ritchie et al., 2017). Although the incidence rate of UC is high in Western countries, its occurrence in East Asian countries has also augmented recently due to increased intake of Westernized food, which is high in protein and fat content, as well as excessive sugar intake with lower fiber consumption (Kondo et al., 2016; Ruemmele, 2016). Also urbanization, which is linked to changes in diet, antibiotic use, hygiene status, microbial exposures, and pollution, has been implicated as a potential risk factor for IBD and MMD. The normal colon mucosa plays an immune, endocrine, and barrier function. Injuries in the intestinal mucosa damage the barrier function; increased intestinal mucosal permeability allows microbes and antigens to invade and excessively stimulate the immune response, triggering intestinal inflammation (Kataoka et al., 2008). The resulting excessive proinflammatory cytokines (tumor necrosis factor α, interleukin 1β, and IL-6) affects colonic damage and ulceration of the colon (Ren et al., 2015). Excessive reactive oxygen species (ROS) are produced, leading to oxidative stress during the inflammatory response, exaggerating inflammatory lesions in the pathogenesis of UC. The efficacy of conventional treatments varies, and the commonly used drugs have long-term side effects (Kim et al., 2010). FRB is a preventive agent that would minimize the inflammation for extended periods of time. FRB may contain prebiotic compounds that selectively enhance the growth of commensal microbiota in the gastrointestinal tract to the host, which have been used for the treatment of IBD (Komiyama et al., 2011). Generally, in the animal model, dextran sodium sulfate (DSS) is used to create experimental colitis. DSS is a water-soluble, negatively charged, sulfated polysaccharide with a molecular weight ranging from 5 to 1400 kDa. DSS damages the epithelial monolayer lining the large intestine and allows the spreading of intestinal contents (e.g., bacteria and their products) into underlying tissue. The DSS colitis model is very popular in IBD study due to its quickness, uncomplicatedness, reproducibility, and controllability. Acute, chronic, and relapsing models of intestinal inflammation can be attained by changing the molecular weight and concentration of DSS and the frequency of administration.

Antidiabetic and antidyslipidemic activities of rice bran have also been reported (Boonloh et al., 2015). To be more applicable of rice bran against MMD and colonic disorders, several technologies as well as fermentation have been used in biotechnological applications to enhance nutrition. It's been previously revealed that rice bran fermented with *Saccharomyces cerevisiae* has antistress and antifatigue effects (Kim et al., 2002). Furthermore, polysaccharide extracts of rice bran fermented with Lentinusedodes showed anticancer and antidefective immune response, and also water extracts of FRB had an antiphotoaging effect (Kim et al., 2010). Ferulic acid and fractionized phenolic compound from rice bran exhibited hypoglycemic effects in a type 2 diabetic mice model (Jung et al., 2007). Moreover, brown rice fermented by *Aspergillus oryzae* has a suppressive effect on the induction of colitis by DSS (Kataoka et al., 2008). Recently, it was found that Driselase-treated rice bran fraction improved glucose and lipid metabolism in SHRSP in a genetic animal model of metabolic syndrome study (Ardiansyah et al., 2006, 2007).

Here, we have focused on raw rice bran compositions and preparation of functional rice bran and their health benefits, particularly the role of FRB in prevention of MMD with DSS-induced colitis and other carcinogen-induced inflammatory and extraintestinal disorders. Considering the several metabolic abnormalities and inflammation-related model, FRB administration followed the following possible mechanisms: (1) management of MMD by reduction of hypertension and lipid abnormalities, (2) management of inflammation by reduction of infiltration of inflammatory cells (Alauddin et al., 2016; Tazawa et al., 2003), (3) suppressing the production of ROS/nitric oxide derived from inflammatory cells (Onuma et al., 2015), and (4) modulating the microbial community by increasing the tight junction barrier integrity (Kataoka et al., 2008; Islam et al., 2008).

2. BASIC COMPOSITION OF RICE BRAN

Rice is the staple food of Bangladesh. Each year, approximately 40 million metric tons of rice are produced with 3 million metric tons of crude rice bran in Bangladesh. Most of the rice bran is either used as animal feed or waste material (BBS, 1999). Previous studies showed rice bran compositional distinctiveness such as carbohydrate, protein, fat, moisture, ash, fiber, amylose contents, phytic acid, minerals, vitamin contents, and the Glycemic Index (GI) (Faria et al., 2012; Bhosale and Vijayalakshmi, 2015). We have previously analyzed rice varieties in a particular region of Bangladesh and found they are composed of protein (7.04%), fat (0.37%), crude fiber (0.26%), and ash (0.58%) in parboiled milled rice (Zubair et al., 2016). A discussion of the chemistry of rice bran oil can be found in the first chapter of this book. Rice bran from different varieties in Bangladesh (BR-5, BR-10, BRRI-28, and BRRI-39 from different automatic rice mills) have been examined and were found to composed of lipid, fatty acid and glyceride (Rahman et al., 2013). Moreover, aromatic rice varieties showed exciting composition such as moisture (11.25%–15.13%), protein (3.23%–6.21%), fat (0.68%–1.45%), and ash (0.88%–1.46%) (Tuncel and Yılmaz, 2011). A recent study showed that cold-treated rice bran (BRRI-28) is useful in many food applications such as food supplements and edible oil extraction (Mohamed, 2014). We uncovered the functional composition of few rice bran varieties in Bangladesh (Table 1).

We especially focused on the functional composition of rice bran (byproduct of the rice milling process), which contains a rich source of bioactive compounds (Abdul-Hamid and Luan, 2000). Rice bran's health benefits and enhanced quality have been reported due to their antioxidant compounds and health benefit. We also extracted rice bran oil from popular rice bran varieties and measured the total fatty acid composition (Table 2).

It's apparent that rice bran is a potential source of high-value antioxidants for use as additives in foods, pharmaceuticals, and cosmetics because of its unsaturated fatty acid as synergists for antioxidants (Lloyd et al., 2000; Rao and Achaya, 1968; Richard et al., 2008). We also focused on the health

Table 1 Bioactive compounds of different rice bran varieties in Bangladesh

Variety name	Total phenolic content (mg GAE/g of dry extract)	Total tannin content (mg TAE/g of dry extract)	Total flavonoid content (mg QE/g of dry extract)	Antioxidant activity (mg AAE/g of dry extract)
BR-11	140.67 ± 0.54*	165.96 ± 0.29*	61.87 ± 0.43*	11.12 ± 0.29
BRRI dhan 28	133.8 ± 0.38	119.79 ± 0.31	33.99 ± 0.41	18.78 ± 0.10*
BRRI dhan 29	138.88 ± 0.25	136.31 ± 0.32	49.69 ± 0.27	16.88 ± 0.22
BRRI dhan 48	134.07 ± 0.27	157.08 ± 0.07*	42.25 ± 0.41	18.69 ± 0.30*
BRRI dhan 49	135.20 ± 0.40	157.03 ± 0.68*	62.43 ± 0.52*	14.04 ± 0.08

Data are represented as mean ± SEM. Each data was analyzed three times and their mean calculated. GAE = Gallic acid equivalent, TAE = Tannic acid equivalent, QE = Quercetin equivalent, AAE = Ascorbic acid equivalent. Significant differences were observed among varieties (*P value $<.05$).

Table 2 Fatty acid composition of different rice bran oils in Bangladesh

Fatty acid %	**BR 11 oil**	**BRRJ Dhan 28 oil**	**BRRI Dhan 29 oil**	**BRRI Dhan 48 oil**	**BRRI Dhan 49 oil**
Laurie acid	ND	0.6617	ND	ND	ND
Myristic acid	0.2381	0.6075	ND	0.4364	1.0042
Palmitic acid	16.4252	18.3878	14.1021	16.7385	17.4408
Stearic acid	1.5208	1.7781	1.7406	1.9147	2.1846
Arachidic acid	0.6980	0.5354	0.7500	0.8262	0.9687
Behenic acid	0.3280	ND	ND	0.3775	ND
Total saturated fatty acids	19.2101	21.9705	16.5927	20.2933	21.5983
Oleic acid	43.1588	44.7921	49.9506	45.3609	39.3869
Eicosenoic acid	0.4380	ND	ND	0.4637	0.4356
Monounsaturated fatty acids	43.5968	44.7921	49.9506	45.8246	39.8225
Linoleic acid	35.9822	32.1265	32.5770	32.6648	37.2559
Linolenic acid	1.2109	1.1109	0.8797	1.2173	1.3233
Polyunsaturated fatty acids	37.1931	33.2374	33.4567	33.8821	38.5792
Total unsaturated	80.7891	78.0295	83.4073	79.7067	78.4017

ND: not detected.

Sample name	Gamma oryzanol content (mg/g)
BR-11 oil	15.62±0.007
BRRI dhan 28 oil	13.38±0.01
BRRI dhan 29 oil	12.67±0.007
BRRI dhan 48 oil	18.86±0.02
BRRI dhan 49 oil	15.95±0.01

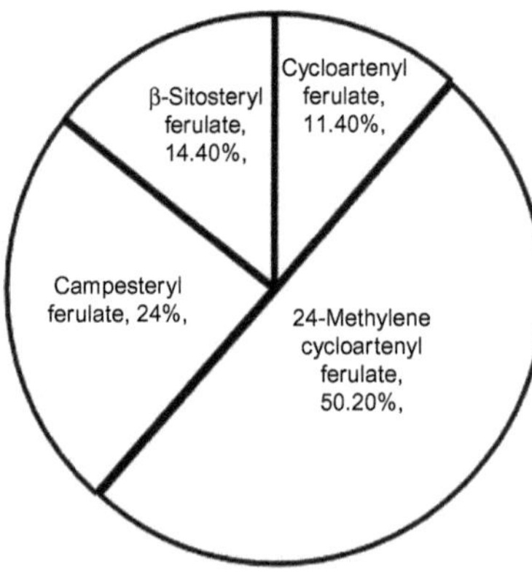

Fig. 1 γ-Oryzanol content in different rice bran oils (left). Ferulates are expressed as the percentages of γ-oryzanol and calculated on the basis of their areas (right).

benefits of γ-oryzanols content and their components in extracted rice bran oil, which was analyzed by gas chromatography (GC) (Fig. 1).

A recent study has quantified the functional compounds in rice bran including γ-oryzanols and vitamin E components in rice bran oil (Rogers et al., 1993), anthocyanin components in red rice (Terahara et al., 1994), and phenolic acids in various rice varieties (Harukaze et al., 1999). Phenolic acids composed of a phenolic ring and an organic carboxylic acid is capable of stabilizing and delocalizing unpaired electrons, which conferred antioxidant properties (Heuberger et al., 2010; Goffman and Bergman, 2004; Chung and Shin, 2007; Goufo et al., 2014). Flavonoids can be synthesized by phenylpropanoid metabolic pathways and are the most commonly encountered compounds in nonpigmented rice varieties (accounting for 77% of all flavonoids) (Goufo et al., 2014). Another important component of fat and oil is the unsaponifiable fraction of rice bran, which contains the antioxidants and micronutrients. Rice bran oil (RBO) has a high percentage of unsaponifiable fractions (4.2%), an excellent source of essential fatty acids, whereas other oils contain less (1%–2%). In addition, RBO has the highest cholesterol-reduction capacity without any allergenic reactions, as does other sources of oil (Rukmini and Raghuram, 1991). However, very few studies to date have examined the nutritional status of nonstabilized, farmer-popular, high-yield rice bran varieties.

3. FERMENTED RICE BRAN PREPARATION AND FUNCTIONAL IMPROVEMENT

FRB preparation is necessary to inactivate lipases and other nutritional inhibitors like field fungi, bacteria, and insects so that their toxicity is reduced without damaging the functional components and protein quality of rice bran. When bran layers are removed from the endosperm during milling, the rice bran must be stabilized using suitable techniques. Specifically, to achieve proper stabilization, each individual bran particle must have the same moisture content, depending on the time and the temperature. Furthermore, to inactivate the enzymes in the rice bran responsible for rancidity, different stabilization methods are used. Among these, microwave energy offers an alternative energy source for stabilization (Kupski et al., 2012; Wataniyakul et al., 2012; Faria et al., 2012). Microwave heating inactivates the enzymes that cause rancidity, such as lipases and lipoxygenases, through the internal heating of particles within the microwave cavity. The cavity makes the dipolar water molecules in the samples excited by the electromagnetic waves, resulting in enhancement of kinetic energy, along with friction, and produces evenly distributed heat through the materials (Malekian et al., 2000). Stabilization fractionation is an important step in industrial processing. It involves the conversion of rice bran into various parts that contain more desirable components than undesirable ones. Subsequently, the different fractions are centrifuged to separate the insoluble fiber fraction (rice bran fiber) from the aqueous dispersible fraction (rice bran soluble). The mixture of both insoluble and soluble extracts is called rice bran balance. Using different technologies, the bran is fully stabilized and the oil is removed. The resultant food grade, defatted rice bran is temporarily stored in food grade silos until it can be used in edible applications. Bleaching of the edible oil typically leaves minor flavor and odor compounds that must be removed by steam distillation before the oil is used. Steam distillation is the final step in the processing of edible oil, whereby any off-flavor and residual free fatty acids left in the oil are removed.

We have produced two types of rice bran fractions, namely the Driselase fraction (DF) and the ethanol fraction (EF). To process the rice bran, 500 g of bran was agitated in 1.0 L of 70% ethanol for 2 h; this yielded two fractions: the solid and filtered fractions (Ardiansyah et al., 2006, 2007). The DF was derived from the solid fraction. The solid fraction of rice bran was dried at room temperature and then suspended in 10 mM acetate buffer (500 mL) containing Driselase (0.2 mg/L, esterase-free commercial plant cell wall-degrading enzyme mixture containing cellulase, xylanase, and laminarinase) from Basidiomycetes spp. The bran was treated overnight at 37°C; the suspension was then filtered and finally lyophilized. As a result, Driselase-treated rice bran had increased quantities of bioactive components that improve glucose and lipid metabolism in the SHRSPs—a genetic animal model of metabolic syndrome (Ardiansyah et al., 2006, 2007).

According to Alauddin et al. (2016), FRB was prepared by two stages of fermentation using fungi and lactic acid bacteria. At the first stage, rice bran was mixed with steam and cooled to about 30°C, after which a spore solution of Aspergillus kawachii was inoculated at an initial concentration of 10^6 spores/g of rice bran. The rice bran was incubated at 30°C in a fermentation chamber for 44 h. The solid-state culture obtained was designated as rice bran koji. Preparation of a mixture of rice bran koji and rice powder (2:1) using a fourfold amount of water at 56°C for 12 h, heated at 85°C for 15 min, and then cooled to about 30°C. At the second stage, the saccharified culture solution was inoculated with a mixture of lactic acid bacteria (*Lactobacillus brevis*, *Lactobacillus rhamnosus*, and *Enterococcus faecium*) at a concentration of 0.01% (w/w). The solution was then incubated at 37°C overnight and then heated at 85°C for 15 min to obtain FRB. The FRB solution was then filtered and lyophilized. The lyophilized powder was kept at 30°C until use. Non-FRB was prepared by the same procedure previously mentioned without the inoculation of the rice bran with fungi or lactic acid bacteria. The macronutrients in FRB and non-FRB were analyzed using conventional methods for food analysis (Fig. 2).

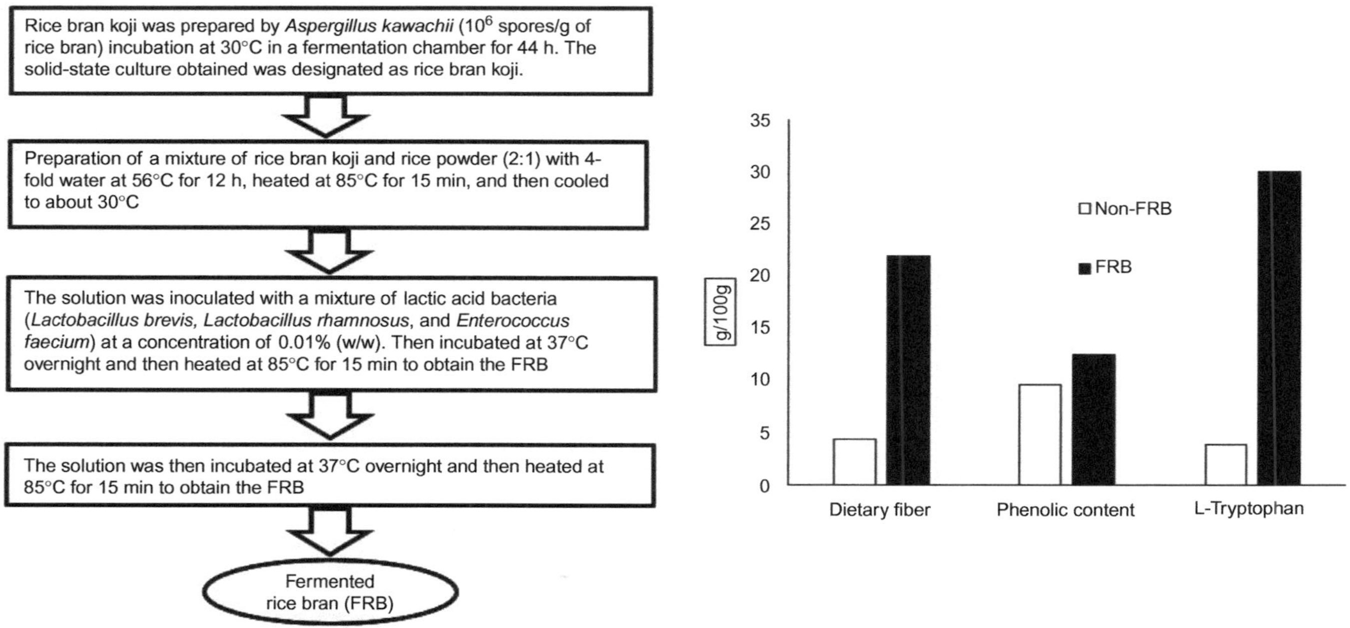

Fig. 2 Fermented rice bran preparation procedure (left) and improvement of its component (right). L-Tryptophan result expressed as (g/100 g) × 1000.

4. RICE BRAN-BASED FUNCTIONAL FOOD, A DRUG ALTERNATIVE

The term "functional food" was first coined in Japan (FOSHU) and then started research on functional food in Europe and America. Insights into functional foods are particularly important because the intake of some physiologically functional constituents of foods is becoming effective in preventing diseases. Technologically, it has become feasible to design and produce physiologically functional foods that are expected to satisfy in whole or in a part today's demand for disease prevention by daily nutrition. Functional food is defined as food products fortified with special constituents that possess advantageous physiological effects (Arai et al., 2001; Saito, 2007). Other government agencies have developed their own definitions for functional food. For example, the United States defines functional food as any amended food or food component that may provide a health advantage beyond its traditional nutrients. Correspondingly, the Institute of Food Technologists (IFT) delineates "ingredients that provide important nutrients for normal growth maintenance and development, and additional biologically active constituents that impart health aids or necessary functional effects." Finally, the American Dietetic Association (ADA) claimed functional foods as "whole, fortified, enriched, or enhanced" that should be consumed commonly to derive health benefits (ADA, 2003; Hasler, 2002).

Precisely, investigators are trying to find out the appropriate food components and their beneficial health effects in this arena. Functional food may improve the health condition and homeostatic behavior, and also some mechanisms through the use of biomarkers or indicators in the body homeostasis. For example, food scientists have determined the health effects and proper or safe functional rice bran that have a positive effect better than synthetic drugs (Martirosyan and Singh, 2015). Thus numerous functional rice bran candidates and >800 plants were found to prevent or reduce metabolic-related disorders and gastrointestinal disorders by assisting the body homeostasis mechanisms. Rice bran foods with an antihypertensive effect have been reported beneficial for human health (Krikorian et al., 2010). Furthermore, FRB, virgin olive oils, olive leaves, pumpkins, corn, and beans reduced diastolic blood pressure and improved insulin secretion or glucose tolerance in a randomized control trial in CVD-risk humans (Brown et al., 2011). Partial replacement of dietary FRB, protein hydrolysate, peptides, and powerful ACE inhibitors decreased the blood pressure in those with noradrenalin-induced hypertension (Mohamed, 2014).

4.1 Anticolitis Effects of Fermented Rice Bran

Kataoka et al. (2008) reported ameliorating effects of *A. oryzae*-mediated fermented brown rice and rice bran (FBRA) on DSS-induced colitis in a rat model. Supplementation of 5% and 10% FBRA on control diet prevented the colitis symptoms and severity, as confirmed by macroscopic and histological findings. Ten percent of the FBRA diet decreased the myeloperoxidase (MPO) activity, which is used as an inflammatory marker of neutrophil infiltration in the colonic mucosa. Also dietary FBRA increased the lactobacilli in the intestine lactobacilli strains, which are important for human health as probiotics and have been reported to suppress IBD. Thus FBRA has a suppressive effect on induction of colitis by DSS and suggests FBRA-mediated modification of colonic microbiota (Kataoka et al., 2008). Islam et al. (2017) investigated the effects of dietary FRB supplementation on UC in a mouse model of DSS-induced UC where RB was dually fermented by Aspergillus kawachii and Lactobacillus sp. Body weight alteration, disease activity index (DAI), histopathology score, tissue MPO activity, cytokine and chemokine transcript levels, and SCFAs and mucin in the colonic tissue were investigated. Based on histopathology scores, DSS caused massive mucosal inflammation with an augmented crypt loss, and inflammatory cell penetration in the control and RB groups, but not in FRB group. MPO activity, thiobarbituric acid-reactive substance levels, and proinflammatory cytokine transcript (Tnf-a, Il-1b, Il-6, and Il-17) levels were significantly reduced in the FRB group ($P < .05$). Besides that, dietary FRB increased the SCFAs, for example, acetic acid (AA), butyric acid (BA), and propionic acid (PA), both before and after DSS given. SCFAs production is intensely linked to the colonic health in human and is produced by the microbiota by breaking down of complex carbohydrates, such as fiber (Islam et al., 2017). SCFAs are the major source of energy for the enteric epithelium. Elevated SCFA levels stimulate colonic epithelial cell proliferation, and increase mucin production and epithelial cell integrity. Specifically, BA contributes to induce colonic regulatory T cells and limits innate immune cell-driven inflammation and prevents autoantibody production. SCFAs play an important role in maintaining tight junction barrier integrity and intestinal homeostasis. Thus FRB could be used as an effective preventive agent for UC (Islam et al., 2008). Kondo et al. (2016) described the protective effects of aqueous extract suspension (AES) isolated from rice bran fermented by *S. cerevisiae* Misaki-1 and *Lactobacillus plantarum* Sanriki-SU8 in DSS-induced IBD model in mice. RB AES showed antioxidant

(DPPH and O^{2-} radical scavenging) and anti-inflammatory effects (inhibition of LPS-induced NO secretion in murine macrophage RAW264.7 cells) in vitro. However, RB AES could not suppress the inflammation in 5% DSS-induced IBD model mice. On the other hand, the RB AES fermented by *L. plantarum* S-SU8 and *S. cerevisiae* Misaki-1 at 30°C for 2 days clearly improved the disease activity index of inflammation in the IBD model mice (Kondo et al., 2016).

4.2 FRB in DSS-Induced Colonic Cancer and Gastrointestinal Disorders

Phutthaphadoong et al. (2010) investigated the suppression of colorectal carcinogenesis in a rat model; processed food prepared by fermented brown rice and rice bran (FBRA) using *A. oryzae* suppresses rat colorectal carcinogenesis. The experimental diets were prepared by mixing 5.0% and 10.0% FBRA with a CE-2 diet. FBRA supplementation in DSS-exposed ApcMin/+ mice significantly inhibited the multiplicity of colon tumors compared to control diet group. FBRA supplementation suppressed the cell proliferative index, which is accompanied by significantly decreased mRNA expressions of Cox2 and iNos in colonic mucosa exposed to DSS ($P<.04$ and .02, respectively). Thus FBRA has chemopreventive properties against inflammation-related tumorigenesis in the colon (Phutthaphadoong et al., 2010). Onuma et al. (2015) investigated FBRA's role in inflammation-related carcinogenesis model in mice, in where regressive QR-32 cells were injected subcutaneously. QR-32 cells are responsible for lethal tumors due to massive infiltration of inflammatory cells. FBRA supplementation of 5% or 10% reduces tumor incidences (35% and 20%, respectively) than in the nontreated group (70%). FBRA did not decrease the formation of 8-hydroxy-2′-deoxyguanine adducts, a marker of oxidative DNA damage in the inflammatory lesions; however, it suppressed TNF-α, Mac-1, CCL3, and CXCL2 gene expression. Thus FRBA inhibited inflammation-related carcinogenesis in mice through inhibition of inflammatory cell infiltration (Onuma et al., 2015). Ochiai et al. (2013) investigated the protective effect of a hydrous ethanol extract of brown rice (ERF) fermented with *A. oryzae* on the methotrexate (MTX)-induced gastrointestinal damage in a rat model. Rats were divided into three groups named control (CON), MTX, and MTX-ERF. CON and MTX groups were fed for 4 weeks on a basal diet, and the MTX-ERF group was fed a 9.16% ERF-containing basal diet; after 3 weeks, MTX were administered. ERF supplementation prevents diarrhea, increased the protein content in small intestinal mucosa, and also improved the survival rate. These results specify that dietary ERF could protect against MTX-induced gastrointestinal damage (Ochiai et al., 2013).

4.3 Preventive Role of FRB on Tumorigenesis

Fermented brown rice and rice bran with *A. oryzae* (FBRA) is also reported to prevent chemically induced carcinogenesis in extraintestinal organs of rodents. Kuno et al. (2016) evaluated the possible chemopreventive effects of FBRA against prostate tumorigenesis in 6-week-old transgenic rats for adenocarcinoma of prostate (TRAP) strain. Rats were fed diets containing 5% or 10% FBRA for 15 weeks. FBRA decreased the occurrence of adenocarcinoma in the lateral prostate and suppressed the development of prostate carcinogenesis. FBRA increased apoptosis and inhibited cell proliferation in histologically high-grade prostatic intraepithelial neoplasias. Also, FBRA supplementation upregulated the phospho-AMP-activated kinase α (Thr172) in the prostate of rats. FBRA limited tumor growth by activating the energy deprivation pathways, which proved that FBRA has translational potential to prevent human prostate cancer (Kuno et al., 2016). Rice bran oil (RBO) extracted from rice bran is an abundant source of bioactive compounds with antioxidant properties such as gamma-oryzanol and phytosteryl ferulate (Islam et al., 2008). Moreover, Islam et al. (2008) investigated the anticolitis effects of RBO in a DSS-induced colitis model. Orally administered RBO diminished the histological results of colitis and inhibited MPO, proinflammatory cytokines, cyclooxygenase-2, and nuclear factor kappa B (NF-κB) (Islam et al., 2008). Pengkumsri et al. (2017) investigated the role of dietary supplementation of Thai black rice bran (RB) extract and yeast beta-glucan (YBG) against DSS-mediated colitis in rat. The protective effect of RB + YBG combinational treatment was higher than that of RB extract and YBG regarding serum antioxidant levels. IL-6, IL-17, and IFN-g levels were significantly reduced by RB + YBG combinational treatment than other tested interventions, which was accompanied by an increase in anti-inflammatory cytokines (IL-10, TGF-b). Supplementation of RB + YBG was a potent alternative nutrient-based therapeutic agent for colitis and to prevent the development of cancer (Pengkumsri et al., 2017).

4.4 Role of FRB on Metabolic Disorders

Medical disorders refer to metabolism-related diseases, including hyperglycemia, hypercholesterolemia, hypertriglyceridemia, and insulin resistance, which usually manifests in the form of type 2 diabetes mellitus, obesity, and cardiovascular diseases (CVD). Rice bran, and its various active components, prevent or ameliorate metabolic disorders. Specifically, a rice bran enzymatic extract-supplemented diet can prevent the adipose and macrophage changes associated with diet-induced obesity in mice (Justo et al., 2016). In addition, the antihyperlipidemic effects (lowering cholesterol and triglyceride) of α-tocopherol have been investigated in F344 rats fed a Western diet (Shibata, et al., 2016). Pigmented rice, which contains anthocyanins and proanthocyanidins concentrated in the bran layer, stimulated glucose uptake by 3T3-L1 adipocytes (a key function in glucose homeostasis). Specifically, basal glucose uptake is increased two- to threefold, whereas mRNA levels of both GLUT1 and GLUT4 are upregulated (Boue, et al., 2016). γ-Oryzanol and FA ester with phytosterols (abundant in rice bran) were reported to prevent high-fat and high-fructose diet (HFFD)-induced metabolic syndrome. Additionally, only γ-oryzanol treatment is more effective than FA in significantly decreasing the liver index and hepatic triglyceride content. Lowered serum levels of C-reactive protein and IL-6, and an increased serum concentration of adiponectin, confirmed that FA and γ-oryzanol can be used as dietary supplements to alleviate the deleterious effects of HFFD (Wang et al., 2015). Adenosine in particular effectively mitigates metabolic syndrome in SHRSP (Ardiansyah et al., 2011). Specifically, single and long-term oral administration of adenosine improves hyperlipidemia and hyperinsulinemia; it also regulated body weight gain and food intake. Studies have shown that enhanced plasma adiponectin levels alleviated hyperinsulinemia, and dietary adenosine can enhance plasma adiponectin and increase insulin sensitivity. Adenosine administration for 3 weeks downregulated mRNA levels of glucose-6-phosphatase, a gene encoding the rate-controlling enzyme of hepatic gluconeogenesis. Adenosine also plays an important role in regulating hepatic mRNA expression of genes involved in β-oxidation, fatty acid synthesis, and AMP-activated protein kinase. In conclusion, various active components of rice bran ameliorate metabolic-related diseases.

5. FERMENTED RICE BRAN MODULATES MULTIFACTORIAL METABOLIC DISEASE AND ITS SENSOR (GLUCOSE, INSULIN, AND TRANSCRIPTION FACTORS)

Common metabolic diseases such as obesity and diabetes are associated with glucose and defective insulin metabolism, which contributes to the development of major medical problems, including hypertension, atherosclerosis, small vessel disease, kidney disease, and blindness. FRB is a potent functional food that can be used for the management of metabolic syndrome by controlling hypertension, insulin resistance, glucose impairment, serum adiponectin level, and AMPK activation. Adiponectin is an important adipokine for insulin sensitization and is involved in some homeostatic functions such as the regulation of glucose and lipid metabolism. FRB improves adiponectin and leptin impairments in SHRSP, which results in an improvement in glucose and lipid metabolism (Esmaili et al., 2014). Plasma adiponectin levels are inversely related to adiposity and directly associated with leptin sensitivity (Nedvidkova et al., 2005). FRB exerts its action on leptin sensitivity and body fat mass via the stimulation of adiponectin secretion even though the exact mechanisms are still unknown. Phenolic compounds and dietary fiber present are used in formulating food products to improve the functionalities and health benefits of such products. Phenolic compounds and dietary fiber produce health benefits by reducing cholesterolemia, modifying glycemic responses, and preventing the development of cardiovascular diseases (Anderson et al., 2009; Kris-Etherton et al., 2002). Microorganisms are used in the brewing and food industries to produce fermented products. They are also used to produce aroma compounds and secondary metabolites for use in processed foods. Thus FRB contains phenolic compounds and dietary fiber as well as good flavor for human use. Chronic supplementation with 5% FRB for 4 weeks increased serum ACE inhibitory activity (Alauddin et al., 2016). This corroborates that fractions of enzyme-treated rice bran improved BP elevation in SHRSP via the inhibition of ACE activity (Ardiansyah et al., 2007). Furthermore, FRB reduces the mRNA levels of enzymes involved in gluconeogenesis (PEPCK and G6PC), which are the rate-limiting enzymes in gluconeogenesis. Insulin can inhibit the transcriptional activity of forkhead box protein O1, which regulates the transcription of PEPCK. The aforementioned mechanisms were therefore involved in the improvements in serum glucose and insulin levels by FRB. Studies have shown that brown rice bran and enzyme-treated rice bran improve glucose tolerance and insulin resistance in mouse and rat models (Ardiansyah et al., 2007; Anderson et al., 2009). Altogether, FRB regulates glucose and lipid metabolism, and contributes effectively to improving hypertension in SHRSP, but some transcription factors like LXRα, SREBP-1c, and ChREBPα mRNA expression levels were downregulated after FRB supplementation in the SHRSP because glucose and insulin coordinate hepatic lipogenesis and the glycolytic gene expression (Alauddin et al., 2016; Kris-Etherton et al., 2002). Moreover, FRB increases the activation of AMPK in the liver, and studies in humans have shown that plasma adiponectin concentration negatively correlates with lipid metabolism (Matsubara et al., 2002). Lipid metabolism disorders are intimately connected to many common lifestyle-related diseases. Cardiovascular and obesity-related diseases, often referred to as the metabolic syndromes, are serious conditions with a clustering of risk factors like dyslipidemia, high blood pressure, and insulin resistance. Triglycerides (TG) are produced in the liver tissue in response to glucose and fatty acids (Pégorier et al., 2004). Glucose and insulin stimulates SREBP-1c either by substrates such as citrate or by increased insulin concentration. Citrate is produced by the action of pyruvate kinase from glucose in a reaction through conversion to pyruvate (Kersten, 2001). This pyruvate is then converted to citrate in the krebs cycle, ultimately producing acetyl coenzyme A, the principal substrate for fatty acid biosynthesis. In addition, glucose-stimulated insulin production induces SREBP-1c gene expression; this elevated level of SREBP-1c promotes the de novo lipogenesis pathway. Glucose can also increase lipogenesis pathways by preventing the release of glucagon from the pancreas. Taken together, these properties may explain the mechanisms of glucose and fat, by which a diet rich in surplus carbohydrates, which rapidly increases serum glucose concentration, subsequently can stimulate lipogenesis pathways in both the liver and adipose tissue. Following the staffing of various transcriptional coactivators and subsequently RNA polymerase II, transcription of the target gene is finally started. Even though SREBP-1c regulates several lipogenic gene expressions, such as acetyl coenzyme-A carboxylase and fatty acid synthase, it is under the regulation of LXR (Pawar et al., 2002). LXR is triggered by binding of its ligands such as oxysterol, which controls excessive cellular cholesterol levels. Nonesterified fatty acids also show LXR ligand binding properties and appear to compete with oxysterol for LXR stimulation. Cellular cholesterol toxicity is controlled by the activation of LXR, which enhances the expression of genes that excite bile acid production (cholesterol 7α-hydroxylase) and cholesterol elimination into bile and inhibit cholesterol immersion. LXR is also an important stimulator of adenosine triphosphate-binding cassette (ABCA1) to stimulate efflux of cholesterol into HDL as well as ABCG5 and ABCG8 to increase cholesterol disposal from hepatic cells into bile and from intestinal cells into the lumen. Ultimately this mechanism markedly reduced cellular cholesterol levels. For a region that is uncertain, LXR stimulation also regulates TG synthesis though enhancing the expression of SREBP-1c. The control of lipids, lipoprotein metabolisms, and glucose homeostasis is a complex process implying numerous genes, the expression of which is regulated in a coordinated manner. When FRB modulates the expression of these genes, the organism can adapt its metabolism to changes in energy requirements. The expression of these genes can be regulated through ligand activated transcription factors, called nuclear transcription factors, which are capable to responding to small lipophilic signaling molecules. Among those nuclear transcription factors, LXR, ChREBP and SREBP-1c play a major role because they are regulated in the metabolic syndrome. FRB supplementation improved the colonic abnormalities by maintaining MPO activity and TBARS levels as well as suppressing the expression of proinflammatory cytokines and chemokines. Furthermore, FRB diet supplementation improved tryptophan, tryptamine, and short-chain fatty acid (SCFA) production that have a positive impact on intestinal barrier function and mucin production. FRB demonstrates the potential of consumption of fermented rice bran as a dietary supplement for preventing intestinal inflammatory disorders (Fig. 3) (Islam et al., 2017).

6. CONCLUSION

FRB is a functional food known to contain abundant phytochemicals with potent health beneficial properties. Evidence from previous in vivo and in vitro studies suggested that these phytochemicals can modulate lifestyle-related disease by reducing hypertension and inhibiting gastrointestinal disorders, potentially through the amelioration of hypertension and oxidative stress, the inhibition of cell proliferation, and the reduction of inflammation. Due to its significant nutritive and therapeutic value, FRB may enhance well-being and health, as well as reduce the risk of disease, providing health benefits and improving quality of life. Thus rice bran can be considered as a super food and/or functional food. In conclusion, there is a strong demand for enrichment of functional bran components in different diet-based approaches that mitigate lifestyle-related disorders and inflammation. Future research efforts should therefore be directed toward the development of effective FRB dietary interventions and the assessment of their effectiveness in reducing the presence of biomarkers indicative of lifestyle-related disease and inflammation.

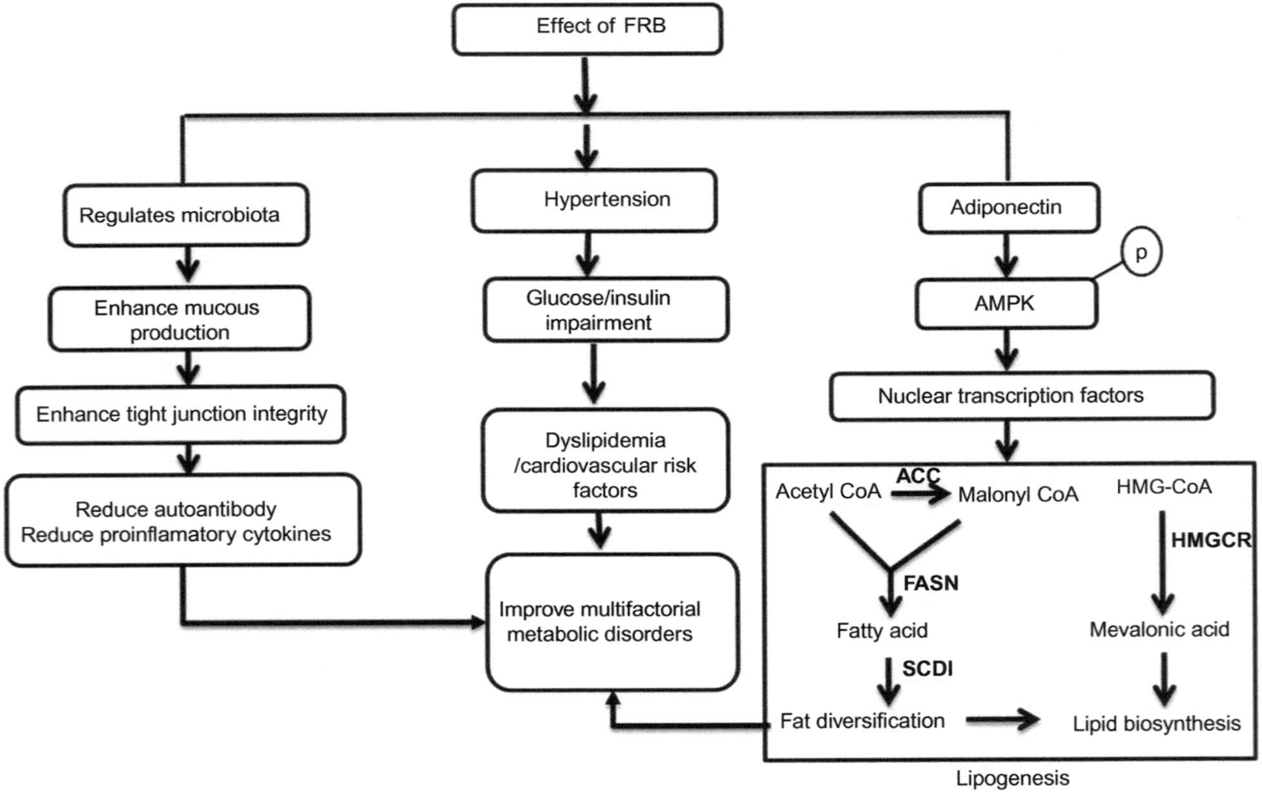

Fig. 3 Proposed mechanism of FRB's effect on multifactorial metabolic disorders.

REFERENCES

Abdul-Hamid, A., Luan, Y.S., 2000. Functional properties of dietary fibre prepared from defatted rice bran. Food Chem. 68 (1), 15–19.

Alauddin, M., Shirakawa, H., Koseki, T., Kijima, N., Budijanto, S., Islam, J., Goto, T., Komai, M., 2016. Fermented rice bran supplementation mitigates metabolic syndrome in stroke-prone spontaneously hypertensive rats. BMC Complement. Altern. Med. 16 (1), 442.

American Dietetic Association, 2003. Position of the American dietetic association and dietitians of Canada: vegetarian diets. J. Acad. Nutr. Diet. 103 (6), 748.

Anderson, J.W., Baird, P., Davis, R.H., Ferreri, S., Knudtson, M., Koraym, A., Waters, V., Williams, C.L., 2009. Health benefits of dietary fiber. Nutr. Rev. 67 (4), 188–205.

Arai, S., Osawa, T., Ohigashi, H., Yoshikawa, M., Kaminogawa, S., Watanabe, M., Ogawa, T., Okubo, K., Watanabe, S., Nishino, H., Shinohara, K., 2001. A mainstay of functional food science in Japan—history, present status, and future outlook. Biosci. Biotechnol. Biochem. 65 (1), 1–13.

Ardiansyah, Shirakawa, H., Koseki, T., Ohinata, K., Hashizume, K., Komai, M., 2006. Rice bran fractions improve blood pressure, lipid profile, and glucose metabolism in stroke-prone spontaneously hypertensive rats. J. Agric. Food Chem. 54 (5), 1914–1920.

Ardiansyah, Shirakawa, H., Koseki, T., Hashizume, K., Komai, M., 2007. The Driselase-treated fraction of rice bran is a more effective dietary factor to improve hypertension, glucose and lipid metabolism in stroke-prone spontaneously hypertensive rats compared to ferulic acid. Br. J. Nutr. 97, 67–76.

Ardiansyah, Shirakawa, H., Shimeno, T., Koseki, T., Shiono, Y., Murayama, T., Hatakeyama, E., Komai, M., 2009. Adenosine, an identified active component from the Driselase-treated fraction of rice bran, is effective at improving metabolic syndrome in stroke-prone spontaneously hypertensive rats. J. Agric. Food Chem. 57 (6), 2558–2564.

Ardiansyah, Shirakawa, H., Koseki, T., Hiwatashi, K., Takahasi, S., Akiyama, Y., Komai, M., 2011. Novel effect of adenosine 5′-monophosphate on ameliorating hypertension and the metabolism of lipids and glucose in stroke-prone spontaneously hypertensive rats. J. Agric. Food Chem. 59 (24), 13238–13245.

Bangladesh Bureau of Statistics, M. O. P., Government Of Bangladesh, Dhaka, 1999. Year Book of Agricultural Statistics BBS. .

Bhosale, S., Vijayalakshmi, D., 2015. Processing and nutritional composition of rice bran. Curr. Res. Nutr. Food Sci. J. 3 (1), 74–80.

Boonloh, K., Kukongviriyapan, V., Kongyingyoes, B., Kukongviriyapan, U., Thawornchinsombut, S., Pannangpetch, P., 2015. Rice bran protein hydrolysates improve insulin resistance and decrease pro-inflammatory cytokine gene expression in rats fed a high carbohydrate-high fat diet. Nutrients 7 (8), 6313–6329.

Boue, S.M., Daigle, K.W., Chen, M.H., Cao, H., Heiman, M.L., 2016. Antidiabetic potential of purple and red rice (*Oryza sativa* L.) bran extracts. J. Agric. Food Chem. 64 (26), 5345–5353.

Brown, A.L., Lane, J., Holyoak, C., Nicol, B., Mayes, A.E., Dadd, T., 2011. Health effects of green tea catechins in overweight and obese men: a randomised controlled cross-over trial. Br. J. Nutr. 106 (12), 1880–1889.

Chung, H.S., Shin, J.C., 2007. Characterization of antioxidant alkaloids and phenolic acids from anthocyanin-pigmented rice (*Oryza sativa* cv. *Heugjinjubyeo*). Food Chem. 104 (4), 1670–1677.

Da Silva, B.C., Lyra, A.C., Rocha, R., Santana, G.O., 2014. Epidemiology, demographic characteristics and prognostic predictors of ulcerative colitis. World J. Gastroenterol.: WJG 20 (28), 9458.

Esmaili, S., Xu, A., George, J., 2014. The multifaceted and controversial immunometabolic actions of adiponectin. Trends Endocrinol. Metab. 25 (9), 444–451.

Faria, S.A.D.S.C., Bassinello, P.Z., Penteado, M.D.V.C., 2012. Nutritional composition of rice bran submitted to different stabilization procedures. Braz. J. Pharm. Sci. 48 (4), 651–657.

Goffman, F.D., Bergman, C.J., 2004. Rice kernel phenolic content and its relationship with antiradical efficiency. J. Sci. Food Agric. 84 (10), 1235–1240.

Goufo, P., Pereira, J., Moutinho-Pereira, J., Correia, C.M., Figueiredo, N., Carranca, C., Rosa, E.A., Trindade, H., 2014. Rice (*Oryza sativa* L.) phenolic compounds under elevated carbon dioxide (CO_2) concentration. Environ. Exp. Bot. 99, 28–37.

Harukaze, A., Murata, M., Homma, S., 1999. Analyses of free and bound phenolics in rice. Food Sci. Technol. Res. 5 (1), 74–79.

Hasler, C.M., 2002. Functional foods: benefits, concerns and challenges—a position paper from the American Council on Science and Health. J. Nutr. 132 (12), 3772–3781.

Heuberger, A.L., Lewis, M.R., Chen, M.H., Brick, M.A., Leach, J.E., Ryan, E.P., 2010. Metabolomic and functional genomic analyses reveal varietal differences in bioactive compounds of cooked rice. PLoS One 5(9), e12915.

Hu, C., Zawistowski, J., Ling, W., Kitts, D.D., 2003. Black rice (Oryza sativa L. indica) pigmented fraction suppresses both reactive oxygen species and nitric oxide in chemical and biological model systems. J. Agric. Food Chem. 51 (18), 5271–5277.

Islam, M.S., Murata, T., Fujisawa, M., Nagasaka, R., Ushio, H., Bari, A.M., Hori, M., Ozaki, H., 2008. Anti-inflammatory effects of phytosteryl ferulates in colitis induced by dextran sulphate sodium in mice. Br. J. Pharmacol. 154 (4), 812–824.

Islam, J., Koseki, T., Watanabe, K., Ardiansyah, Budijanto, S., Oikawa, A., Alauddin, M., Goto, T., Aso, H., Komai, M., Shirakawa, H., 2017. Dietary supplementation of fermented rice bran effectively alleviates dextran sodium sulfateinducedcolitis in mice. Nutrients 9, 747. https://doi.org/10.3390/nu9070747.

Jung, E.H., Ran Kim, S., Hwang, I.K., Youl Ha, T., 2007. Hypoglycemic effects of a phenolic acid fraction of rice bran and ferulic acid in C57BL/KsJ-db/db mice. J. Agric. Food Chem. 55 (24), 9800–9804.

Justo, M.L., Claro, C., Zeyda, M., Stulnig, T.M., Herrera, M.D., Rodríguez-Rodríguez, R., 2016. Rice bran prevents high-fat diet-induced inflammation and macrophage content in adipose tissue. Eur. J. Nutr. 55 (6), 2011–2019.

Kahlon, T.S., Chow, F.I., Sayre, R.N., Betschart, A.A., 1992. Cholesterol-lowering in hamsters fed rice bran at various levels, defatted rice bran and rice bran oil. J. Nutr. 122 (3), 513–519.

Kataoka, K., Ogasa, S., Kuwahara, T., Bando, Y., Hagiwara, M., Arimochi, H., Nakanishi, S., Iwasaki, T., Ohnishi, Y., 2008. Inhibitory effects of fermented brown rice on induction of acute colitis by dextran sulfate sodium in rats. Dig. Dis. Sci. 53 (6), 1601–1608.

Kersten, S., 2001. Mechanisms of nutritional and hormonal regulation of lipogenesis. EMBO Rep. 2 (4), 282–286.

Kim, K.M., Yu, K.W., Kang, D.H., Suh, H.J., 2002. Anti-stress and anti-fatigue effect of fermented rice bran. Phytother. Res. 16 (7), 700–702.

Kim, C.J., Kovacs-Nolan, J.A., Yang, C., Archbold, T., Fan, M.Z., Mine, Y., 2010. L-tryptophan exhibits therapeutic function in a porcine model of dextran sodium sulfate (DSS)-induced colitis. J. Nutr. Biochem. 21 (6), 468–475.

Komiyama, Y., Andoh, A., Fujiwara, D., Ohmae, H., Araki, Y., Fujiyama, Y., Mitsuyama, K., Kanauchi, O., 2011. New prebiotics from rice bran ameliorate inflammation in murine colitis models through the modulation of intestinal homeostasis and the mucosal immune system. Scand. J. Gastroenterol. 46 (1), 40–52.

Kondo, S., Kuda, T., Nemoto, M., Usami, Y., Takahashi, H., Kimura, B., 2016. Protective effects of rice bran fermented by Saccharomyces cerevisiae Misaki-1 and Lactobacillus plantarum Sanriki-SU8 in dextran sodium sulphate-induced inflammatory bowel disease model mice. Food Biosci. 16, 44–49.

Krikorian, R., Nash, T.A., Shidler, M.D., Shukitt-Hale, B., Joseph, J.A., 2010. Concord grape juice supplementation improves memory function in older adults with mild cognitive impairment. Br. J. Nutr. 103 (5), 730–734.

Kris-Etherton, P.M., Hecker, K.D., Bonanome, A., Coval, S.M., Binkoski, A.E., Hilpert, K.F., Griel, A.E., Etherton, T.D., 2002. Bioactive compounds in foods: their role in the prevention of cardiovascular disease and cancer. Am. J. Med. 113 (9), 71–88.

Kuno, T., Nagano, A., Mori, Y., Kato, H., Nagayasu, Y., Naiki-Ito, A., Suzuki, S., Mori, H., Takahashi, S., 2016. Preventive effects of fermented brown rice and rice bran against prostate carcinogenesis in TRAP rats. Nutrients 8 (7), 421.

Kupski, L., Cipolatti, E., Rocha, M.D., Oliveira, M.D.S., Souza-Soares, L.D.A., Badiale-Furlong, E., 2012. Solid-state fermentation for the enrichment and extraction of proteins and antioxidant compounds in rice bran by *Rhizopus oryzae*. Braz. Arch. Biol. Technol. 55 (6), 937–942.

Lloyd, B.J., Siebenmorgen, T.J., Beers, K.W., 2000. Effects of commercial processing on antioxidants in rice bran. Cereal Chem. 77 (5), 551–555.

Malekian, F., Rao, R.M., Prinyawiwatkul, W., Marshall, W.E., Windhauser, M., Ahmedna, M., 2000. Lipase and Lipoxygnase Activity, Functionality, and Nutrient Losses in Rice Bran During Storage. Louisiana Agricultural Experiment Station, Louisiana State University Agricultural Center Bulletin Number 870.

Martirosyan, D.M., Singh, J., 2015. A new definition of functional food by FFC: what makes a new definition unique? Funct. Foods Health Dis. 5 (6), 209–223.

Matsubara, M., Maruoka, S., Katayose, S., 2002. Decreased plasma adiponectin concentrations in women with dyslipidemia. J. Clin. Endocrinol. Metabol. 87 (6), 2764–2769.

Min, B., Chen, M.H., Green, B.W., 2009. Antioxidant activities of purple rice bran extract and its effect on the quality of low-NaCl, phosphate-free patties made from channel catfish (*Ictalurus punctatus*) belly flap meat. J. Food Sci. 74(3).

Mohamed, S., 2014. Functional foods against metabolic syndrome (obesity, diabetes, hypertension and dyslipidemia) and cardiovasular disease. Trends Food Sci. Technol. 35 (2), 114–128.

Mottillo, S., Filion, K.B., Genest, J., Joseph, L., Pilote, L., Poirier, P., Rinfret, S., Schiffrin, E.L., Eisenberg, M.J., 2010. The metabolic syndrome and cardiovascular risk: a systematic review and meta-analysis. J. Am. Coll. Cardiol. 56 (14), 1113–1132.

Nakornriab, M., Sriseadka, T., Wongpornchai, S., 2008. Quantification of carotenoid and flavonoid components in brans of some thai black rice cultivars using supercritical fluid extraction and high-performance liquid chromatography-mass spectrometry. J. Food Lipids 15 (4), 488–503.

Nedvidkova, J., Smitka, K., Kopský, V., Hainer, V., 2005. Adiponectin, an adipocyte-derived protein. Physiol. Res. 54 (2), 133.

Nicolosi, R.J., Austrian, L.M., Hegsted, D.M., 1991. Rice bran oil lowers serum total and low density lipoprotein cholesterol and apo B levels in nonhuman primates. Atherosclerosis 88 (2–3), 133–142.

Ochiai, M., Shiomi, S., Ushikubo, S., Inai, R., Matsuo, T., 2013. Effect of a fermented brown rice extract on the gastrointestinal function in methotrexate-treated rats. Biosci. Biotechnol. Biochem. 77, 243–248.

Onuma, K., Kanda, Y., Suzuki Ikeda, S., Sakaki, R., Nonomura, T., Kobayashi, M., Osaki, M., Shikanai, M., Kobayashi, H., Okada, F., 2015. Fermented brown rice and rice bran with aspergillus oryzae (FBRA) prevents inflammation-related carcinogenesis in mice, through inhibition of inflammatory cell infiltration. Nutrients 7 (12), 10237–10250.

Palou, M., Sánchez, J., García-Carrizo, F., Palou, A., Picó, C., 2015. Pectin supplementation in rats mitigates age-related impairment in insulin and leptin sensitivity independently of reducing food intake. Mol. Nutr. Food Res. 59 (10), 2022–2033.

Pawar, A., Xu, J., Jerks, E., Mangelsdorf, D.J., Jump, D.B., 2002. Fatty acid regulation of liver X receptors (LXR) and peroxisome proliferator-activated receptor α (PPARα) in HEK293 cells. J. Biol. Chem. 277 (42), 39243–39250.

Pengkumsri, N., Sundaram Sivamaruthi, B., Sirilun, S., Suwannalert, P., Rodboon, T., Prasitpuriprecha, C., Peerajan, S., Butrungrode, W., Chaiyasut, C., 2017. Dietary supplementation of Thai black rice bran extract and yeast beta-glucan protects the dextran sodium sulphate mediated colitis induced rat. RSC Adv. 7, 396–402.

Pégorier, J.P., Le May, C., Girard, J., 2004. Control of gene expression by fatty acids. J. Nutr. 134 (9), 2444S–2449S.

Phutthaphadoong, S., Yamada, Y., Hirata, A., Tomita, H., Hara, A., Limtrakul, P., Iwasaki, T., Kobayashi, H., Mori, H., 2010. Chemopreventive effect of fermented brown rice and rice bran (FBRA) on the inflammation-related colorectal carcinogenesis in ApcMin/+ mice. Oncol. Rep. 23 (1), 53–59.

Quirós-Sauceda, A.E., Palafox-Carlos, H., Sáyago-Ayerdi, S.G., Ayala-Zavala, J.F., Bello-Perez, L.A., Alvarez-Parrilla, E., De La Rosa, L.A., González-Córdova, A.F., Gonzalez-Aguilar, G.A., 2014. Dietary fiber and phenolic compounds as functional ingredients: interaction and possible effect after ingestion. Food Funct. 5 (6), 1063–1072.

Rahman, M.L., Absar, N., Mondal, M.I.H., Hossan, M.K., Khatun, S., Khan, G.A.M., Islam, M.J., 2013. Lipid, fatty acid and glyceride composition of five different varieties of rice bran oil. Bangladesh Res. Publ. J. 9 (3), 111–115.

Rao, M.K., Achaya, K.T., 1968. Unsaturated fatty acids as synergists for antioxidants. Eur. J. Lipid Sci. Technol. 70 (4), 231–234.

Ren, T., Tian, T., Feng, X., Ye, S., Wang, H., Wu, W., Qiu, Y., Yu, C., He, Y., Zeng, J., Cen, J., 2015. An adenosine A3 receptor agonist inhibits DSS-induced colitis in mice through modulation of the NF-κB signaling pathway. Sci. Rep. 5, 9047.

Richard, D., Kefi, K., Barbe, U., Bausero, P., Visioli, F., 2008. Polyunsaturated fatty acids as antioxidants. Pharmacol. Res. 57 (6), 451–455.

Ritchie, L.E., Taddeo, S.S., Weeks, B.R., Carroll, R.J., Dykes, L., Rooney, L.W., Turner, N.D., 2017. Impact of novel sorghum bran diets on DSS-induced colitis. Nutrients 9 (4), 330.

Rogers, E.J., Rice, S.M., Nicolosi, R.J., Carpenter, D.R., McClelland, C.A., Romanczyk, L.J., 1993. Identification and quantitation of γ-oryzanol components and simultaneous assessment of tocols in rice bran oil. J. Am. Oil Chem. Soc. 70 (3), 301–307.

Ruemmele, F.M., 2016. Role of diet in inflammatory bowel disease. Ann. Nutr. Metab. 68 (Suppl. 1), 32–41.

Rukmini, C., Raghuram, T.C., 1991. Nutritional and biochemical aspects of the hypolipidemic action of rice bran oil: a review. J. Am. Coll. Nutr. 10 (6), 593–601.

Ryan, E.P., Heuberger, A.L., Weir, T.L., Barnett, B., Broeckling, C.D., Prenni, J.E., 2011. Rice bran fermented with Saccharomyces boulardii generates novel metabolite profiles with bioactivity. J. Agric. Food Chem. 59 (5), 1862–1870.

Saenjum, C., Chaiyasut, C., Chansakaow, S., Suttajit, M., Sirithunyalug, B., 2012. Antioxidant and anti-inflammatory activities of gamma-oryzanol rich extracts from Thai purple rice bran. J. Med. Plant Res. 6 (6), 1070–1077.

Saito, M., 2007. Role of FOSHU (food for specified health uses) for healthier life. Yakugaku zasshi: J. Pharm. Soc. Jpn. 127 (3), 407–416.

Seril, D.N., Liao, J., Yang, G.Y., Yang, C.S., 2003. Oxidative stress and ulcerative colitis-associated carcinogenesis: studies in humans and animal models. Carcinogenesis 24 (3), 353–362.

Shibata, A., Kawakami, Y., Kimura, T., Miyazawa, T., Nakagawa, K., 2016. α-Tocopherol attenuates the triglyceride-and cholesterol-lowering effects of rice bran tocotrienol in rats fed a western diet. J. Agric. Food Chem. 64 (26), 5361–5366.

Tazawa, H., Okada, F., Kobayashi, T., Tada, M., Mori, Y., Une, Y., Sendo, F., Kobayashi, M., Hosokawa, M., 2003. Infiltration of neutrophils is required for acquisition of metastatic phenotype of benign murine fibrosarcoma cells: implication of inflammation-associated carcinogenesis and tumor progression. Am. J. Pathol. 163 (6), 2221–2232.

Terahara, N., Saigusa, N., Ohba, R., Ueda, S., 1994. Composition of anthocyanin pigments in aromatic red rice and its wine. Nippon Shokuhin Kogyo Gakkaishi 41 (7), 519–522.

Tuncel, N.B., Yılmaz, N., 2011. Gamma-oryzanol content, phenolic acid profiles and antioxidant activity of rice milling fractions. Eur. Food Res. Technol. 233 (4), 577.

Wang, O., Liu, J., Cheng, Q., Guo, X., Wang, Y., Zhao, L., Zhou, F., Ji, B., 2015. Effects of ferulic acid and γ-oryzanol on high-fat and high-fructose diet-induced metabolic syndrome in rats. PLoS One 10(2), e0118135.

Wataniyakul, P., Pavasant, P., Goto, M., Shotipruk, A., 2012. Microwave pretreatment of defatted rice bran for enhanced recovery of total phenolic compounds extracted by subcritical water. Bioresour. Technol. 124, 18–22.

Zubair, M.A., Rahman, M.S., Islam, M.S., Abedin, M.Z., Sikder, M.A., 2016. A comparative study of the proximate composition of selected Rice varieties in Tangail, Bangladesh. J. Environ. Sci. Nat. Resour. 8 (2), 97–102.

FURTHER READING

DeFronzo, R.A., Bonadonna, R.C., Ferrannini, E., 1992. Pathogenesis of NIDDM: a balanced overview. Diabetes Care 15 (3), 318–368.

Foufelle, F., Ferré, P., 2002. New perspectives in the regulation of hepatic glycolytic and lipogenic genes by insulin and glucose: a role for the transcription factor sterol regulatory element binding protein-1c. Biochem. J. 366 (2), 377–391.

Harris, R.A., 1992. Carbohydrate metabolism I: major metabolic pathways and their control. In: Textbook of Biochemistry With Clinical Correlations. pp. 291–358, CiNii.

Hartmann, H., Probst, I., Jungermann, K., Creutzfeldt, W., 1987. Inhibition of glycogenolysis and glycogen phosphorylase by insulin and proinsulin in rat hepatocyte cultures. Diabetes 36 (5), 551–555.

Hu, C., Zawistowski, J., Ling, W., Kitts, D.D., 2003. Black rice (*Oryza sativa* L. indica) pigmented fraction suppresses both reactive oxygen species and nitric oxide in chemical and biological model systems. J. Agric. Food Chem. 51 (18), 5271–5277.

Iizuka, K., Horikawa, Y., 2008. ChREBP: a glucose-activated transcription factor involved in the development of metabolic syndrome. Endocr. J. 55 (4), 617–624.

Kelley, D., Mitrakou, A., Marsh, H., Schwenk, F., Benn, J., Sonnenberg, G., Arcangeli, M., Aoki, T., Sorensen, J., Berger, M., 1988. Skeletal muscle glycolysis, oxidation, and storage of an oral glucose load. J. Clin. Investig. 81 (5), 1563.

Kim, D., Han, G.D., 2011. Ameliorating effects of fermented rice bran extract on oxidative stress induced by high glucose and hydrogen peroxide in 3T3-L1 adipocytes. Plant Foods Hum. Nutr. 66 (3), 285.

Kozuka, C., Yabiku, K., Sunagawa, S., Ueda, R., Taira, S.I., Ohshiro, H., Ikema, T., Yamakawa, K., Higa, M., Tanaka, H., Takayama, C., 2012. Brown rice and its component, γ-oryzanol, attenuate the preference for high-fat diet by decreasing hypothalamic endoplasmic reticulum stress in mice. Diabetes 61 (12), 3084–3093.

Landau, B.R., Wahren, J., Chandramouli, V., Schumann, W.C., Ekberg, K., Kalhan, S.C., 1996. Contributions of gluconeogenesis to glucose production in the fasted state. J. Clin. Investig. 98 (2), 378.

Okada, F., Kobayashi, M., Tanaka, H., Kobayashi, T., Tazawa, H., Iuchi, Y., Onuma, K., Hosokawa, M., Dinauer, M.C., Hunt, N.H., 2006. The role of nicotinamide adenine dinucleotide phosphate oxidase-derived reactive oxygen species in the acquisition of metastatic ability of tumor cells. Am. J. Pathol. 169 (1), 294–302.

Parrado, J., Miramontes, E., Jover, M., Gutierrez, J.F., de Terán, L.C., Bautista, J., 2006. Preparation of a rice bran enzymatic extract with potential use as functional food. Food Chem. 98 (4), 742–748.

Petersen, K.F., Laurent, D., Rothman, D.L., Cline, G.W., Shulman, G.I., 1998. Mechanism by which glucose and insulin inhibit net hepatic glycogenolysis in humans. J. Clin. Investig. 101 (6), 1203.

Pilkis, S.J., Granner, D.K., 1992. Molecular physiology of the regulation of hepatic gluconeogenesis and glycolysis. Annu. Rev. Physiol. 54 (1), 885–909.

Satter, M.A., Ara, H., Jabin, S.A., Abedin, N., Azad, A.K., Hossain, A., Ara, U., 2014. Nutritional composition and stabilization of local variety rice bran BRRI-28. Int. J. Sci. Technol. 3 (5), 306–313.

Shepherd, P.R., Kahn, B.B., 1999. Glucose transporters and insulin action—implications for insulin resistance and diabetes mellitus. N. Engl. J. Med. 341 (4), 248–257.

Suzuki, T., Yoshida, S., Hara, H., 2008. Physiological concentrations of short-chain fatty acids immediately suppress colonic epithelial permeability. Br. J. Nutr. 100 (2), 297–305.

Van Eijk, H.M., Bloemen, J.G., Dejong, C.H., 2009. Application of liquid chromatography-mass spectrometry to measure short chain fatty acids in blood. J. Chromatogr. B 877 (8), 719–724.

CHAPTER 9

Rice Husk, Rice Husk Ash and Their Applications

Yanping Zou, Tiankui Yang
Wilmar Global Research and Development Center, Shanghai, China

1. INTRODUCTION

Rice is one of the oldest ancient crops, which was initially planted thousands of years ago. Nowadays, rice is cultivated over 100 countries and consumed as staple food by more than half of the world's population. In the past 20 years, the output of rice increased by almost 50%. The world's production and area harvested for paddy rice is shown in Fig. 1. In 2014, the worldwide cultivation area for rice was about 162.72 million ha, and approximately 741.48 million tons of rice was produced. >90% of rice is produced in Asia, in which China, India, and Indonesia contribute 27.85%, 21.20%, and 9.55% shares of the total output, respectively (FAOSTAT, 2014).

The kernel of rice mainly consists of endosperm, husk, bran, and germ, in which the endosperm accounts for 70%, rice husk (RH) for 20–21%, rice bran for 6–8%, and rice germ for 1%, respectively, of the total seed weight. During production of milled rice, large quantities of RH are produced as byproducts. For example, the output of rice paddy in China is 208.22 million tons in 2015 (National Bureau of Statistics of China, 2017). Theoretically, about 41.64 million tons of RH are generated in China alone, which is a major issue for the rice milling industry. Currently, most RH is underutilized or left unused. It is difficult to use RH efficiently due to the intrinsic properties of RH such as hard surface, poor nutritive value, high silicon content, low bulk density, and difficult to decompose with bacteria. Previous treatment of RH including onsite burning to produce steam or electricity, open dumping, or land-filling also led to serious environmental pollution including smog, dust, and a greenhouse effect (Kuan et al., 2012; Soltani et al., 2015). On the other hand, when RH is burned, vast amounts of rice husk ash (RHA) are produced, which could be another source of pollution. Therefore it is important to utilize RH/RHA comprehensively and efficiently.

Rice Bran and Rice Bran Oil
https://doi.org/10.1016/B978-0-12-812828-2.00009-3

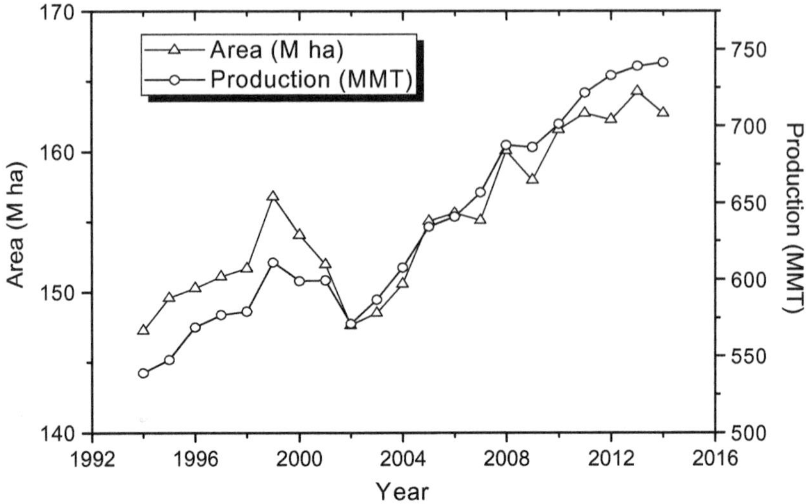

Fig. 1 World production and cultivation area of rice from 1994 to 2014.

RH is a potential material, either in its raw form or in ash form, for production of many value-added products. This chapter introduces the utilization of RH/RHA as a bioadsorbent. Firstly, the physicochemical characterizations of RH/RHA and how the properties of RH/RHA affects their final utilization are presented. Secondly, we present a summary of how silica and silica aerogel are produced from RHA by different methods. Last, we present a glimpse of the application of RHA as a bioadsorbent in vegetable oil refining and removal of heavy metals.

2. CHARACTERIZATIONS OF RICE HUSK/RICE HUSK ASH

2.1 Characterizations of RH

RH is the hard shell covering paddy rice seed, which provides nutrients and metabolite accumulations during grain development, and protects seeds from physical damage and attacks by pathogens, insects, and pests. RH comprises two major, modified, leaf-like structures called the lemma and palea, which completely encase the caryopsis. The structured layers of RH are divided into four categories, namely (1) the rough outer epidermis with surface hairs, where the silica is highly concentrated; (2) sclerenchyma; (3) spongy parenchyma cells; and (4) inner epidermis, whose surface is relatively smooth and free of hair (Champagne et al., 2004).

A study using back-scattered electrons and X-ray images of the RH showed silica was distributed mostly in the husk's outer surface (Stroeven et al., 1999), whereas the midregion and inner epidermis contained less silica. A typical morphology of RH determined by scanning electron micrograph (SEM) showed that the outer surface of RH is relatively globular, which is well organized and has a corrugate structure. The epidermal cells of lemma are arranged in linear ridges and furrows, and the ridges are punctuated with prominent globular protrusions. The relatively stable Si—O carcass and biomass assembled around it formed the tight structure of RH, and thus the surface of RH is relatively nonporous (Jiang, 2010). Although the silicon atoms are presented all over the husk, they are concentrated in the protuberances and hairs (trichomes) on the outer and inner epidermis (Genieva et al., 2008). The side section of RH by SEM showed that an interlayer exists between the inner and outer surfaces, and the interlayer is composed of interlaced plates and sheets, which is loose and honeycombed, and contains a large number of holes with the dimension of 10 μm (Jiang, 2010).

Typical dimensions of RH are about 8–10 mm in length, 2–3 mm in width, and 0.2 mm in thickness (Fang et al., 2004). The bulk density of RH ranges from 100 to 160 kg/m^3, and a true density ranges from 670 to 740 kg/m^3, whereas RH can be only compressed to 400 kg/m^3. RH contains about 80% organic substance and 20% inorganic materials. Crude protein and fat are very low, ranging from 2.0% to 2.8%, and 0.3% to 0.8%, respectively, whereas crude fiber ranges from 34.5% to 45.9% (Champagne et al., 2004), mainly including hemicellulose from 28.6% to 41.5%, hemicellulose from 14.0% to 28.6%, and lignin from 20.4% to 33.7% (Quispe et al., 2017).

Components of RH from different countries by proximate analysis are shown in Table 1. The chemical components of RH are found to vary in different samples due to the different locations, varieties, climate, soil, and fertilizer used during the growth of rice. Moisture content varied from 4.55% to 10.57%, volatile matter varied from 58.22% to 71.24%, and ash varied from 9.29% to 30.18%. Elements such as C, H, O, N, and S are mostly concerned, in which C, O, and H are the major elemental constituents. Ultimate analysis showed that RH contained C 29.98%–50.455%, H 4.40%–6.58%, O 35.20%–59.46%, N 0.05%–4.26%, and S 0.00%–0.64%. As reported by Olupot et al. (2016), 10 selected RH in Uganda from one geographical region exhibited bulk density of 88.82–124.26 kg/m^3, moisture content 9.16%–11.21% (wet-based, wb), volatile matter contents 58.78%–66.37% wb, ash contents 15.87%–25.56% (dry-based, db), fixed carbon 14.8%–17.8% db, and carbon contents 30% 34.5% db.

Table 1 Proximate and ultimate value of rice husks from different countries

	Proximate analysis (%)				Elemental analysis (%)					
Location	Moisture	Volatile matter	Ash	Fixed carbon	C	H	O	N	S	Reference
Uganda	10.57	61.68	22.93	15.40	29.98	4.46	42.31	0.42	0.005	Olupot et al. (2016)
Brazil	8.00	71.24	12.50	16.27	35.86	4.40	59.46	0.28	–	Rambo et al. (2015)
China	4.55	61.78	30.18	8.04	37.65	5.13	55.40	1.63	0.18	Ma et al. (2015)
China	8.38	76.85[a]	14.77	–	43.06	6.08	46.60	4.26	–	Liu et al. (2013)
Malaysia	6.73	62.61	17.06	14.96	38.22	5.88	–	0.68	0.07	Alias et al. (2014)
Ghana	8.59	58.22	24.71	8.48	34.90	5.15	59.00	0.31	0.64	Titiloye et al. (2013)
India	4.65	68.89	9.29	17.17	43.10	5.33	42.27	0.00	0.00	Singh et al. (2013)
India	9.45	70.60	17.09	2.97	50.45	6.58	41.46	1.49	0.23	Natarajan and Ganapathy (2009)
Thai	6.65	60.90	19.11	13.34	38.00	4.73	50.2	0.37	0.09	Garivait et al. (2006)
Vietnam	–	70.20	15.70	14.10	43.50	5.50	35.20	0.05	0.02	Simonov et al. (2003)

[a]Including volatile matter and fixed carbon.
–, not determined.

2.2 Characterizations of RHA

RHA is a general term describing all types of ash produced by combustion of RH. When RH is incinerated, it produces 17%–20% of RHA, which is a lightweight, bulky, and highly porous material with a density of around 180–200 kg/m^3. There are two types of RHA, that is, white rice husk ash (WRHA) and black rice husk ash (BRHA), depending on whether the combustion is complete or incomplete (Ugheoke and Mamat, 2012a). The controlled combustion of RH in the atmosphere can lead to production of WRHA containing almost pure silica (>95%) in a hydrated amorphous form with high porosity and reactivity (Vlaev et al., 2003). The controlled pyrolysis of RH in nitrogen or inert atmosphere results in production of BRHA containing different amounts of carbon and silica (Ghaly and Mansaray, 1999).

SEM photograph of RHA obtained by combustion of RH in an electric oven at 600°C for 2 h is shown in Fig. 2 (Xu et al., 2012). After combustion in air, the major parts retained their original shape, but small parts of RHA suffered structural damage (Fig. 2A). Both the external and internal surfaces

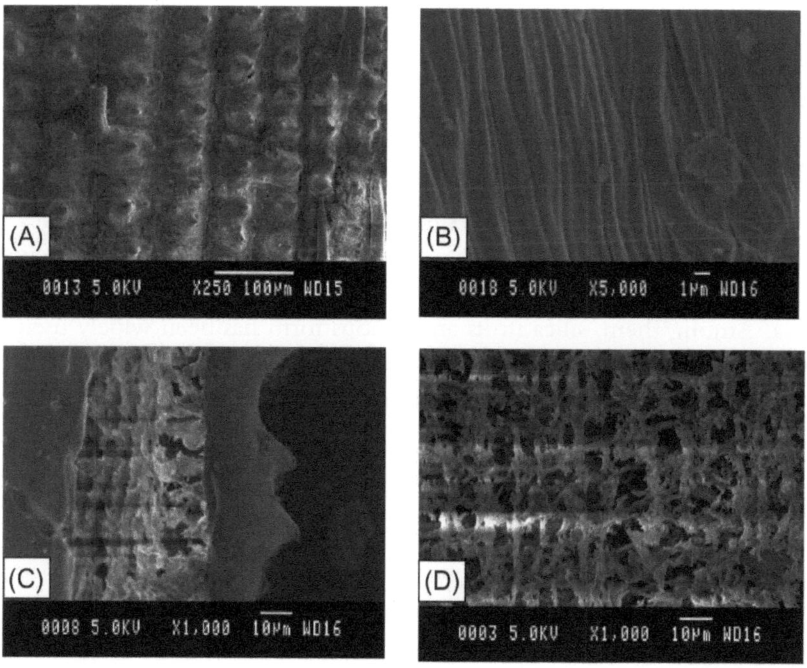

Fig. 2 SEM photograph of the outer surface (A), inner surface (B), side section (C), and interlayer (D) of rice husk ash.

of RHA have a dense structure, suggesting that the exterior and interior surface of RHA is covered with a compact membrane without any micropores (Fig. 2A and B). Cross-section of RHA shows that the exterior surface of RHA is thicker as compared to the interior surface, and there is an interlayer that consisted of a crisscross mesh of chips between the two surfaces (Fig. 2C). The chips are arranged in loose honeycombed fashion and contain a large number of holes. The SEM image of an interlayer of RHA (Fig. 2D) verifies the loose honeycomb-shape structure. Many nanosized pores ranging from several nanometers to several microns are distributed in the interlayer. These pores contribute to the huge specific surface area and high reactivity of RHA when it is ground to powder.

Using SEM and transmission electron microscope (TEM) analysis to investigate the microstructure of RHA, Ouyang and Chen (2003) pointed out the three-layer model of the microstructure of RHA. Two kinds of pores exist in RHA; one is microsized pores (about 10 μm), which are formed by interlacing of the fiber sheet dependent on the structure of RH but independent on the combustion conditions, and the other is nanosized pores (<50 nm), which are formed by nano SiO_2 particles and dependent on the combustion conditions. The nano SiO_2 particles and nanopores are the basic factors to specific surface area and high activity of RHA.

2.3 Characterization of Silica in RH/RHA

In nature, the polymorphs of silica are quartz, cristobalite, tridymite, coestite, stishovite, lechatelerite, and silica gel (Bronzeoak Ltd., 2003). Silica or silicon dioxide (SiO_2) usually exists in two forms, amorphous and crystalline. Processing of silica of a specific quality results in several types of specialty silica. Among them, silica in its amorphous form has been widely used in industry and is more reactive than crystalline silica.

Based on temperature range and duration of combustion of the husk, crystalline and amorphous forms of silica are obtained (Stroeven et al., 1999; Asavapisit and Ruengrit, 2005). It is generally accepted that amorphous silica is formed below 800°C, whereas crystalline silica occurs above 900°C (Chandrasekhar et al., 2006). The crystalline and amorphous forms of silica have different properties, and it is significant to produce ash with correct specifications for specific end use.

The crystal forms of silica in RHA are usually determined by X-ray diffraction (XRD). XRD patterns of RHA obtained by combustion of RH at 300, 500, 700, and 900°C for 10 h in a muffle furnace are shown in Fig. 3

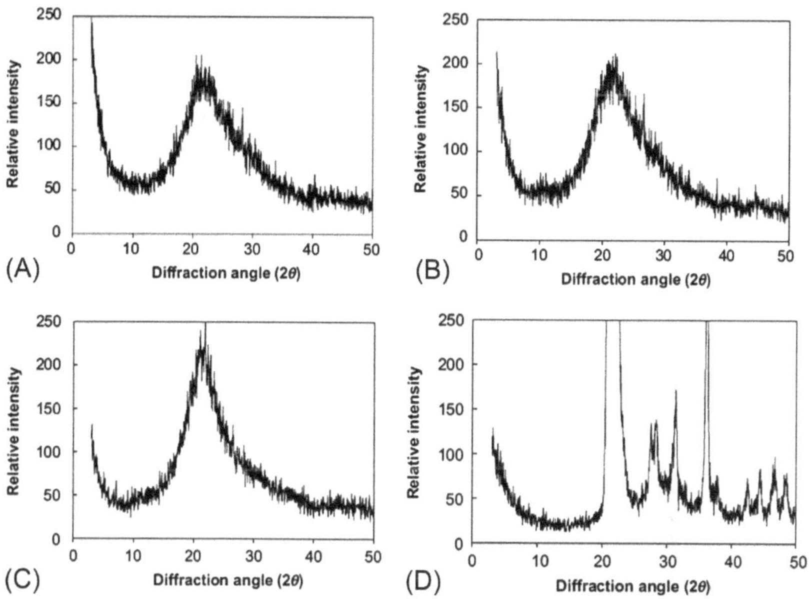

Fig. 3 XRD patterns of rice husk ash obtained at 300°C (A), 500°C (B), 700°C (C), and 900°C (D).

(Kim et al., 2008). The broad smooth hump between 15° 2θ and 35° 2θ in the diffractogram is a characteristic feature of amorphous material, indicating that pyrolysis converted the crystalline cellulose structure to amorphous, random, disordered structure. When the temperature is raised from 500 to 700°C, there is a little increase of peak intensity at 22° 2θ. Heating to 900°C increases the intensity of peaks at 15° 2θ and 36° 2θ. Additional peaks were observed at 43° 2θ, 45° 2θ, 47° 2θ, and 49° 2θ, which are typical of crystalline phase of silica. Many researchers also state that increasing the temperature up to 900°C leads to crystallization of the ash from amorphous form into cristobalite or tridymite. Furthermore, formation of crystalline phase is accelerated by the presence of metallic impurities such as potassium oxide in the RH (Krishnarao et al., 2001; Haslinawati et al., 2009).

2.4 Factors Affecting the Properties of RHA

The properties of RHA such as color, activity, content of impurities, and SiO_2 are influenced by various factors such as incinerating conditions (temperature and duration), rate of heating, burning equipment, pretreatment (acid leaching or not), geographical location and origin, crop variety, and fertilizer used. The physicochemical properties of RHA by different incinerating conditions and pretreatments are listed in Table 2. In general, RHA consists of SiO_2, C, K_2O, P_2O_5, CaO, and minor amounts of Mg, Fe, and Na. Although the composition of RH may depend on several factors, the percentage of silica (SiO_2) in the ash ranges between 80% and 99%. All the other constituents of RHA, except potassium and calcium, are <1%. Silica is the most abundant element in RHA, and no other plant except RH is able to retain such a huge proportion of silica. Trace elements most commonly found in RHA are Na, K, Ca, Mg, Fe, Cu, Mn, and Zn, and differences in the composition may be due to geographical factors, year of harvest, sample preparation, and analysis methods (Della et al., 2002). The levels of impurities such as K_2O and Na_2O were reported to be related to soil type and the amount of fertilizer used during the process of paddy growth (Moraes et al., 2014).

Incinerating conditions, including combustion system and temperature, are key factors affecting the properties of RHA. Large-scale combustion of RH can be achieved by cyclonic furnace, inclined grate furnace, rotary kiln, and fixed bed and fluidized bed furnace. The fluidized bed process is more practical in view of completion of mixing and heat transformation during the combustion of RH, and flexible in production of different grades of RHA (Soltani et al., 2015; Rozainee et al., 2008). RHA obtained from a fluidized bed reactor showed lower levels of ignition and carbon loss, and higher levels of silica compared with other burning conditions (Nehdi et al., 2003). A study by Fernandes et al. (2016) reviewed the chemical composition, loss on ignition (LOI), total carbon, specific weight, and specific surface area of RHA from different combustion technologies such as grate furnace, fluidized bed, and suspension/entrained combustion. The main characteristics affected by these combustion techniques are surface area, LOI, SiO_2, and total carbon content, and the structure of silica. The morphology, specific weight, and other metal oxides of all RHA types are almost the same. Their results implied that ash generated from different combustion technologies is suitable for different applications.

Structure and properties of RHA is also dependent on combustion temperature of RH. RHA obtained under various combustion temperatures (400–900°C) exhibited a strong temperature dependence of the specific surface area and pore volume (Xiong et al., 2009). The specific surface area and pore volume of RHA decreased with increasing combustion temperature when solid combustibles are mostly burned off. Densification of the amorphous silica structure of RHA becomes lower with increased temperature, whereas the Si—O—Si bond angle increases with temperature. Therefore a

Table 2 Physicochemical properties of RHA

Samples	1	2	3	4	5	6	7	8
Location	Brazil	Malaysia	China	Egypt	Iran	Japan	Iran	Malaysia
Burning conditions	Fluidized bed	Ferro-cement furnace, < 690°C	Downdraft gasifier, 387–815°C	Torbed reactor, 750°C	Muffle furnace, 10° C/min until 700°C	600–1200°C	Electric furnace, 600°C	Muffle furnace, 500–900°C
Treatment	Untreated	Untreated	Untreated	Untreated	Acid leaching	Acid leaching	Acid leaching	Acid leaching
Chemical composition (%)								
SiO_2	96.71	88.32	94.36	94.60	97.89	99.14	95.55	99.58
Al_2O_3	0.09	0.46	0.26	0.30	0.02	0.03	0.13	0.17
Fe_2O_3	0.01	0.67	0.23	0.30	0.16	0.03	0.03	0.03
K_2O	0.69	2.91	0.65	1.30	0.18	0.12	0.05	0.02
CaO	–	0.67	1.56	0.40	0.27	0.16	0.56	0.04
MgO	–	0.44	0.86	0.30	0.09	0.08	0.09	0.02
Na_2O	–	–	0.39	0.20	0.18	0.06	–	–
P_2O_5	0.23	–	–	0.30	0.13	0.29	0.45	0.11
Physical properties								
Loss on ignition	2.96	5.81	–	1.80	1.23	–	2.75	–
Specific gravity (g/cm^3)	2.22	2.11	–	2.05	–	–	–	–
Mean particle size (μm)	19.56	63.8	–	7.15	1.14	–	–	0.52
BET surface area (m^2/g)	–		142		284.3	–	272.96	218
Reference	Fernandes et al. (2016)	Habeeb and Mahmud (2010)	Ma et al. (2015)	Nehdi et al. (2003)	Vayghan et al. (2013)	Umeda and Kondoh (2008)	Ghorbani et al. (2015)	Bakar et al. (2016)

decrease in burning temperature is desirable for applications of RHA in catalysis, absorption, and chemical synthesis as a Si source. Taku et al. (2016) reported that RHA calcined at temperatures between 400°C and 800°C contain >70% silica, and the content of silica only varied slightly with different calcination temperature, but the specific gravity of RHA decreases with increasing calcination temperature. Behak and Núñez (2013) showed that the LOI decreases when the combustion temperature increased up to 650°C, and the organic content of RHA linearly decreases with the increase of incineration temperature. The color of RHA changes with temperature, from gray with black points in RHA combusted at 500°C to whitish gray with a pinkish tone in RHA combusted at 650, 800, and 900°C. The greater the combustion temperature, the thinner and more brittle the produced RHA, with an intense pinkish tone. However, Bakar et al. (2016) found that the effect of combustion at different temperatures between 500 and 900°C on the silica production is very small, particularly at temperature above 600°C.

As an effective method to reduce the amount of metallic impurities in RHA, chemical pretreatment of RH with acid leaching is employed by many researchers. Acids include mineral acids such as HCl, H_2SO_4, H_3PO_4, and HNO_3, and organic acids such as acetic, citric, and oxalic acid (Chandrasekhar et al., 2005; Mahmud et al., 2016; Fernandes et al., 2016; Vayghan et al., 2013; Bakar et al., 2016). As shown in Table 2, pretreatment of RH with acid leaching resulted in production of RHA/silica with high specific surface area and less amount of metal contaminants such as K_2O.

Among all acids studied, HCl seems to be the most effective and extensively used agent in removing metallic impurities from RH. Chakraverty et al. (1988) compared the effect of HCl, HNO_3, and H_2SO_4 with different concentrations on removal of metallic impurities. They found that 1 N HCl is effective in removing most of the metallic impurities, and the ash obtained by this treatment is completely white in color following complete combustion of RH. The order of efficiency is $HCl > H_2SO_4 > HNO_3$. However, Umeda and coworkers obtained very high purity of silica ashes with 99.14% and 99.3% when using 5% citric acid and 1%–5% H_2SO_4, respectively (Umeda and Kondoh, 2008; Umeda et al., 2007).

Thermogravimetric (TG) analysis of untreated and acid-leached RH revealed three distinct stages of weight loss. As shown in Fig. 4 (Bakar et al., 2016), the first stage is the removal of moisture, which took place at a temperature range of 50–150°C. The loss of water is 1%–2% irrespective

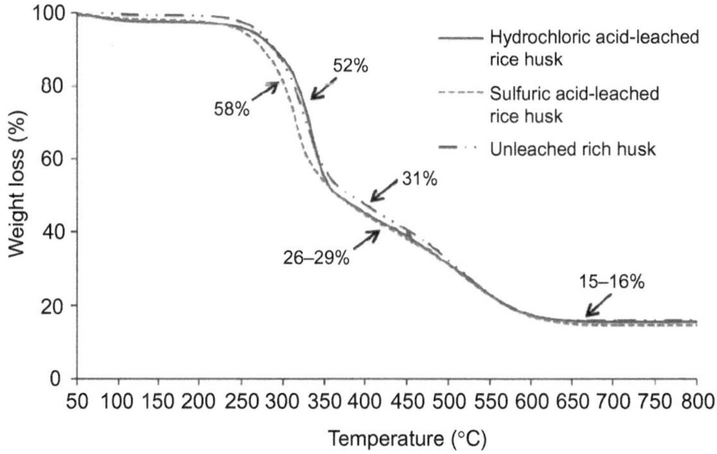

Fig. 4 Thermogravimetry (TG) curves of unleached and acid-leached RH.

of acid leaching. The second stage is release of volatile matter by thermal decomposition of hemicellulose and cellulose at 240–360°C. RH with acid-leaching showed lower thermal stability compared to unleached RH. The third stage is the combustion of combustible materials due to lignin, and the weight loss is 26%–31%. A similar phenomenon was also reported by Chakraverty et al. (1985). All these results suggested that acid treatment of RH could decrease the combustion temperature.

SEM showed acid-leached RH contained a more even and less rough outer surface than the unleached RH, which might be attributed to hydrolysis of some organic components by the acids (Bakar et al., 2016). Leaching of RH with 0.01 N HCl for 1 h, and then followed by combustion at 10°C/min until the temperature was 700°C, resulted in porous microstructure of acid-leached ashes. Meanwhile, surface of nontreated ashes was almost melted. Consequently, when ash from acid-leached RH was ground, finer particles with homogeneous size distribution, higher surface area, and larger nanoporosity could be obtained (Vayghan et al., 2013).

Acid leaching of RH is usually conducted before the combustion, which would produce silica powder of high specific surface area and better quality. Real et al. (1996) compared the effect of acid leaching before and after combustion. It was found that the specific surface area decreases from 260 m^2/g in silica obtained by acid leaching before combustion to 1 m^2/g in silica obtained by directly acid leaching of combusted ashes. This is due to strong interaction between the silica and the potassium contained in the RH, which leads to dramatic decrease of the specific surface area if K^+ cations are not removed prior to heat treatment.

Apart from acid leaching, other pretreatment methods using ionic liquid, alkaline solution, and deionized water were reported. Because acid leaching does not improve the amorphicity of the RHA produced, Javed et al. (2009) treated the RH with 0.1 N NaOH at 60°C for 3 h and found that the amorphicity of produced silica was improved by reducing the O—Si—O bond strength. Lee et al. (2017) reported that when RH was incubated at 130°C for 36 h with 100% ionic liquid (1-butyl-3-methylimidazolium hydrogen sulfate), significantly increased amount of ash was obtained after the RH was pyrolyzed at 800°C for 48 h. Furthermore, the ash contains 99.5% silica with 1.9 times the surface area and 2.4 times the pore volume of the silica from untreated ash. Using deionized water to leach RH is not effective enough to remove metallic impurities or to decompose the organic matter. Water-leached RH only has a specific surface area of 1.45 m^2/g, much smaller than that from acid-leached RH (10.39 m^2/g) (Liou, 2004).

3. PRODUCTION OF SILICA FROM RICE HUSK ASH

As an important raw material, silica finds many applications in ceramics, rubbers, plastics, microelectronics, food, pharmaceuticals, personal care, and structural and adsorptive materials (Sun and Gong, 2001). Traditionally, silica is produced from quartz sand or quartz rock, which is sourced from mining. The industrial process generally involves the reaction between sodium carbonate and smelted quartz sand at 1300–1500°C to produce sodium silicate, and the reaction between sodium silicate and sulfuric acid to precipitate silica. To produce 1 ton of silica, 0.51 ton of sulfuric acid and 0.53 ton of sodium carbonate is needed. At the same time, 0.23 ton of carbon dioxide, 0.74 ton of sodium sulfate, and 20 tons of waste water are produced (Soltani et al., 2015). Therefore the traditional way to produce silica is high-energy consumption and high levels of pollution, which will limit its large scale commercial applications with much concern on mineral resources and sustainable development.

RHA, which is a byproduct of combustion of RH to generate electricity, contains >65% of silica in amorphous form and exhibits high activity. RHA could be a cheaper and economical raw material for preparing silica gels and powders. Silica from RAH is green and renewable compared with that from quartz sand, and it can be really claimed as bio-silica.

3.1 Alkaline Extraction and Acid Precipitation of Silica

It is well known that solubility of amorphous silica at pH 10 and below is very low, whereas amorphous silica has increased solubility in solutions with pH values above 10. This unique solubility behavior makes silica extractable in a pure form of silica gel by solubilizing it under alkaline conditions, subsequently precipitating it at a low pH. The extraction of silica gel, sometimes also known as white carbon black, is based on the following reactions (1) and (2):

$$SiO_2 + 2NaOH = Na_2SiO_3 + H_2O \tag{1}$$

$$Na_2SiO_3 + H_2SO_4 = SiO_2 + Na_2SO_4 + H_2O \tag{2}$$

When amorphous silica contained in the RHA is dissolved in alkaline solution, a silicate solution forms that has a solubility of 876 mg/L in water at 25°C. After the sodium silicate is acidified, a supersaturated solution of silica gel is produced by means of a polymerization process, which is divided into three phases: monomer polymerization to form particles; particle growth; and particle union in ramified chains that extend throughout the solution (De Lima et al., 2011).

A typical commercial production of silica gel and activated carbon from RHA using alkaline extraction and acid precipitation is illustrated in Fig. 5. After combustion of RH in a stokerfeed boiler at 500–600°C for 20–30 min, the ash is cooled down, screened, and sieved to remove some impurities like unseparated rice. It is then transported to a pressure reactor, where 10%–12% sodium hydroxide solution (0.4–0.5:1, v/w) is added. The reaction takes place at pressure ranging from 3 to 4 bar and temperature ranging from 130–140°C for 3–5 h. The obtained resultant is filtered to get sodium silicate solution, which is a viscous, transparent, colorless to pale yellow solution, and filter cake, which can be further processed to make activated carbon. The sodium silicate solution is neutralized with 8%–10% H_2SO_4 at 70–80°C for 2–3 h to precipitate the silica. Addition of the sulfuric acid must be slow until an acidic condition is reached. After filtration and water washing, the filter cake with moisture content being usually about 75%–80% is homogenized by agitation, and then the homogenate is dried in a spray drier. Different characteristics of silica used as additives for rubber, feed, and food can be produced by this process. Characteristics of several commercial products of silica are shown in Table 3.

The reaction between NaOH and RHA can also be conducted at higher temperatures in the range of 180–200°C and higher pressure ranging from 6

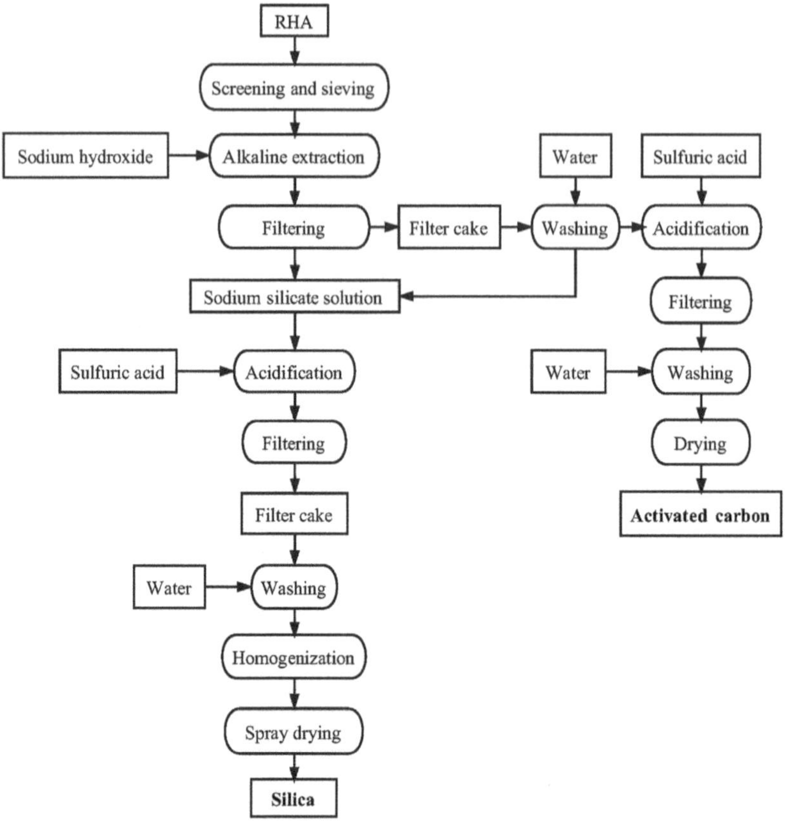

Fig. 5 Process flow chart for production of silica and activated carbon.

Table 3 Characteristics of some commercial products of silica

Samples	**1**	**2**	**3**
Appearance	Micropearl	Micropearl	Powder
SiO_2 content (%)	≥92	≥90	≥92
Loss on heating (%)	4.0–8.0	≤6.5	≤6.5
Loss on ignition (%)	≤7.0	≤7.0	≤7.0
pH(10% solution)	6.0–7.5	6.0–7.5	6.0–7.5
Soluble salts (Na_2SO_4, %)	≤2.0	≤1.5	≤0.7
B.E.T. (m^2/g)	140–180	–	165–195
Tap density (g/cm^3)	0.26–0.33	–	–
Bulk density (g/cm^3)	–	0.19–0.21	0.19–0.28
Application	Rubber additive	Feed additive	Rubber additive

to 8 atm (Todkar et al., 2016). But high reaction temperature and pressure can be avoided if ash is obtained by burning RH at 650°C, because RHA is mostly amorphous silica, which is more reactive.

Foletto et al. (2006) studied the effect of time, temperature of reaction, and composition of the reaction mixture (expressed in terms of molar ratios of $NaOH/SiO_2$ and H_2O/SiO_2) on the extraction of silica in an open and closed system. It is found that the increase of the H_2O/SiO_2 ratio from 11 to 22 practically does not modify the values of the conversion for the different assayed $NaOH/SiO_2$ ratios, and the increase in the conversion is small with the rise of the $NaOH/SiO_2$ ratio. The silica conversion into silicate presents a similar behavior for the two types of systems. The influence of reaction temperature on silica conversion is significant. It is observed that the conversion increases with the temperature rise, reaching 92% at 200°C in only 20 min of reaction. In another study, Liu et al. (2016) optimized the parameters of extraction of silica from acid-leaching RHA. It was observed that silica dissolution rate increased significantly with the increase of ratio of acid-leaching RHA and NaOH solution, the maximum value of the dissolution rate of silica was at solid/liquid ratio of 1:8. With the reaction time and the reaction temperature augmenting, the silica dissolution rate increased continuously, and it reached the maximum value at 2.5 h and 90°C. The silica dissolution rate climbed and then declined with the increase of the concentration of NaOH aqueous solution. The silica dissolution rate was highest when 10% NaOH (wt/wt) aqueous solution was used (Liu et al., 2016).

Metallic impurities can adversely affect the properties of silica from RHA and its application, so it is important to purify silica with fewer impurities to improve its characteristics. Kalapathy et al. (2000) produced precipitated silica with >4% of sodium as contaminant, which needed an extra washing step and drying step to lower the concentration of sodium below 0.1%. Later, they proposed an improved method for production of silica with lower sodium (Kalapathy et al., 2002). After alkaline extraction from RHA, the final pH 4.0 of precipitation environment was achieved by addition of sodium silicate into hydrochloric acid, oxalic acid, or citric acid solution instead of the normal procedure, in which the acid solution was added into sodium silicate until pH 7. Compared with the normal procedure, the silica prepared at pH 4.0 using oxalic acid and citric acid contained less content of sodium and carbon. The reason might be that gelation at pH 4.0 was slower, and hence sodium ions diffused readily out of gel matrix. This improved method did not require an additional washing step to prepare silica and could be an alternative to the current method that involves high-energy sand smelting.

3.2 Recyclable Routes for Production of Silica

Subbukrishna et al. (2007) patented a novel process, which was confirmed from laboratory scale to pilot plant, for silica precipitation where the chemicals used are regenerated. The process is characterized by extraction of silica gel with NaOH and precipitation of silica with CO_2, and regeneration of NaOH and CO_2 with a fresh calcium hydroxide. The involved reactions include Eqs. (3)–(7), which can enable recycling of NaOH and CO_2. Silica produced by this process possessed specific surface area of 150–200 m^2/g and bulk density of 120–200 g/L. The conversion of silica in the ash is about 80%–90%, and the purity of silica can be as high as 98%.

$$\text{Ash} + \text{NaOH} = \text{Na}_2\text{O} \cdot n\text{SiO}_2 + \text{Undigested ash} \tag{3}$$

$$\text{Na}_2\text{O} \cdot n\text{SiO}_2 + \text{CO}_2 = n\text{SiO}_2 + \text{Na}_2\text{CO}_3 \tag{4}$$

$$\text{Na}_2\text{CO}_3 + \text{Ca(OH)}_2 = \text{CaCO}_3 + 2\text{HaOH} \tag{5}$$

$$\text{CaCO}_3 \underset{\Delta}{\underline{\underline{\text{calcined}}}} \text{CaO} + \text{CO}_2 \tag{6}$$

$$\text{CaO} + \text{H}_2\text{O} = \text{Ca(OH)}_2 \tag{7}$$

Another recyclable route (Fig. 6) for preparation of silica powder using RHA and NH_4F was proposed by Ma et al. (2012). By dissolving silica into 4–5 mol/L NH_4F solution at a reaction temperature of 110°C for 2–3 h $(NH4)_2SiF_6$ and NH_3 are produced. Addition of $(NH4)_2SiF_6$ solution to

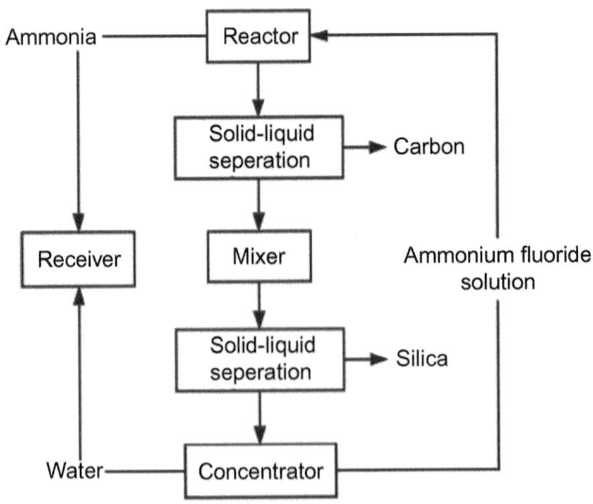

Fig. 6 Recyclable route for production of silica using NH_4F.

NH_3H_2O caused precipitation of high purity spherical silica powder with a diameter of 50–60 nm and the yield of 94.6%. After recycling of NH_4F solution four times, the yield of silica did not obviously decrease, which confirmed the reactant is recyclable. The process of silica dissolution and precipitation is described by reactions (8) and (9), respectively.

$$6NH_4F + SiO_2 = (NH_4)_2SiF_6 + 4NH_3 + 2H_2O \quad (8)$$

$$(NH_4)_2SiF_6 + 4NH_3 + (n+2)H_2O + = 6NH_4F + SiO\downarrow + nH_2O \quad (9)$$

3.3 Sodium Carbonate Activation

As illustrated in Fig. 7, a green and sustainable process for simultaneous production of silica and activated carbon (AC) has been developed by Liu et al. (2011). The procedure mainly included three steps: (1) activation stage: RHA was activated with Na_2CO_3 powder, and CO_2 released from this process could be reused to precipitate silica; (2) dissolved stage: the activated RHA was continuously boiled for some time with large amount of water and then filtered; the AC was prepared by thoroughly washing solid residue to neutral with distilled water and then dried; (3) carbonization stage: the filtrate was neutralized by CO_2 and then precipitated to separate SiO_2. The reactions involved in this process are listed in Eqs. (10)–(13), respectively.

$$Na_2CO_3 \underset{\Delta}{\underline{\underline{900°C}}} Na_2O + CO_2\uparrow \quad (10)$$

$$Na_2O + nSiO_2 = Na_2O \cdot nSiO_2 \quad (11)$$

$$Na_2O \cdot nSiO_2 + mH_2O = Na_2O \cdot nSiO_2 \cdot mH_2O \quad (12)$$

$$Na_2O \cdot nSiO_2 \cdot mH_2O + CO_2 = Na_2CO_3 + nSiO_2 \cdot mH_2O \quad (13)$$

The authors optimized experimental conditions as following: RHA and Na_2CO_3 with a ratio of 1:1.75 was heated at 900°C for 45 min, and then sodium silicate was extracted by reflux in 350 mL water for 2 h, silica was precipitated from sodium silicate by CO_2 released from the activation process. The particle size of prepared silica was approximately 40–50 nm, and the leaching rate was about 72%–98%. The capacitance value, iodine adsorption capacity, and surface area of AC could achieve 180 F/g, 1708 mg/g, 570 m^2/g, respectively. In addition, the authors found that K_2CO_3 owned more superiority than Na_2CO_3 in view of dosage, the hole structure, and prodigious surface area as well as the adsorption capability for the obtained silica (Liu et al., 2012). The average pore size and the surface

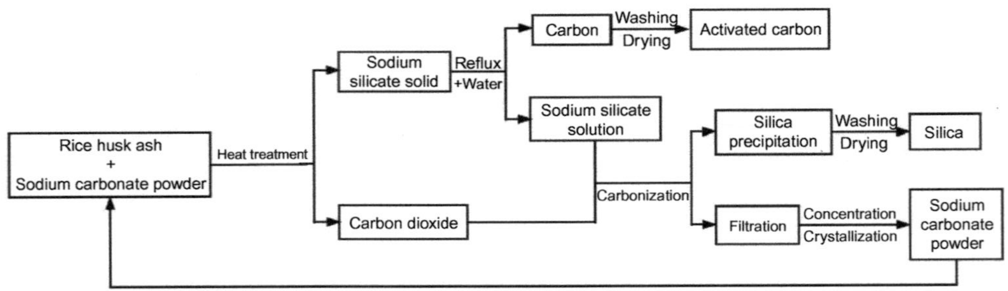

Fig. 7 Procedure for the preparation of silica and activated carbon simultaneously from RHA.

area of the AC prepared with the activation of K_2CO_3 at 1000°C were 4 nm and 1713 m^2/g. The capacitance value was up to 190 F/g, and the maximum adsorption capacity of methylene blue for AC reached 210 mg/g. The particle size of silica was 40–50 nm, but the leaching rate could be up to 96.84%. In this synthetic procedure, Na_2CO_3 and K_2CO_3 powders could be recycled and reused as the reactant to activate RHA, which stated that inexpensive, sustainable, and environmentally friendly were the dominant features for this method (Liu et al., 2016).

3.4 Other Methods

Microwave heating has been successfully used in chemical reactions. Extraction of silica gel from RHA by microwave heating is reported by Rungrodnimitchai et al. (2009). The experiment was carried out by heating RHA in sodium hydroxide solution with different concentrations in the microwave oven for 5 or 10 min. The best condition for silica gel production was reaction with 2.0 M sodium hydroxide at microwave power of 800 W for 10 min. However, silica gel prepared by low concentration of sodium hydroxide solution had the highest ability of water adsorption. This microwave-heating method required less energy and gives higher reaction efficiency, although it still needed some other resources such as sodium hydroxide and sulfuric acid.

Sodium silicate was prepared by reacting RHA with 10 M sodium hydroxide (Kumchompooa et al., 2017) and microwaving the resulting mixture at 400, 600, and 800 watts for 5 and 10 min. It was found that heating at 600 watts for 5 and 10 min corresponded with the stoichiometry of Na and Si of Na_2SiO_3, which agreed with vibration of (Na)O—Si—O(Na) in the FT-IR spectra. While heating at 400 watts, the formation of sodium hydroxide and silica of RHA was incomplete, and at 800 watts, the hydrolysis of sodium silicate occurred. After drying, the sodium silicate powder could be used as a catalyst in the production of biodiesel from palm oil.

Noncatalytic high-temperature and high-pressure water treatment processes in autoclave and steam-explosion systems were used to treat RH (Mochidzuki et al., 2001). The RHA obtained by this method exhibited some properties that are different from those obtained by conventional techniques. The results showed that the hydrothermal reaction hardly affected the local silicate structures within the temperature range tested (<240°C). The RH silica contained fewer metallic impurities, which was comparable to that obtained with the hydrochloric acid-leaching method. It is suggested that the hot-water-treatment residue is applicable to the preparation of water-glass-like materials required in some liquid-phase syntheses of inorganic materials, for example, an advanced production system of mesoporous silicate.

A novel method, the hydrothermobaric process, was developed by Ugheoke and Mamat (2012b). Hydrothermobaric purification refers to a process that utilizes single or heterogeneous phase reactions in aqueous media at high temperature ($T > 243°C$) and pressure ($P > 3$ MPa) to cause leaching or solutionizing of oxide impurities, as well as degradation of organic compounds of RH. The RH was pretreated with water (1:10, wt/v) in a hydrothermobaric condition at 300°C for 10 or 30 min. Then a nano-silica product was obtained by placing the RH residue in zirconia crucibles inside a box furnace with static air and the temperature raised by 10°C per minute until 650°C, and then kept for 4 h. It was found that most of the impurities existing in the RH could be removed after 30 min, and only phosphorous oxide that tends to remain in the solid phase in large quantities. The nano-silica retained amorphous structure, possessed higher surface area and total pore volume, and smaller average pore diameter, though the product had some residual carbon adsorbed onto its surface. If RH was pretreated for 45 min, nano-silica with purity of 98.9% and yield of 92.3% could be obtained (Ugheoke et al., 2013). This method is claimed as inexpensive, fast, commercially scalable and viable, and environmental friendly.

Subcritical water, defined as liquid water in the temperature range of boiling point to critical point (100–374°C) or near critical point, has low viscosity, low dielectric constant, high solubility of organic substances, and can be used as media for various synthesis reactions. RH was hydrothermally processed with 30% nitric acid solution to prepare nano-silica (Tolba et al., 2015). At subcritical water conditions (160°C, 2 h), organic compounds can be decomposed, and trace metal impurities can be turned into soluble ions as nitrate salts. Therefore almost 100% silica can be obtained with yield of 81%, and the particle size distribution of silica was 10–50 nm.

4. PRODUCTION OF SILICA AEROGEL FROM RICE HUSK ASH

Silica gel is a rigid three-dimensional network of colloidal silica and is classified as aquagel (pores are filled with water), xerogel (aqueous phase in the pores is removed by evaporation), or aerogel (solvent is removed by supercritical extraction), depending on how they are made (Kalapathy et al., 2000). Aerogel is a structure-controllable, mesoporous or nanoporous, light solid material with high specific surface area (500–1200 m^2/g), high porosity (80%–99.8%), low bulk density (~0.003 g/cm^3), high thermal insulation value (0.005 W/m K), ultralow dielectric constant (k = 1.0–2.0), and low index of refraction (~1.05). Silica aerogel has many commercial applications in fields such as fillers for paints and varnishes, thermal and acoustic insulation materials, adsorbents and catalyst supports, and electronic materials (Dorcheh and Abbasi, 2008; Tang and Wang, 2005).

Silica aerogel is usually produced by a sol-gel method, which includes three key steps: gel preparation, aging, and drying of the gel. First, the sol is produced from a silica source solution, and the gelation occurs by addition of a catalyst. Second, the prepared gel is aged in its mother liquids for different times. The purpose is to strengthen the gel to prevent it from shrinking during the drying step. Third, the gel is dried under special conditions such as supercritical drying, ambient pressure drying, or freeze-drying to make the pore free of any liquid. The three-dimensional porous structure of the aerogels is influenced by precursor, acid type, methylation agent, water/precursor ratio, pH of the medium, aging time, and drying method (Dorcheh and Abbasi, 2008; Temel et al., 2017).

Compared with organoalkoxysilanes precursors, such as tetramethylorthosilicate (TMOS), tetraethylorthosilicate (TEOS), and polyethoxydisiloxane (PEDS), RHA is a cheap and abundant raw material for production of silica aerogel. The process flow chart is illustrated in Fig. 8 (Nayak and Bera, 2009; Cui et al., 2015). After preparation of silica gel by alkaline extraction and acid precipitation, the gel is aged, and the water in the gel is replaced by solvents such as ethanol or heptane, then the gel is dried to produce kinds of silica aerogel with different properties.

Silica aerogel was first prepared from RHA by Tang and Wang (2005) using a sol-gel method and supercritical CO_2 drying. The extracted sodium silicate solution was neutralized using sulfuric acid solution to obtain the silica gel. The gel was then washed with water and the solvent exchanged with ethanol. The aged alcogel was subsequently dried using supercritical carbon dioxide drying. The specific surface area of the aerogel was 597.7 m^2/g, and bulk density, 38.0 kg/m^3. The pore diameter was between 10 and 60 nm. The silica aerogel was comparable to that from TEOS in densities and pore volumes, but the specific surface area was smaller. In addition, they found if the alcogel was dried at ambient pressure, it produced a silica xerogel instead of an aerogel. However, when a small amount of TEOS was added into the gel, the pore surface of the gel was partially substituted with the group —OSi $(OC_2H_5)_3$ during the gelation. Therefore a silica aerogel with specific

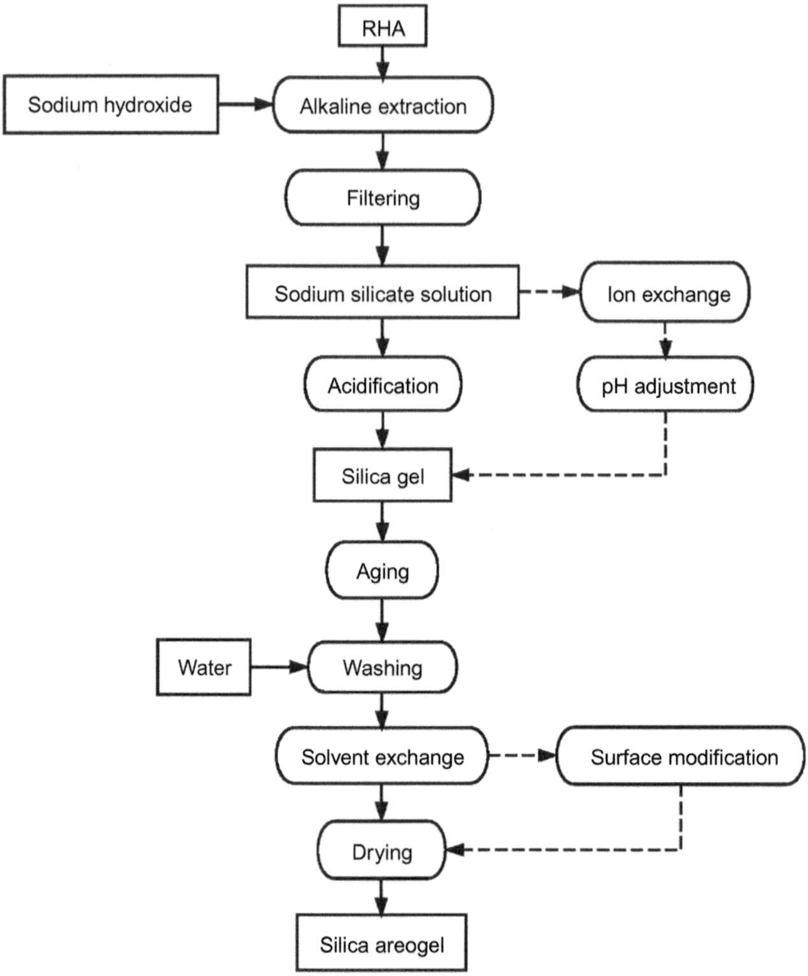

Fig. 8 Process flow chart for production of silica aerogel from RHA.

surface area 500 m^2/g and a bulk density of 0.33 g/cm^3 was prepared even by ambient pressure drying at 40°C for 10 h (Li and Wang, 2008). Nayak and Bera (2009) also prepared silica aerogel using TEOS as surface modifier during the gelation. After exchange of the pore water of the gel by ethanol, the surface modification was carried out by aging the alcogel in TEOS/ethanol solution at 70°C for 24 h. The solvent was replaced by n-heptane then dried at ambient pressure. The n-heptane solution added, due to its low surface tension, ensured that shrinkage in the gel network was greatly reduced. The aerogel obtained was crack-free with specific surface area of 273 m^2/g

and bulk density of 0.67 g/cm^3. Tadjarodi et al. (2012) prepared and characterized nanoporous silica aerogel from RH by drying at atmospheric pressure. The prepared gel was structurally strengthened by TEOS during gelation and aged at room temperature for 24 h. The water was then replaced by ethanol and then dried directly at atmospheric pressure at 40°C for 10 h to yield a typical silica aerogel with specific surface area of 315 m^2/g and bulk density of 0.32 g/cm^3.

Using ion exchange method (Cui et al., 2015), a silicic acid with pH of 2.1–3.1 was prepared from RHA sodium silicate solution over a Na-type 732 cation exchange resin. The pH of silicic acid solution was adjusted to certain value with 1 N of NaOH to form gel. After aging in H_2O/ethanol or TEOS/ethanol, the silica aerogel was obtained by supercritical ethanol drying at 10 MPa and 270°C. It was found that the highest quality of silica aerogel was produced at the gel pH 5 and SiO_2 concentration of 6% and 8%. Although aging at TEOS/ethanol solution improved aerogel properties such as shrinkage rate, density, and specific surface area, the effect was not significant. Therefore it was recommended to use H_2O/ethanol for aging with regard to the cost and further modification of silica aerogel. In subsequent research (Cui et al., 2017), an amine-modified silica aerogel was prepared using 3-(aminopropyl)triethoxysilane (APTES) as the modification agent. The ethanol-exchanged gel was further modified by amino in APTES/ethanol solution at 50°C for 7 days. After drying by ethanol supercritical fluid, silica aerogel with surface area of 593.45 cm^2/g, pore volume of 2.14 cm^3/g, and average pore size of 12.26 nm was produced. Due to physical adsorption and chemical adsorption, the CO_2 adsorption capacity of this modified aerogel was much higher than that of unmodified aerogel, in which the adsorption of CO_2 was only based on physical adsorption.

The effects of gelation acid type (acetic, hydrochloric, nitric, oxalic, and sulfuric acid), dryer type (air, freeze, oven, and vacuum), and the addition of TEOS on the structural and physical properties of aerogels produced from RHA was systematically studied by Temel et al. (2017). Table 4 shows the textural and physical properties of silica aerogel as affected by these parameters. It can be concluded that silica aerogels dried by air dryer had the largest specific surface area and pore size due to a relatively lower drying shrinkage, which resulted from the uniform heating and high contacting efficiency between aerogel and hot air in the air dryer. Therefore air drier might to be the most suitable dryer to prepare silica aerogel. With regard to the acid used to prepare gel, the size of formed sodium salt during gelation influenced the properties of aerogel. The sodium oxalate ($Na_2C_2O_4$), sodium sulfate

Table 4 Effect of dryer, acid, and addition of TEOS on the properties of silica aerogel obtained from RHA

	SBET (m^2/g)	BJH pore size (nm)	V_{meso} (cm^3/g)	Density (g/cm^3)	Porosity (%)
Dryer type					
Air	287.7	11.19	0.923	0.29	87
Freeze	213.1	8.11	0.345	0.3	86
Oven	234	7.33	0.333	0.18	91
Vacuum oven	285.4	7.79	0.881	0.43	80
Acid type[a]					
Acetic	294.4	10.7	1.022	0.37	83
Hydrocholoric	268	11.79	0.998	0.38	83
Nitric	287.7	11.19	0.923	0.29	87
Oxilic	322.5	10.85	1.048	0.21	90
Sulfuric	294.9	10.33	1.044	0.39	82
Additive					
Without TEOS[b]	140.7	12.05	0.418	0.38	82
With TEOS[b]	234	7.33	0.333	0.19	91
without TEOS[a]	247.8	5.38	0.524	0.71	68
With TEOS[a]	287.8	11.19	0.923	0.29	87

[a] Aerogel was air-dried.
[b] Aerogel was oven-dried.

(Na_2SO_4), and sodium acetate (NaOAc) from gelation was much easier to be eliminated by water washing than sodium nitrate ($NaNO_3$) and sodium chloride (NaCl). Sodium chloride may have agglomerated in the gel network due to its smaller size, and hence may block the pores. Thus aerogel prepared with oxalic acid had the largest BET specific surface area and porosity, and the lowest density. The addition of TEOS was favorable for increasing the specific surface area and porosity, and reduction of the tap density of silica aerogel.

Rajanna et al. (2015) prepared silica aerogel microparticles (SAMs) from RHA for drug delivery application using water-in-mineral oil emulsion technique. The big difference of this process from the previously mentioned process is that the gelation occurred in an emulsion mixture of mineral oil (namely kerosene) and dual surfactants (mixture of Span 80 and Tween 80, HLB value of 6.3). Gelation parameters including speed of agitation, sol-to-oil ratio, and surfactant concentration affected shape, textural properties, size, and size distribution of the aerogel microparticles. These parameters were optimized by Taguchi design of experiments and found to be agitation speed of 1200 rpm, 1:3 sol-to-oil ratio, and a surfactant concentration of 5 wt %, respectively. The prepared aerogel was found to have a total porosity of 99.34%, BET-specific surface area of $652\,m^2/g$, and pore volume of $3.0\,cm^3/g$. High holdings of ibuprofen and eugenol by SAMs were found at a relatively lower SC-CO_2 pressure of 150 bar. About 80% of adsorbed ibuprofen was released within half an hour, and 100% of eugenol was released over a period of 17 days, suggesting that SAMs is valid for drug delivery applications.

5. APPLICATION OF RICE HUSK/RICE HUSK ASH AS BIOADSORBENT

After combustion of RH, the obtained RHA mainly contains SiO_2 and carbon, and still retains a cellular structure skeleton with large specific surface area and high porosity. RHA is insoluble in effluent, and exhibits good chemical stability and high durability, therefore it has tremendous potential as a bioadsorbent for removal of fatty acids and pigments in the vegetable oil refining process, and heavy metals, dyes, pesticides, and other organic pollutants from waste water.

5.1 Adsorbent in Vegetable Oils Refining

The majority of vegetable oils are triglycerides; the minor nontriglycerides must be removed to produce oils with acceptable quality. Minor nontriglycerides such as phospholipids, free fatty acids (FFA), and pigments such as carotenoid and chlorophyll, which are usually removed in degumming, deacidification, and decoloration of vegetable oils by hydration, NaOH neutralization, activated clay bleaching, respectively, are also reported to be adsorbed and removed by RHA.

Proctor and Palaniappan tested the adsorptive capacity of two kinds of RHA, the "alkaline ash" that was prepared by further heat treatment of partially combusted RHA from a local company (pH 8.7 as determined from a 4% suspension of this material in deionized water), and H_2SO_4-activated RHA (pH 6.6), in soybean oil/hexane miscellas at room temperature. It was found that activated RHA had similar adsorptive capacity as a commercial bleaching earth for lutein, and both were superior to that of the "alkaline ash". Acid activation of RHA enhanced adsorption of lutein, whereas heat treatment above 600°C reduced it. Optimum conditions for processing RHA were a combustion temperature of 500°C and 5% acid activation (Proctor and Palaniappan, 1989). However, the "alkaline ash" was more effective in reducing FFA content than that of acid-activated RHA, and the difference became more obvious when a large amount of ash was used, indicating that the mode of adsorption differed from that of lutein. Adsorption of FFA followed Freundlich isotherm, and addition of isopropanol promoted the adsorption of FFA (Proctor and Palaniappan, 1990). Structural study using XRD and SEM showed almost the same structure between these two materials except for a slight decrease in particle size in acid-activated RHA. Trace alkali oxide was removed during the activation process, which attributed to the enhanced lutein binding effect of activated RHA, but presence of these oxides enhanced binding of FFA. Alkali oxide may possibly compete with lutein for binding sites and modify adsorption sites to favor FFA retention. Heating changed the crystal forms of silica, thus resulting in the change of adsorption performance of RHA (Proctor, 1990). Under a simulated commercial temperature and pressure conditions (100°C, 2 mmHg pressure, 30 min) of bleaching of alkali-refined soy oil, RHA is effective in adsorption of phospholipid on a surface area basis but ineffective for adsorbing lutein, free fatty acids, and peroxides (Proctor et al., 1995). In contrast, Liew et al. (1993) found that rice hull ash obtained by heat treatment and acid activated followed by washing was not effective as an adsorbent for carotene in palm oil. Unwashed acid-activated ash prepared by heat treatment of RH below 300°C and drying of acid-activated ash below 200°C had much higher activity, and about 90% of the carotene in the palm oil hexane miscella was removed. It is suggested that the removal of carotene was caused by chemical interactions involving the adsorbed acid and the carotene.

Processing conditions of RHA including heating temperature and time may affect the adsorptive ability of the ash. Chang et al. (2001) studied the bleaching efficiency of RHA produced in the range of 300–1000°C for 6–120 min in the flow of nitrogen gas and found that the specific surface area and pore size increased to a maximum value at 500 and 700°C, respectively. The specific surface area decreased yet pore size enlarged to a plateau after 30 min of heating at 500°C. The maximum bleaching efficiency in sesame oil was obtained using RHA by combustion of RH at 500°C for 60 min. If the pore diameter of RHA was <50 Å, the specific surface area had little effect on bleaching efficiency. It was also reported that RHA prepared by carbonization of RH in air in a muffle furnace at 500°C for 10 h was the most effective in reducing residual FFA in degummed soy oil. Furthermore, higher adsorption efficiency was observed in oil-hexane miscella system than in degummed oil, because better contact between oil and adsorbents was attained due to reduced viscosity and specific gravity of miscella, but acid activation for RHA did not improve the FFA adsorption efficiency (Yoon et al., 2011).

RHA could be an appropriate adsorbent to improve the quality of used cooking oil. The RHA prepared from acid treatment and untreated one had their own specific capacity to recovery quality of used palm oil. The improvement effect of HCl- and HNO_3-treated RHA was not significant, but acid-treated RHA had significantly higher abilities to reduce total polar compounds and made brighter colors compared to the untreated RHA. In agreement with the study of Proctor and Palaniappan (1990), the untreated ash seemed to be significantly better in reducing PV and FFA contents (Nattaporn and Porjai, 2015).

When waste frying oil (WFO) was pretreated with RHA, the quality of WFO was regenerated, and thus the yield, and physical and chemical properties of obtained biodiesel was improved. The highest yield of biodiesel was 87.8% when WFO was treated with 3% RHA. The acid value, iodine value, peroxide value, and refractive index significantly decreased compared with that from untreated WFO, suggesting that treated biodiesel were more resistant to oxidation than those without purification treatment (Ismail and Ali, 2015). Biodiesel contains traces of alcohol, catalyst, glycerol, water, and unreacted glycerides. The traditional way to purify biodiesel is to wash with water or slightly acid solution, which produces high volumes of waste water. RHA can be an effective alternative adsorbent for biodiesel purification (Saengprachum et al., 2013; Manique et al., 2012). Addition of RHA into biodiesel could effectively remove organic and inorganic impurities such as free and total glycerin. The effect was comparable to a commercial adsorbent Magnesol and 1% aqueous H_3PO_4 solution.

5.2 Adsorbent for Removal of Heavy Metals

Contamination of air, water, and soil with heavy metal ions is hazardous to the Earth. Many efforts have been made to use RH/RHA as a cheap and low-cost adsorbent in the removal of heavy metals from the aqueous environment. The most investigated metals include Cd, Pb, Zn, Cu, Co, Ni, and Hg from waste water. The capability and rate of absorption of RH/RHA are dependent on various parameters such as the effect of pH, initial concentration, agitation rate, sorbent dosage, contacting time, and temperature.

5.2.1 Removal of Heavy Metals by RH

Using central composite face-centered experimental design in response surface methodology (RSM), Zulkali et al. (2006) investigated the effects of initial concentration of lead, temperature, RH loading, and pH for an optimized condition of lead uptake from the aqueous solution. The initial concentration of 50.0 mg/L, temperature of 60°C, biomass loading of 0.2 g, and pH of 5.0 had been found to be the optimum conditions for the maximum uptake of lead ions (98.11%) in batch mode. Under the optimum conditions, the lead uptake was attained to be circa 8.60 mg/g. The removal of lead from waste water by RH was also investigated by Abdel-Ghani et al. (2007). The adsorption capacity of RH was compared with maize cobs and sawdust, and it was found RH was the most effective, for which the removal of Pb (II) reached 98.15% at room temperature. The adsorption efficiencies were reported to be pH-dependent, increasing with increased solution pH from 2.5 to 6.5. The potential of raw RH was also assessed for the adsorption of copper from simulated waste water (Gandhimathi et al., 2008). The efficiency of RH with particle size of 150–300 μm for Cu (II) removal was 74% at 5 g/L of adsorbent dose at 30°C. Isotherm study showed the adsorption fit better with Freundlich isotherm than Langmuir isotherm, and a kinetics study showed that the pseudo-second order model provided a better fit for RH than the pseudo-first order model.

Raw RH was first boiled with distilled water for 5 h to make it free from colored compounds (BRH) or treated with 1% formaldehyde at room temperature to immobilize the color and water-soluble substances (FRH), and then they were utilized for the removal of hexavalent chromium from synthetic waste waters (Bansala et al., 2009). Effect of various process parameters, namely pH, adsorbent dose, initial chromium concentration, and contact time, has been studied in batch systems. The removal of Cr (VI) was dependent on the physicochemical characteristics of the adsorbent, adsorbate concentration, and other studied process parameters. Maximum removal was reported at 71.0% and 76.5% for BRH and FRH, respectively, using 20 g/L of adsorbent dose at pH 2.0. Isotherm study showed Freundlich and Dubinin-Radushkevich (D-R) isotherm models fitted well. The SEM micrographs obtained before and after Cr (VI) adsorption indicated the presence of new shiny bulky particles over the surface of metal-loaded BRH and FRH, and the FTIR spectra confirmed the complexation of Cr (VI) with functional groups (—OH and C=O) present in the adsorbents. There was not much difference in the adsorption capacity of BRH and FRH at equilibrium time, suggesting BRH could be an attractive option for removal of low concentration of Cr (VI) in waste water in small-scale industries. However, in another study, the removal of Cr (VI) by untreated RH was reported to be 99%–100% under optimal operating conditions (Sarkar et al., 2013).

Zhang et al. (2013) investigated the potential of RH and NaOH-treated RH for the biosorption of zinc from aqueous solution. The rate of adsorption was fast in the first 10 min, and the solution metal concentration reached equilibrium within 30 min. Maximum adsorption capacity of zinc onto untreated RH was 12.41 mg/g at adsorbent dosage of 1 g/L at 25°C. After RH was maintained in 0.1 N NaOH for 24 h, the adsorption capacity of this chemically treated RH increased to 20.08 mg/g, because the physical and chemical property of RH was improved by dissolving lipid, protein, and soluble polysaccharide in the RH, so the loosened cells in the surface are conductive to the uptake of more metal ions. The principal mechanism for Zn (II) biosorption by using NaOH-treated RH was ion exchange with carboxyl, amino, and hydroxyl groups.

The ability of RH to remove chromium, lead, and cadmium from waste water has been investigated in batch experiments and continuous fluidized bed column experiment (Al-Baidhani and Al-Salihy, 2016). Under optimal conditions, removal of the Cr (II), Pb (II), and Cd (II) ions from aqueous solution was 84%, 90%, and 97.9%, respectively. The adsorption isotherms of Cd (II) adsorption onto the RH fitted better with Freundlich model than for Langmuir model. The adsorption of Cd (II) in continuous fluidized bed was faster and more efficient due to increasing the adsorption surface area of the adsorbent, which leads to reduced dead zones between the particles.

Raw or pristine raw rice husk (PRH) was reported as an effective adsorber for Cd (II), Cu (II), Pb (II), and Zn (II) cations (Alexander et al., 2017). The percent adsorption of trace metals in aqueous solutions by PRH was in the order of Pb > Cd > Cu > Zn under competitive equilibration conditions. Adsorption isotherms of Cd (II), Pb (II), and Zn (II) fitted well with the Freundlich model than Langmuir model except for Cu (II), which fitted well with the Langmuir model. Cu and Zn adsorption was endothermic process with positive enthalpies (ΔH^0) 6.38 and 11.96 kJ/mol, whereas Cd and Pb adsorption was exothermic process with negative enthalpies (ΔH^0) −7.85 and − 6.12 kJ/mol, respectively. Spatial elemental distribution plots obtained by laser ablation inductively coupled plasma-mass spectrometry (LA-ICP-MS) on the PRH, demonstrating that metals were adsorbed on the silica-rich areas of the RH. The authors also stated that PRH effectively adsorbed Cd, Cu, Fe, Pb, and Zn from contaminated soils and acid mine waters.

5.2.2 Removal of Heavy Metals by RHA

Feng et al. (2004) tested the adsorption capacity of RHA for the removal of lead and mercury from aqueous water. The study was carried out as a function of contact times, ionic strength, particle size, and pH. RHA was prepared by immersing RH in 1 N HCl aqueous solution for 4 h, and then heat treating RH at 700°C for 4 h, subsequently grinding and dry sieving to obtain fractions with different particle sizes. The adsorption of lead and mercury ions by RHA was found to be much more rapid than in many other methods and attained a constant value after 10 min. The finer the RHA particles, the higher the pH of the solution, and the lower the concentration of the supporting electrolyte, potassium nitrate solution, the greater amount of Pb and Hg ions absorbed on RHA. The adsorption of Pb (II) ions from aqueous solution using RHA was investigated by Wang and Lin (2008); it was found that the rate of removal of Pb (II) and the removal of Pb (II) at equilibrium were increased upon increasing the initial lead concentration, pH, stroke speed, or adsorption temperature. The data of the adsorption kinetics indicated that the process was physisorption-controlled, and the pseudo-second order rate equation suitably interpreted the overall process.

In another study, the adsorption of Pb (II) ions from aqueous solution was investigated on RHA, which was collected from a local rice mill in India. Optimum conditions for the removal of Pb (II) ions were reported to be pH 5, an adsorbent dosage of 5 g/L of solution, and an equilibrium time of 1 h. The adsorption capacity of RHA for Pb (II) ions was reported to be 91.74 mg/g. The change of entropy (ΔS^0) and enthalpy (ΔH^0) were 0.132 kJ/(mol K) and 28.923 kJ/mol, respectively. The value of the adsorption energy (E), using the Dubinin-Radushkevich isotherm, was 9.901 kJ/mol, indicated that the adsorption process was chemical in nature (Naiya et al., 2009). Furthermore, the RHA was used to remove Pb (II) from an effluent sample obtained in a battery manufacturing unit, and it was found that the adsorption of Pb (II) on RHA was 96.83%, which meets the IS 10500 of 1992 norms for discharge water.

Krishnani et al. (2008) prepared a biomatrix (RHA) from RH and studied the adsorption effect of nine heavy metal ions as a function of pH and metal concentrations in single and mixed solutions. Raw RH was subjected to 1.5% alkali treatment and then autoclaved at 121°C for 30 min to remove the low molecular weight lignin compounds. Batch and column adsorption studies were applied to show mechanistic aspects, especially the role of calcium and magnesium present in the biomatrix in an ion exchange mechanism. The metal-binding capacity of RHA is strongly pH dependent with more metal cations bound at higher pH, and the maximum uptake of metal ion took place at pH 5.5–6. The increasing order of adsorption capacity obtained from the Langmuir isotherm in mmol/g was: Ni (0.094), Zn (0.124), Cd (0.149), Mn (0.151), Co (0.162), Cu (0.172), Hg (0.18), and Pb (0.28), whereas the sorption of Cr (III) onto RHA at pH 2 was 1.0 mmol/g. The RHA biomatrix has adsorption capacity comparable or greater to other reported sorbents and can be regenerated by treatment with HCl or HNO_3.

Akhtar et al. (2010) used RHA for the removal of Pb (II), Cd (II), Zn (II) and Cu (II) divalent metal ions from aqueous solutions over pH range $(1-10)$ via batch adsorption technique. RH was first treated with 0.1 M HNO_3 and 1 M K_2CO_3, and then it was thermally treated at a heating rate of 10 K/min for 8 h under nitrogen flow 500 mL/min to increase the surface area of the RH. The adsorption equilibrium was well described by Freundlich, Langmuir, and Dubinin-Radushkevish (D-R) isotherm models at equilibrium time of 20 min at pH 6 and using 0.2 g of sorbent. The chemical and thermal activation of RH increased the removal efficiency of all the metal ions. The numerical values of thermodynamic parameters indicated the exothermic nature, spontaneity, and feasibility of the sorption process. The desorption study of metal components from RHA surface could be performed with 0.1 M HCl. The sorption mechanism developed illustrates the strong interactions of sorbates with the active sites of the sorbent coupled with efficient and environmentally clean exploitation of rice waste product.

More recently, RHA is also reported to be an effective adsorbent for the removal of Cr (VI) metal ions from aqueous solution. Georgieva et al. (2015) reported the most favorable conditions for removing Cr (VI) from aqueous solutions via adsorption onto black rice husk ash (BRHA) were found to be pH 2, adsorbent dosage level of 15 g/L in the concentration range 25–200 mg/L, temperature interval 10°C-30°C, and contact time 30–120 min. Although there was a tendency to reduce the adsorption capacity of BRHA with increasing temperature, the percent of removal of Cr (VI) ions at initial concentration 50 mg/L even at 30°C was 98.46%. The kinetic data of Cr (VI) sorption are in a good agreement with pseudo-second order kinetic model at all studied temperatures and initial concentrations. The value of activation energy (41.57 kJ/mol) is on limit between physical and chemical adsorption, but the physisorption is the predominant adsorption mechanism for Cr (VI) removal by BRHA. However, when RHA was treated with 5 M NaOH to remove all silica completely, activated carbon (AC) with high surface area (750 m^2/g) was prepared, and the obtained AC showed the highest rate of adsorption. With a decrease in pH from 4.4 to 2, the adsorption capacity increased from 3 to 25.2 mg/g. The adsorption of Cr (VI) followed pseudo-second order behavior. The changes in Gibbs free energy, enthalpy, and entropy affected by thermodynamic parameters were found to be negative, which confirmed that the adsorption of Cr (VI) on AC was spontaneous, exothermic, and favored low temperatures (Mukri et al., 2016).

The adsorption capacities of metals by RH and RHA are summarized in Tables 5 and 6, respectively. According to most studies presented, modified

Table 5 RH used as adsorbent for the removal of heavy metals

Adsorbate	Adsorption conditions	Maximum adsorption capacity (mg/g) or removal percentage (%)	Isotherm models or equations	Reference
Pb (II)	Cs: 2 g/L; Cm: 50 mg/L; pH 5; T: 60°C	8.60		Zulkali et al. (2006)
Cu (II)	Cs: 5 g/L; Cm: 10 mg/L; T: 30°C; t: 90 min; agitation: 150 rpm	74%	F, P	Gandhimathi et al. (2008)
Cr (VI)	Cs: 20 g/L; Cm: 100 mg/L; pH 5; T: 25°C; t: 90 min; agitation: 150 rpm	71.0% (BRH) 76.5% (FRH)	F, D-R, P	Bansala et al. (2009)
Cr (VI)	Cs: 40 g/L; Cm: 50 mg/L; pH 2.2; T: 30°C; t: 30 min; agitation: 160 rpm	99–100%	L	Sarkar et al. (2013)
Zn (II)	Cs: 1 g/L; Cm: 10–100 mg/L; pH 3.5; T: 30°C; t:	12.41 20.08 (NRH)	L, P	Zhang et al. (2013)

Table 5 RH used as adsorbent for the removal of heavy metals—cont'd

Adsorbate	Adsorption conditions	Maximum adsorption capacity (mg/g) or removal percentage (%)	Isotherm models or equations	Reference
	240 min; agitation: 150 rpm			
Pb (II) Cd (II) Cr (II)	Cs: 30 g/L; Cm: 25 mg/L; pH 6; T: 25°C; t: 180 min; agitation: 150 rpm	90.0% 97.9% 84.0%	F, L	Al-Baidhani and Al-Salihy (2016)
Cd (II) Cu (II) Pb (II) Zn (II)	Cs: 100 g/L; Cm: 10 mg/L; pH 6.9; Room temperature; t: 16 h; agitation: 170 rpm	– – 96.2% 65.8%	F (Cd, Pb, Zn) L (Cu)	Alexander et al. (2017)
Cu (II)	Cs: 4 g/L; Cm: 150 ppm; pH 4; T: 26°C; t: 60 min; agitation: 180 rpm	25.6	F, L, P	Adekola et al. (2016)

BRH: boiled rice husk; FRH: formaldehyde-treated rice husk; NRH: NaOH-treated rice husk; Cs: concentration of sorbent; Cm: initial concentration of heavy metal; T: adsorption temperature; t: adsorption time; F, L, and D-R represent Freundlich, Langmuir, and Dubinin-Radushkevich isotherm models, respectively. P: pseudo-second order rate equation.

Table 6 RHA used as adsorbent for the removal of heavy metals

Adsorbate	Adsorption conditions	Maximum adsorption capacity (mg/g) or removal percentage (%)	Isotherm models or equations	Reference
Pb (II) Hg (II)	pH 5.6–5.8; T: 15°C	10.86 3.23	F, L	Feng et al. (2004)
Pb (II)	Cs: 3 g/L; Cm: 700 mg/L; pH 4.2; T: 30°C; Speed: 200 stroke/min	207.5	P	Wang and Lin (2008)

Continued

Table 6 RHA used as adsorbent for the removal of heavy metals—cont'd

Adsorbate	Adsorption conditions	Maximum adsorption capacity (mg/g) or removal percentage (%)	Isotherm models or equations	Reference
Ni (II) Zn (II) Mn (II) Co (II) Cu (II) Cd (II) Hg (II) Pb (II)	Cs: 3 g/L; Cm: 50–200 mg/L; pH 6 (except for Hg and Cu at pH 5.5); T: 32°C; t: 180 min; agitation: 200 rpm.	5.52 8.14 8.30 9.57 10.9 16.7 36.1 58.1	F, L	Krishnani et al. (2008)
Pb (II)	Cs: 5 g/L; Cm: 3–100 mg/L; pH 5; T: 30°C; t: 60 min;	99.3%	F, D-R, P	Naiya et al. (2009)
Pb (II) Cd (II) Cu (II) Zn (II)	Particle size of RHA: 100 μm; Cs: 2–20 g/L; Cm: 4.8–157 $\times 10^{-5}$ M; pH 6; T: 30°C; t: 10–50 min; agitation: 100 rpm.	99% 97% 96% 95%	F, L, D-R	Akhtar et al. (2010)
Cr (VI)	Cs: 15 g/L; Cm: 50 mg/L; pH 2; T: 30°C; t: 60 min;	98.46	P	Georgieva et al. (2015)
Cr (VI)	pH 2; Room temperature; t: 90 min;	25.2	P	Mukri et al. (2016)
Mn (II) Fe (II)	Cs: 5 g/L; Cm: 100 mg/L; pH 5; T: 30°C; t: 420 min;	3.21 18.84	F, L, P (Fe) L, P (Mn)	Adekola et al. (2016)

Cs: concentration of sorbent; Cm: initial concentration of heavy metal; T: adsorption temperature; t: adsorption time; F, L, and D-R represent Freundlich, Langmuir, and Dubinin-Radushkevich isotherm models, respectively. P: pseudo-second order rate equation.

RH or RHA show higher potential for removing heavy metals compared with raw or untreated RH. It is observed that the RHA has properties that justify its use as an adsorbent. In most of the cases, the adsorption onto RH/RHA is well represented by Langmuir and Freundlich isotherms, and the adsorption usually follows pseudo-second order kinetics. The adsorption is a surface phenomenon, and the surface is easily accessible to the ions in solution. The adsorption of heavy metals from aqueous solution is influenced by various physical and chemical parameters like pH, temperature, initial heavy metal concentration, amount of adsorbent and adsorbate, particle size of adsorbent, etc. These parameters determine the overall adsorption by affecting the selectivity and amount of heavy metals removed.

6. CONCLUSION

RH is an important byproduct produced in huge amounts from the milling process of paddy rice. It is a major challenge faced by the rice milling industry. In this chapter, an attempt is made to demonstrate the physicochemical characterizations of RH/RHA ash and their potential use in production of silica gel and silica aerogel, which can be used as a bioadsorbent in the vegetable oil refining process and in removing of heavy metals. Silica exists abundantly in RH in amorphous form, which is distributed mostly in the husk's outer surface. The amorphous form of silica doesn't change when RH is incinerated below 800°C, but its content is further increased to >80%, which is beneficial in the following extraction of silica gel and subsequently preparation of silicon-based materials. Although a method using alkaline extraction and acid precipitation is practicable and employed extensively in an industrial scale, new environment-friendly and recyclable technologies such as sodium carbonate activation in gas phase are emerging. Silica aerogel could be produced from RHA by sol-gel method, with applications in many fields such as drug delivery systems. As a low-cost bioadsorbent, RH/RHA could effectively remove metals including Cd, Pb, Zn, Cu, Co, Ni, and Hg from waste water. Moreover, RHA has been used in vegetable oil refining process for adsorption of free fatty acid, phospholipids, and pigments.

Nowadays, the use of RH and its thermal degradation product (RHA) is still a hot topic in many studies, but most of them have been performed on a laboratory scale. Compared with the huge amount of RH produced worldwide each year, the comprehensive applications of RH/RHA are relatively underutilized. Therefore the opportunities for RH/RHA in incorporation into silica and silicon-based materials, building materials, bioenergies, bioadsorbents, and even foods are vast. It could be assumed that the production and usage of RH/RHA and its constituents will continue to competitively increase. As a natural, sustainable, and renewable biomass resource, RH/RHA could become a potential precursor for the production of high value-added silica or silicon-based materials for practical applications. The comprehensive utilization of RH/RHA not only facilitates utilization of an abundantly available agro-waste to value-added products but also helps to reduce the environmental pollution.

REFERENCES

Abdel-Ghani, N.T., Hefny, M., El-Chaghaby, G.A.F., 2007. Removal of lead from aqueous solution using low cost abundantly available adsorbents. Int. J. Environ. Sci. Technol. 4, 67–73.

Adekola, F.A., Hodonou, D.S.S., Adegoke, H.I., 2016. Thermodynamic and kinetic studies of biosorption of iron and manganese from aqueous medium using rice husk ash. Appl Water Sci 6, 319–330.

Akhtar, M., Iqbal, S., Kausar, A., Bhanger, M.I., Shaheen, M.A., 2010. An economically viable method for the removal of selected divalent metal ions from aqueous solutions using activated rice husk. Colloids Surf. B: Biointerfaces 75, 149–155.

Al-Baidhani, J.H., Al-Salihy, S.T., 2016. Removal of heavy metals from aqueous solution by using low cost rice husk in batch and continuous fluidized experiments. Int. J. Chem. Eng. Appl. 7, 6–10.

Alexander, D., Ellerby, R., Hernandez, A., Wu, F., Amarasiriwardena, D., 2017. Investigation of simultaneous adsorption properties of Cd, Cu, Pb and Zn by pristine rice husks using ICP-AES and LA-ICP-MS analysis. Microchem. J. 135, 129–139.

Alias, N., Ibrahim, N., Hamid, M.K., Hasbullah, H., Ali, R.R., Sadikin, A.N., Asli, U.M., 2014. Thermogravimetric analysis of rice husk and coconut pulp for potential biofuel production by flash pyrolysis. Malaysian J. Anal. Sci. 18, 705–710.

Asavapisit, S., Ruengrit, N., 2005. The role of RHA-blended cement in stabilizing metal-containing wastes. Cem. Concr. Compos. 27, 782–787.

Bakar, R.A., Yahya, A., Gana, S.N., 2016. Production of high purity amorphous silica from rice husk. Procedia Chem. 19, 189–195.

Bansala, M., Gargb, U., Singha, D., Garg, V.K., 2009. Removal of Cr (VI) from aqueous solutions using pre-consumer processing agricultural waste: a case study of rice husk. J. Hazard. Mater. 162, 312–320.

Behak, L.B., Núñez, W.P., 2013. Effect of burning temperature on alkaline reactivity of rice husk ash with lime. Road Mater. Pavement Des. 14, 570–585.

Bronzeoak Ltd., 2003. Rice Husk Ash Market Study. EXP 129, DTI/Pub URN. 03/668.

Chakraverty, A., Mishra, P., Banerjee, H.D., 1985. Investigation of thermal decomposition of rice husk. Thermochim. Acta 94, 267–275.

Chakraverty, A., Mishra, P., Banerjee, H.D., 1988. Investigation of combustion of raw and acid-leached rice husk for production of pure amorphous white silica. J. Mater. Sci. 23, 21–24.

Champagne, E.T., Wood, D.F., Juliano, B.O., Bechtel, D.B., 2004. The rice grain and its gross compositions. In: Champagne, E.T. (Ed.), Rice Chemistry and Technology. third ed. Am. Assoc. Cereal Chem., St Paul, MN, pp. 77–108.

Chandrasekhar, S., Pramada, P.N., Praveen, L., 2005. Effect of organic acid treatment on the properties of rice husk silica. J. Mater. Sci. 40, 6535–6544.

Chandrasekhar, S., Pramada, P., Majeed, J., 2006. Effect of calcination temperature and heating rate on the optical properties and reactivity of rice husk ash. J. Mater. Sci. 41, 7926–7933.

Chang, Y.Y., Lina, C.I., Chen, H.K., 2001. Rice hull ash structure and bleaching performance produced by ashing at various times and temperatures. J. Am. Oil Chem. Soc. 78, 657–660.

Cui, S., Yu, S.W., Lin, B.L., Shen, X.D., Gu, D.M., 2015. Preparation of SiO_2 aerogel from rice husk ash. RSC Adv. 5, 65818–65826.

Cui, S., Yu, S.W., Lin, B.L., Shen, X.D., Zhang, X., Gu, D.M., 2017. Preparation of amine-modified SiO_2 aerogel from rice husk ash for CO_2 adsorption. J. Porous. Mater. 24, 455–461.

De Lima, S., De Vasconcelos, R.P., Paiva, O.A., Cordeiro, G.C., Chaves, M., Toledo, R.D., Fairbairn, E., 2011. Production of silica gel from residual rice husk ash. Quim Nova 34, 71–75.
Della, V.P., Kuhn, I., Hotza, D., 2002. Rice husk ash as an alternate source for active silica production. Mater. Lett. 57, 818–821.
Dorcheh, A.S., Abbasi, M.H., 2008. Silica aerogel; synthesis, properties and characterization. J. Mater. Process. Technol. 199, 10–26.
Fang, M., Yang, L., Chen, G., Shi, Z., Luo, Z., Cen, K., 2004. Experimental study on rice husk combustion in a circulating fluidized bed. Fuel Process. Technol. 85, 1273–1282.
FAOSTAT, 2014. http://www.fao.org/faostat/en/#data/QC/visualize. (Accessed Aug., 2017).
Feng, Q., Lin, Q., Gong, F., Sugita, S., Shoya, M., 2004. Adsorption of lead and mercury by rice husk ash. J. Colloid Interface Sci. 278, 1–8.
Fernandes, I.J., Calheiro, D., Kieling, A.G., Moraes, C.A.M., Rocha, T.L.A.C., Brehm, F.A., Modolo, R.C.E., 2016. Characterization of rice husk ash produced using different biomass combustion techniques for energy. Fuel 165, 351–359.
Foletto, E.L., Gratieri, E., De Oliveira, L.H., Jahn, S.L., 2006. Conversion of rice hull ash into soluble sodium silicate. Mater. Res. 9, 335–338.
Gandhimathi, R., Ramesh, S.T., Arun Praveeth, E., 2008. Adsorptive removal of copper from aqueous solution onto raw rice husk: kinetics and isotherms. Nat. Environ. Pollut. Technol. 7, 763–768.
Garivait, S., Chaiyo1, U., Patumsawad, S., Deakhuntod, J., 2006. Physical and chemical properties of Thai biomass fuels from agricultural residues. The 2nd Joint International Conference on "Sustainable Energy and Environment (SEE 2006)" 1–23, November, 2006, Bangkok, Thailand.
Genieva, S.D., Turmanova, S.C., Dimitrova, A.S., Vlaev, L.T., 2008. Characterization of rice husks and the productions of its thermal degradation in air or nitrogen atmosphere. J. Therm. Anal. Calorim. 93, 387–396.
Georgieva, V.G., Tavlieva, M.P., Genieva, S.D., Vlaev, L.T., 2015. Adsorption kinetics of Cr(VI) ions from aqueous solutions onto black rice husk ash. J. Mol. Liq. 208, 219–226.
Ghaly, A.E., Mansaray, K.G., 1999. Comparative study on the thermal degradation of rice husk in various atmospheres. Energy Sources 21, 867–881.
Ghorbani, F., Sanati, A.M., Maleki, M., 2015. Production of silica nanoparticles from rice husk as agricultural waste by environmental friendly technique. Environ. Stud. Persian Gulf 2, 56–65.
Habeeb, G.A., Mahmud, H.B., 2010. Study on properties of rice husk ash and its use as cement replacement material. Mater. Res. 13, 185–190.
Haslinawati, M., Matori, K., Wahab, Z., Sidek, H., Zainal, A., 2009. Effect of temperature on ceramic from rice husk ash. Int. J. Basic Appl. Sci. 9, 111–117.
Ismail, S.A., Ali, R.F., 2015. Physico-chemical properties of biodiesel manufactured from waste frying oil using domestic adsorbents. Sci. Technol. Adv. Mater. 16, 1–9.
Javed, S.H., Tajwar, S., Shafaq, M., Zafar, M., Kazmi, M., 2009. Characterization of silica from sodium hydroxide treated rice husk. J. Pak. Inst. Chem. Eng. 37, 97–101.
Jiang, X.H., 2010. The Research on Application of the Rice Husk Ash. Master Dissertation, Harbin Institute of Technology, People's Republic of China.
Kalapathy, U., Proctor, A., Schultz, J., 2000. A simple method for production of pure silica from rice hull ash. Bioresour. Technol. 73, 257–262.
Kalapathy, U., Proctor, A., Schultz, J., 2002. An improved method for production of silica from rice hull ash. Bioresour. Technol. 85, 285–289.
Kim, M.J., Yoon, S.H., Choi, E., Gil, B., 2008. Comparison of the adsorbent performance between rice hull ash and rice hull silica gel according to their structural differences. LWT Food Sci. Technol. 41, 701–706.

Krishnani, K.K., Meng, X., Christodoulatos, C., Boddu, V.M., 2008. Biosorption mechanism of nine different heavy metals onto biomatrix from rice husk. J. Hazard. Mater. 153, 1222–1234.

Krishnarao, R., Subrahmanyam, J., Kumar, T.J., 2001. Studies on the formation of black particles in rice husk silica ash. J. Eur. Ceram. Soc. 21, 99–104.

Kuan, C.Y., Yuen, K.H., Liong, M.T., 2012. Physical, chemical and physicochemical characterization of rice husk. Br. Food J. 114, 853–867.

Kumchompooa, J., Wongwaib, W., Puntharod, R., 2017. Microwave-assisted preparation of sodium silicate as biodiesel catalyst from rice husk ash. Key Eng. Mater. 761, 461–466.

Lee, J.H., Kwon, J.H., Lee, J.W., Lee, H.S., Chang, J.H., Sang, B.I., 2017. Preparation of high purity silica originated from rice husks by chemically removing metallic impurities. J. Ind. Eng. Chem. 50, 79–85.

Li, P., Wang, T., 2008. Preparation of silica aerogel from rice hull ash by drying at atmospheric pressure. Mater. Chem. Phys. 112, 398–401.

Liew, K.Y., Yee, A.H., Nordin, M.R., 1993. Adsorption of carotene from palm oil by acid-treated rice hull ash. J. Am. Oil Chem. Soc. 70, 539–541.

Liou, T.H., 2004. Evolution of chemistry and morphology during the carbonization and combustion of rice husk. Carbon 42, 785–794.

Liu, Y., Guo, Y.P., Zhu, Y.C., An, D.M., Gao, W., Wang, Z., Ma, Y.J., Wang, Z.C., 2011. A sustainable route for the preparation of activated carbon and silica from rice husk ash. J. Hazard. Mater. 186, 1314–1319.

Liu, Y., Guo, Y.P., Gao, W., Wang, Z., Ma, Y.J., Wang, Z.C., 2012. Simultaneous preparation of silica and activated carbon from rice husk ash. J. Clean. Prod. 32, 204–209.

Liu, Y., Yuan, X.Z., Huang, H.J., Wang, X.L., Wang, H., Zeng, G.M., 2013. Thermochemical liquefaction of rice husk for bio-oil production in mixed solvent (ethanol–water). Fuel Process. Technol. 112, 93–99.

Liu, X., Chen, X., Yang, L., Chen, H., Tian, Y., Wang, Z., 2016. A review on recent advances in the comprehensive application of rice husk ash. Res. Chem. Intermed. 42, 893–913.

Ma, X.Y., Zhou, B., Gao, W., Qu, Y.N., Wang, L.L., Wang, Z.C., Zhu, Y.C., 2012. A recyclable method for production of pure silica from rice hull ash. Powder Technol. 217, 497–501.

Ma, Z., Ye, J., Zhao, C., Zhang, Q., 2015. Gasification of rice husk in a downdraft gasifier. Bioresources 10, 2888–2902.

Mahmud, A., Megat-Yusoff, P.S.M., Ahmad, F., Farezzuan, A.A., 2016. Acid leaching as efficient chemical treatment for rice husk in production of amorphous silica nanoparticles. ARPN J. Eng. Appl. Sci. 11, 13384–13388.

Manique, M.C., Faccini, C.S., Onorevoli, B., Benvenutti, E.V., Caramão, E.B., 2012. Rice husk ash as an adsorbent for purifying biodiesel from waste frying oil. Fuel 92, 56–61.

Mochidzuki, K., Sakoda, A., Suzuki, M., Izumi, J., Tomonaga, N., 2001. Structural behavior of rice husk silica in pressurized hot-water treatment processes. Ind. Eng. Chem. Res. 40, 5705–5709.

Moraes, C.A.M., Fernandes, I.J., Calheiro, D., Kieling, A.G., Brehm, F.A., Rigon, M.R., Filho, J.A.B., Schneider, I., Osorio, E., 2014. Review of the rice production cycle: byproducts and the main applications focusing on rice husk combustion and ash recycling. Waste Manag. Res. 32, 1034–1048.

Mukri, B.D., Chowdhury, K.K.A., Suryakala, D., Subrahmanyam, C., 2016. Alkali-treated carbonized rice husk for the removal of aqueous Cr (VI). Bioresources 11, 9175–9189.

Naiya, T.K., Bhattacharya, A.K., Mandal, S., Das, S.K., 2009. The sorption of lead (II) ions on rice husk ash. J. Hazard. Mater. 163, 1254–1264.

Natarajan, E., Ganapathy, S.E., 2009. Pyrolysis of rice husk in a fixed bed reactor. Int. J. Mech. Aerosp. Indust., Mechat. Manuf. Eng. 3, 959–963.

National Bureau of Statistics of China, 2017. http://www.stats.gov.cn/tjsj/ndsj/2016/indexch.htm. (accessed Aug., 2017).
Nattaporn, S., Porjai, T., 2015. Recovery of used frying palm oil by acidified ash from rice husk. J. Food Sci. Agric. Technol. 1, 193–196.
Nayak, J.P., Bera, J., 2009. Preparation of silica aerogel by ambient pressure using rice husk as raw materials. Trans. Indian Ceram. Soc. 62, 1–4.
Nehdi, M., Duquette, J., El Damatty, A., 2003. Performance of rice husk ash produced using a new technology as mineral admixtures in concrete. Cem. Concr. Res. 33, 1203–1210.
Olupot, P.W., Candia, A., Menya, E., Walozi, R., 2016. Characterization of rice husk varieties in Ugandafor biofuels and their techno-economic feasibility in gasification. Chem. Eng. Res. Des. 107, 63–72.
Ouyang, D., Chen, K., 2003. SEM/TEM study on the microstructure of rice husk ash and nano-SiO_2 in it. J. Chin. Electron Microsc. Soc. 22, 390–394.
Proctor, A., 1990. X-ray diffraction and scanning electron microscope studies of processed rice hull silica. J. Am. Oil Chem. Soc. 67, 576–584.
Proctor, A., Palaniappan, S., 1989. Soy oil lutein adsorption by rice hull ash. J. Am. Oil Chem. Soc. 66, 1618–1621.
Proctor, A., Palaniappan, S., 1990. Adsorption of soy oil free fatty acids by rice hull ash. J. Am. Oil Chem. Soc. 67, 15–16.
Proctor, A., Clark, P.K., Parker, C.A., 1995. Rice hull ash adsorbent performance under commercial soy oil bleaching conditions. J. Am. Oil Chem. Soc. 72, 459–462.
Quispe, I., Navia, R., Kahhat, R., 2017. Energy potential from rice husk through direct combustion and fast pyrolysis: a review. Waste Manag. 59, 200–210.
Rajanna, S.K., Kumar, D., Vinjamur, M., Mukhopadhyay, M., 2015. Silica aerogel microparticles from rice husk ash for drug delivery. Ind. Eng. Chem. Res. 54, 949–956.
Rambo, M.K.D., Schmidt, F.L., Ferreira, M.M.C., 2015. Analysis of the lignocellulosic components of biomass residues for biorefinery opportunities. Talanta 144, 696–703.
Real, C., Alcala, M.D., Criado, J.M., 1996. Preparation of silica from rice husks. J. Am. Ceram. Soc. 79, 2012–2016.
Rozainee, M., Ngo, S.P., Salema, A.A., Tan, K.G., 2008. Fluidized bed combustion of rice husk to produce amorphous siliceous ash. Energy Sustain. Dev. 12, 33–42.
Rungrodnimitchai, S., Phokhanusai, W., Sungkhaho, N., 2009. Preparation of silica gel from rice husk ash using microwave heating. J. Met. Mater. Miner. 19, 45–50.
Saengprachum, N., Poothongkam, J., Pengprecha, S., 2013. Glycerin removal in biodiesel purification process by adsorbent from rice husk. Int. J. Sci. Eng. Technol. (2), 474–478.
Sarkar, D., Das, K., Bandyopadhyay, A., 2013. Analysis of bio-sorption of Cr (VI) onto raw rice husk by a hybrid theoretical model using results of batch experiments. Adsorpt. Sci. Technol. 31, 747–765.
Simonov, A.D., Mishenko, T.I., Yazykov, N.A., Parmon, V.N., 2003. Combustion and processing of rice husk in the vibrofluidized bed of catalyst or inert material. Chem. Sustain. Dev. 11, 277–283.
Singh, H., Sapra, P.K., Sidhu, B.S., 2013. Evaluation and characterization of different biomass residues through proximate & ultimate analysis and heating value. Asian J. Eng. Appl. Technol. (2), 6–10.
Soltani, N., Bahrami, A., Pech-Canul, M.I., Gonzalez, L.A., 2015. Review on the physicochemical treatments of rice husk for production of advanced materials. Chem. Eng. J. 264, 899–935.
Stroeven, P., Bui, D.D., Sabuni, E., 1999. Ash of vegetable waste used for economic production of low to high strength hydraulic binders. Fuel 78, 153–159.
Subbukrishna, D.N., Suresh, K.C., Paul, P.J., Dasappa, S., Rajan, N.K.S., 2007. Precipitated silica from rice husk ash by IPSIT process. 15th European Biomass Conference and Exhibition, Berlin, Germany.

Sun, L., Gong, K., 2001. Silicon-based materials from rice husks and their applications. Ind. Eng. Chem. Res. 40, 5861–5877.
Tadjarodi, A., Haghverdi, M., Mohammadi, V., 2012. Preparation and characterization of nano-porous silica aerogel from rice husk ash by drying at atmospheric pressure. Mater. Res. Bull. 47, 2584–2589.
Taku, J.K., Amartey, Y.D., Kassar, T., 2016. Comparative elemental analysis of rice husk ash calcined at different temperatures using X-ray flourescence (XRF) technique. Am. J. Civil Eng. Archit. 4, 28–31.
Tang, Q., Wang, T., 2005. Preparation of silica aerogel from rice hull ash by supercritical carbon dioxide drying. J. Supercrit. Fluids 35, 91–94.
Temel, T.M., İkizler, B.K., Terzioğlu, P., Yücel, S., Yeliz Başaran Elalmış, Y.B., 2017. The effect of process variables on the properties of nanoporous silica aerogels: an approach to prepare silica aerogels from biosilica. J. Sol-Gel Sci. Technol. 84, 51–59. https://doi.org/10.1007/s10971-017-4469-x.
Titiloye, J.O., Bakar, M.S., Odetoye, T.E., 2013. Thermochemical characterisation of agricultural wastes from West Africa. Ind. Crop. Prod. 47, 199–203.
Todkar, B.S., Deorukhkar, O.A., Deshmukh, S.M., 2016. Extraction of silica from rice husk. Int. J. Eng. Res. Dev. 12, 69–74.
Tolba, G.M.K., Barakat, N.A.M., Bastaweesy, A.M., Ashour, E.A., Abdelmoez, W., El-Newehy, M.H., Al-Deyab, S.S., Kim, H.Y., 2015. Effective and highly recyclable nano-silica produced from the rice husk for effective removal of organic dyes. J. Ind. Eng. Chem. 29, 134–145.
Ugheoke, I.B., Mamat, O., 2012a. A critical assessment and new research directions of rice husk silica processing methods and properties. Maejo Int. J. Sci. Technol. 6, 430–448.
Ugheoke, B.I., Mamat, O., 2012b. Hydro thermo-baric processing and properties of nano silica from rice husk. Appl. Mech. Mater. 152, 177–182.
Ugheoke, B.I., Mamat, O., Ari-Wahjoedi, B., 2013. A direct comparison of processing methods of high purity rice husk silica. Asian J. Sci. Res. 6, 573–580.
Umeda, J., Kondoh, K., 2008. High-purity amorphous silica originated in rice husks via carboxylic acid leaching process. J. Mater. Sci. 43, 7084–7090.
Umeda, J., Kondoh, K., Michiura, Y., 2007. Process parameters optimization in preparing high-purity amorphous silica originated from rice husks. Mater. Trans. 48, 3095–3100.
Vayghan, A.G., Khaloo, A.R., Rajabipour, F., 2013. The effects of a hydrochloric acid pre-treatment on the physicochemical properties and pozzolanic performance of rice husk ash. Cem. Concr. Compos. 39, 131–140.
Vlaev, L.T., Markovska, I.G., Lyubchev, L.A., 2003. Non-isothermal kinetics of pyrolysis of rice husk. Therochim. Acta 406, 1–7.
Wang, L.H., Lin, C.I., 2008. Adsorption of lead (II) ion from aqueous solution using rice hull ash. Ind. Eng. Chem. Res. 47, 4891–4897.
Xiong, L., Sekia, E.H., Sujaridworakun, P., Wada, S., Saito, K., 2009. Burning temperature dependence of rice husk ashes in structure and property. J. Metals Mater. Miner. 19, 95–99.
Xu, W.T., Lo, T.Y., Memon, S.A., 2012. Microstructure and reactivity of rich husk ash. Constr. Build. Mater. 29, 541–547.
Yoon, S.H., Kim, M., Gil, B., 2011. Deacidification effects of rice hull-based adsorbents as affected by thermal and acid treatment. LWT Food Sci. Technol. 44, 1572–1576.
Zhang, Y., Zheng, R., Zhao, J., Zhang, Y., Wong, P.K., Ma, F., 2013. Biosorption of zinc from aqueous solution using chemically treated rice husk. Biomed. Res. Int. https://doi.org/10.1155/2013/365163.
Zulkali, M.M.D., Ahmad, A.L., Norulakmal, N.H., 2006. *Oryza sativa* L. husk as heavy metal adsorbent: optimization with lead as model solution. Bioresour. Technol. 97, 21–25.

CHAPTER 10

Nutritional Ingredients and Active Compositions of Defatted Rice Bran

Xuhui Zhuang, Tie Yin, Wei Han, Xiaolin Zhang
Academy of State Administration of Grain, Beijing, China

1. INTRODUCTION

Rice bran is one of the main byproducts in the process of the rice milling. It is the outer brown layer of brown rice and is separated during the milling process. All over the world, >63 million tons of rice bran is produced each year (Webber et al., 2014). Rice bran contains lipid (rice bran contains 15–20% oil), and high activities of lipases and lipoxygenases result in easy rancidity of the lipid due to lack of economic stabilization methods (Lakkakula et al., 2004). Therefore rice bran is mainly used as livestock feed or boiler fuel, and only a small amount is applied to extraction and preparation rice bran oil (Webber et al., 2014). To improve utilization of rice bran, development of feasible stabilization methods and the value-added processes of the nutrients and active compositions are necessary. Oil, proteins, and carbohydrates are the main nutrients in rice bran. There are also plenty of other nutrients and active compositions, such as nonstarch polysaccharides, phenolic acids, flavonoids, tocopherols, tocotrienols, c-oryzanol, and phytic acid. Most of these compounds have antioxidant and anticancer activities (Goufo and Trindade, 2014, Henderson et al., 2012). Defatted rice bran (DRB), also called rice bran meal, is the main byproduct in the extraction process of the rice bran oil from full fat rice bran (FRB) (Kaur et al., 2012). Most of the water-soluble chemicals are in the DRB, including most carbohydrates, proteins, and phytic acid. In this chapter, areas of research related to the nutrition and active compositions in the DRB are highlighted. The outline of this chapter is as follows: nutritional ingredients in the FRB and DRB; the starch and its application in the DRB; structures and applications of the nonstarch polysaccharides in the DRB; brief introduction to the rice bran protein; other active phytochemicals in the DRB; and the development prospect of the DRB.

Rice Bran and Rice Bran Oil
https://doi.org/10.1016/B978-0-12-812828-2.00010-X

2. NUTRITIONAL INGREDIENTS AND THEIR CONTENTS IN RICE BRAN AND DEFATTED RICE BRAN

There are an abundance of nutrients in rice bran. In the fatty phase, the oil is most important. However, other nutritional ingredients such as γ-oryzanol, vitamins, and phenolics are also of great significance to human health (Goufo and Trindade, 2014). The main nutrients in DRB are proteins, carbohydrates, lignins, and phytic acid (Henderson et al., 2012; Fig. 1).

According to the analysis of the extruded stabilized FRB samples and defatted ones from different areas in China (Cao, 2015), the content of ingredients in rice bran samples vary in accordance with species, producing areas and processing means. The content of crude fat in FRB from the Heilongjiang province is about 19%, yet it is only 13% in Beijing. The starch content in the former is about 24%, whereas in the latter it's about 33%. According to a report by Beloshapka et al. (2016), DRB contains 17% crude protein, 24.4% dietary fiber (including 22.5% insoluble and 1.7% soluble), and 34.2% starch (including 29.8% digestible starch and 4.4% resistant starch). In the bran from the parboiled rough rice, the contents of the protein, fat, ash, fiber, and starch are 11.4%, 21.2%, 21.3%, 13.9% and 32.3%, respectively (Gopala Krishna et al., 2012). Hashimoto et al. (1987) reported that the contents of protein, ash, and oil are 13.3–16.3%, 10.00–10.16%, and 19.1–25.5%, respectively, in parboiled rice bran, compared to 12.7–15.1%, 10.05–11.89%, and 18.0–21.2%, respectively in rice bran. The contents of the nutrients in stabilized rice bran of Tarom cultivar by

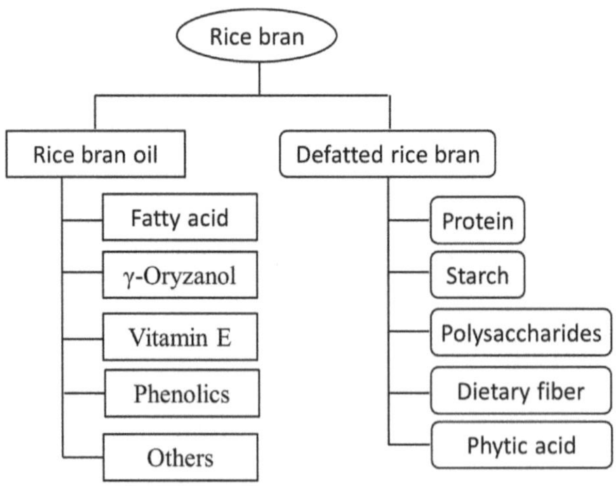

Fig. 1 The main nutrients in rice bran.

extruding are very different from those in unstabilized ones (Rafe et al., 2017). The protein contents in the former and the latter are about 8.50% and 15.00%, respectively, whereas the contents of the digestible carbohydrates are about 28.55% and 16.25%. In the unstabilized rice bran, the content of the phytate is about 23.34 mg/g, whereas it is almost undetected in the extruding treated samples (Table 1).

3. STARCH IN RICE AND RICE BRAN

3.1 Properties of Starch in Rice and Rice Bran

Starch is the major carbohydrate storage in cereal grains with the main structure of α-D-linked glucose. The starch in rice bran is the remnant endosperm starch in the rice milling process. The starch granules in rice are in the range from 3 to 10 μm (Dang and Copeland, 2004, Delcour et al., 2010). In contrast to most other cereals, it contains 20–60 individual granules in one rice amyloplast, which makes a formation of compound granules with a diameter up to 150 μm (Bechtel and Pomeranz, 1978).

Generally, starch is composed of amylose and amylopectin. They are all polymer molecules of D-glucosyl units joined together through glycosidic linkages. Amylose is a linear one, made up of R-1,4 glycosidic linkages, and amylopectin is a highly branched one through R,1-4 and R,1-6 glycosidic linkages (Akoh et al., 2008). The molecular weight of amylose is in the range of 10^5–10^6 Da, and amylopectin is in the range of 10^7–10^8 Da. In DRB starch, the polymerization values of the amylose is in the range of 920–1110 glucose units with average chain lengths of 250–370 (Takeda et al., 1986). Polymerization of rice amylopectin ranges from 2700 to 12,900 (Takeda et al., 2003), and the average chain lengths were 18–22, 15–18, and 17–20, respectively, for three rice species of Indica, Japonica and wax (wax is a starch with an amylose content $<2\%$) (Lu et al., 1997). There are also other minor compositions in rice starch (Delcour et al., 2010). Most of them are lipids (0.1–1.3%) (Morrison and Azudin, 1987) and proteins (0.25%) (Baldwin, 2001).

Amylose has beneficial health effects including slower digestion and acting as prebiotics (Topping et al., 2003), whereas amylopectin possesses good baking properties (Fox and Manley, 2014). Amylose:amylopectin (AM:AP) ratio in rice bran starch may affect its applications. Feed with different AM: AP ratio starch additions were applied to evaluate growth properties of juvenile obscure puffer fish (Liu et al., 2014). According to this study, fish fed with a diet containing AM:AP at a ratio of 6:19 showed best growth, feed

Table 1 Nutrients in rice bran and defatted rice bran (Cao, 2015)

Sample	Dry matter (%)	Crude protein of DW (%)	Ash of DW (%)	Crude fiber of DW (%)	Crude fat of DW (%)	Starch of DW (g/100 g)	Carbohydrate of DW (%)	Calories (Cal/g)
FRB-1	91.29 ± 0.03	15.82 ± 0.29	10.02 ± 0.06	8.69 ± 0.30	18.89 ± 0.92	23.91 ± 1.91	55.29	4739.45
FRB-2	89.20 ± 0.06	15.23 ± 0.40	8.72 ± 0.00	7.90 ± 0.25	12.43 ± 0.97	33.30 ± 0.49	63.34	4587.61
DRB-1	89.44 ± 0.04	20.24 ± 0.07	12.65 ± 0.01	10.23 ± 0.07	1.25 ± 0.55	31.57 ± 2.06	65.51	3765.96
DRB-2	87.24 ± 0.05	17.87 ± 0.03	10.78 ± 0.78	11.43 ± 0.25	3.53 ± 2.11	29.92 ± 0.32	67.60	3686.59

DRB-1, defatted rice bran samples from Heilongjiang; *DRB-2*, defatted rice bran samples from Beijing; *DW*, dry weight; *FRB-1*, Full-fat rice bran samples from Heilongjiang; *FRB-2*, full-fat rice bran samples from Beijing.

efficiency, and protein efficiency ratio. In the fish fed the highest AM:AP (12:13) diet, the hepatosomatic index, plasma total cholesterol concentration, liver glycogen and lipid content, and gluconokinase, pyruvate kinase, and fructose-1,6-bisphosphatase activities were lower than that in fish fed with low-amylose diets. Activities of liver trypsin, intestinal trypsin, liver amylase, intestinal amylase, intestinal lipase, and starch digestibility in fish fed AM:AP ratios of 9:16 and 12:13 were lower than those fed a lower (3:12 and 6:19) diet. Postprandial plasma glucose and triglyceride peak value of fish fed a diet of AM:AP ratio 12:13 were lower than in fish fed the low-amylose diets, and the peak time of plasma glucose was also later.

Table 2 shows classification of rice starches according to amylose contents (Delcour et al., 2010; Juliano et al., 1981).

Gelatinization temperature of rice starch is distributed in the range from 62°C to 79°C (Tester and Morrison, 1990; Vandeputte et al., 2003b; Delcour et al., 2010). It is related to the amylose content in rice starch. Waxy and lower amylose starches have a low peak gelatinization temperature (Vandeputte et al., 2003b). The starch has a better swelling power with an increase in gelatinization temperature. Rice starches have higher swelling power with an increase of the amylose content (Lii et al., 1996; Noosuk et al., 2003; Vandeputte et al., 2003a). Waxy starches demonstrated higher swelling power than that of nonwaxy ones (Delcour et al., 2010). Rice starches with higher amylose content also show better characteristics in terms of gel texture, hardness, and firmness (Tan and Corke, 2002; Vandeputte et al., 2003c). For the starches of the high amylose contents with the same gelatinization temperature range, the gel softness highly depends on the chain length distribution (Villareal et al., 1997). In the cooking process, starches with higher amylose contents indicate more rigid features (Rani and Bhattacharya, 1989). Starches with lower amylose contents are weak and

Table 2 The classification of rice starches according to their amylose contents (Delcour et al., 2010; Juliano et al., 1981)

Grade	Amylose content
Waxy	0% to 2%
Very low	5% to 12%
Low	12% to 20%
Intermediate	20% to 25%
High	25% to 33%
Mutants	35% to 40%

fragile (Rani and Bhattacharya, 1989), and thus the rice containing this kind of starch is soft and sticky after cooking (Reddy et al., 1994). Besides amylose content, amylopectin structure and protein also contribute to the taste and texture of the rice after cooking. Rice with long B-chains of amylopectin has a firmer texture (Reddy et al., 1993; Ramesh et al., 1999). Rice with higher protein contents is more nutritious and has a special texture, as protein can form a barrier to decrease leaching of rice components during cooking (Hamaker and Griffin, 1990; Derycke et al., 2005a). After cooking, rice will be hardened and its stickiness will decrease during the cooling and storage procedure. Firmness of cooked rice is suggested to be related to starch retrogradation during storage at low temperature. It is proven that starch retrogradation and cooked rice firmness are linearly related at different storage temperatures (Perdon et al., 1999). Starch retrogradation is slow when storage temperature decreased from 60°C to 40°C. However, as storage temperature further dropped to 20°C, the starch became significantly firmed (Philpot et al., 2006). Apart from temperature, protein and lipids in the starch also significantly influence the retrogradation. Removal of proteins will softened the cooked rice, but it will not contribute to changes in firmness upon cooling (Philpot et al., 2006). Meanwhile, removal of lipids can significantly increase the rate of retrogradation and firmness of gels (Philpot et al., 2006).

In the parboiling of the rice, a three-step hydrothermal treatment process that consists of soaking, steaming, and drying is needed by using paddy or brown rice as a resource. The parboiled rice is obtained after dehulling and milling. Of all the changes upon parboiling process, the most obvious one is starch gelatinization. It is proven that the degree of starch gelatinization is positively related to the hardness of cooked parboiled rice (Delcour et al., 2010). After steaming, drying, and cooling, starch gelatinization and retrogradation increased the rice firmness (Priestley, 1976). Apart from starch gelatinization, the complexation generated by rice lipids and amylose during heating and cooling restricts the leaching of rice components and swelling of starch granules during cooking (Biliaderis et al., 1993; Derycke et al., 2005a,b).

3.2 Digestibility of Starch

Starch digestion is consisted of oral digestion, esophageal transit, gastric digestion, small intestinal digestion, and large intestinal fermentation (Bornhorst and Paul, 2014). Starch is transformed into glucose after chewing and enzymolysis. The digestibility in ileum is commonly considered as a measure of the starch availability for absorption during digestion (Bornhorst and Paul, 2014). In a study of feeding different specific varieties of cooked rice to pigs, it is proven that medium-grain (Amaroo variety) rice and waxy rice had higher ileal digestibilities (96.2% and 99.1%, respectively) (Pluske et al., 2007). Long grain rice (Doongara variety) showed significantly lower ileal starch digestion (88.6%). This is mainly due to higher content of amylose in long grain rice (Pluske et al., 2007).

3.3 Applications of Rice Starch

Rice starch is widely used in industrial products and functional foods. For example, starch can be converted to bioethanol as biofuel to partly replace petroleum or fossil fuel energy (Badarudin et al., 2014). On the other hand, rice starch also can be developed into valuable functional ingredients, such as fructose, maltose syrups, wine, glucose, and trehalose (Akoh et al., 2008). Generally, there are two ways to achieve the conversion process by biomethod. One is fermentation using microorganisms, and the other is biocatalyst using the amylase superfamily (Akoh et al., 2008).

Another application of rice starch is production of rice wine. Rice wine, also called Chinese rice wine or yellow wine, has been popular in China, Japan, Korea, and other Asian countries for thousands of years, and is one of three of the most ancient alcoholic beverages in the world (Cheng et al., 2012). Traditionally, rice wines are usually fermented from rice, wheat, and some Chinese medicines by yeast, oryzae, or other microbes. The starch in rice and wheat offer the microbes an abundance of carbon and nitrogen for fermentation. Besides ethanol, fermentation technology and the addition of traditional herbs also give the rice wine a bright brown color, subtle enjoyable flavor, and functional ingredients for the prevention of age and diseases (Seo et al., 2009; Fei and Pan, 2006; Guo et al., 2007). Therefore, rice wine has health-enhancing properties (Wei et al., 2017). In the National Standard of China GB13662-2008, Chinese rice wine can be classified into four categories according to the concentrations of the total sugars: dry-type (total sugars ≤15 g/L), semidry-type (15.1 g/L ≤ total sugars ≤ 40 g/L), semisweet-type (40.1 g/L total sugars ≤ 100 g/L), and sweet-type wines (total sugars >100 g/L). Among the four types, semidry-type and semisweet-type are dominant, and the sweet-type is produced in the smallest amounts.

Chen et al. (2014) summarized the rice wine process. The main production procedure of the rice wine is shown in Fig. 2.

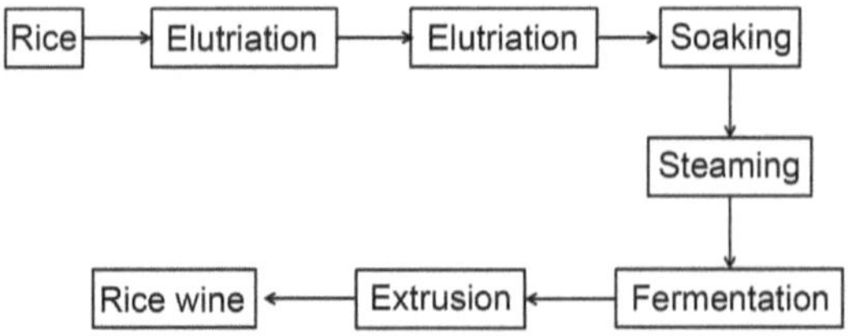

Fig. 2 The flow diagram showing the brewing process of rice wine.

Traditionally, the aging time of rice wine is a major character to determine its price. The reason is that rice wines have characteristic flavor and taste features with different aging times. Determinations of the flavor features can be performed by sensory panel (Zhong et al., 2012), gas chromatography mass spectrometer (Yang et al., 2017), or electronic nose (Wei et al., 2017).

In conclusion, starch is a major energy resource for human food and animal feed. Apart from it, amylose also acts as a prebiotic and can influence intestinal bacteria. Ratio of the amylose and amylopectin affect the taste, flavor, and digestibility. In the energy and health aspects, starch also can be used as resources for biofuels, functional ingredients, and wine. With the development of biotechnology, starch will be widely used in other fields.

4. NONSTARCH POLYSACCHARIDES

In rice bran, there are also plenty of nonstarch carbohydrates, such as monosaccharide, oligosaccharide, and polysaccharide. Nonstarch polysaccharides (NSPs), resistant starch, and oligosaccharides compose the major ingredients of total dietary fiber in most species of grain. In general, NSPs are considered to have antinutritional effects due to their viscous properties, which interfere with digestion and absorption of nutrients (Choct and Annison, 1992). The negative correlation between NSP and its nutritive value has been well studied in poultry (Antoniou et al., 1981; Choct and Annison, 1990; Bedford et al., 1991), pigs (King and Taverner, 1975), dogs, and cats (Earle et al., 1998). However, it has a healthy function as dietary fiber. Dietary fiber can reduce risk of serious diseases including colorectal cancer, cardiovascular disease, and diabetes (Collins et al., 2010). The major NSP in cereal grains are cellulose, β-glucan, α-glucan, arabinoxylan, etc. Their contents vary with the types of grains. The types and levels of NSP in different cereal grains have been reported in numerous references, and the content in rice is less than most other cereal grains, such as wheat, barley, rye, and maize

Table 3 The types and levels of NSP in rice (% dry matter)

NSP	Soluble	Insoluble	Total
Cellulose	Trace	0.3	0.3
β-glucan	0.1	–	0.1
Arabinoxylan	Trace	0.2	0.2
Galactan	0.1	Trace	0.1
Uronic acid	0.1	Trace	0.1
Total	0.3	0.5	0.8

(Choct, 2002; Englyst, 1989). The different polysaccharides in rice are summarized in Table 3 (Choct, 2002; Englyst, 1989). Rice bran contains 0.6 g arabinoxylan and 0.8 g β-glucan per kilogram dry matter (Mathlouthi et al., 2002).

4.1 Different Kinds of Nonstarch Polysaccharides

4.1.1 Cellulose

Cellulose is an indispensable component of cell walls and is the world's most abundant biopolymer (Limayem and Ricke, 2012). Cellulose is constituted of linear homopolymer of β-(1-4)-linked glucose units with molecular weights of over 10^6 Da (as shown in Fig. 3A). Strong hydrogen bonds exist between cellulose molecules, giving the polymers poor water solubility. Furthermore, due to the absence of cellulase, humans cannot digest cellulose. Therefore it does not contribute directly to our nutrition. However, cellulose can be hydrolyzed by 72% sulfuric acid, hot *N*-methylmorpholino *N*-oxide, and cellulase into soluble oligosaccharides, which are good sources of prebiotics (Kumar et al., 2012).

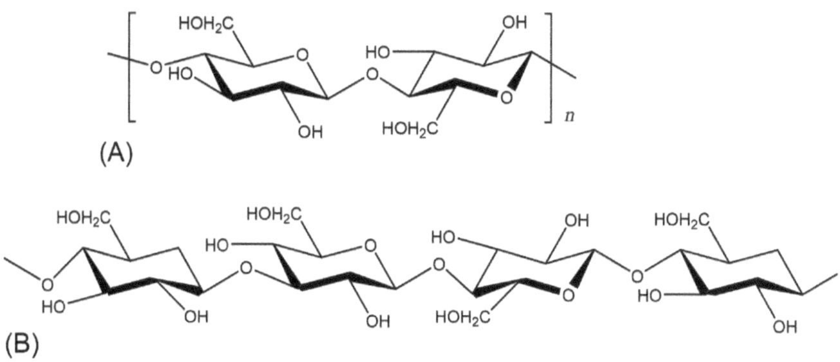

Fig. 3 Chemical structures of cellulose (A) and mixed-linked β-glucans (B).

4.1.2 Mixed-Linked β-Glucans

Besides cellulose, there is another group of polysaccharides formed by β-glucose units called mixed-linked β-glucans. Mixed-linked β-glucans are composed of glucose units in a linear chain by both β-(1→3) and β-(1→4) linkages (Fig. 3B) (Bengtsson et al., 1990). The (1→3)- and (1→4)-linked glucopyranosyl residues are present in a ratio of about 1:2.5, and the presence of 1→3 linkages in the chain results in irregularities in the molecules of mixed-linked β-glucans (Bengtsson et al., 1990). Mixed-linked β-glucans have better water solubility compared to cellulose (Bamforth, 1982), and the water-soluble β-glucans have molecular weights in the range of 200,000 and 300,000 Da, which depend on the degrees of polymerization of 1200–1850 monomers (Woodward et al., 1983). The role of water-soluble β-glucans as dietary fibers has attracted wide attention. Clinical studies indicated that β-glucans could reduce both serum cholesterol levels and postprandial blood glucose response (Tiwari and Cummins, 2011). Another report (Klopfenstein and Hoseney, 1987) revealed that mixed-linked β-glucan-enriched bread is useful in diets to control blood cholesterol levels. Klopfenstein (1988) also discussed the effects of mixed-linked β-glucans on blood glucose levels, hormone responses, colon cancer, and vitamin and mineral bioavailability in detail.

4.1.3 Mixed-Linked α-Glucans

In rice bran, one typical water-soluble polysaccharide is α-glucan. An antitumor dextran-like α-glucan named RON was obtained from DRB by diethylaminoethyl-Sepharose CL-6B separation (Takeo et al., 1988). It is composed mainly of α-(1→6)-glucosidic linkages with a small amount of C-3 branches at a molecular weight over 10^6 Da. RON has potential antitumor activities against syngeneic tumors, Meth-A fibrosacoma, and Lewis lung carcinoma not only via intraperitoneal administration but also by oral administration, with optimum doses around 30 mg/kg (Takeo et al., 1988). It is one of the few reports on the antitumor activities of α-glucan. Tanigami et al. (1991) attempted to hydrolyze RON by formic acid and ultrasonic irradiation. Molecular weight of RON is degraded to 10^4 Da from 10^6 Da (Tanigami et al., 1991). The hydrolyzed polysaccharide fractions also have strong antitumor and immunoregulation activities (Tanigami et al., 1991). In vitro, the extracted polysaccharides have been proven to enhance cytotoxic effects in tumor cells by stimulating nitric oxide (NO) production and tumor necrosis factor-α (TNF-α) secretion in a dose-dependent manner (Wang et al., 2016b). Biomodification of the rice bran polysaccharides might be a good way to enhance their activities, and new bioactivities could be found. The modified rice bran polysaccharides by *Grifola frondosa*, an edible fungus, have antioxidant activities as well as two-way adjusting effects on the production of NO in macrophages (Liu et al., 2017).

4.1.4 Arabinoxylan

The content of arabinoxylan in rice is highest among all polysaccharides except starch (Table 3). In general, the structure of cereal arabinoxylans consists of a chain linked with xylose by linear (1 → 4)-β-D-xylan backbone and substituted an L-arabino as residue attached through 2-O and 3-O atoms (Annison et al., 1992). In most cases, the 5-O-trans-feruloyl larabinofuranosyl substituents existed in the molecules. In rice bran, the arabinoxylan content is 4.84%–5.11% but can only be directly extracted by water at a concentration of 0.35%–0.77% (Zhang et al., 2014). Biomodification of the arabinoxylans not only improves its solubility but also can enhance the bioactivities including antitumor and immunoregulation. Among the biomodified arabinoxylans, the most representative one is BioBran/MGN-3 arabinoxylan, also called MGN-3. MGN-3 is a natural blend of hemicelluloses derived from partially hydrolyzed rice bran hemicellulose B with shiitake mushroom enzymes (Lentinus edodes mycelia extract) (Ooi et al., 2017). It is developed and manufactured in Japan by Daiwa Pharmaceutical Co, Ltd. collaborated with Prof. Mamdooh Ghoneum at the University of California. MGN-3 as a nontoxic food supplement has different brand names such as BioBran (Globally), Lentin Plus (Japan/Asia), Ribraxx (Australia/New Zealand), BRM4 (United States), and others (https://biobran.org/overview). The main chemical structure of MGN-3 is an arabinoxylan with a xylose in its main chain and an arabinose in its side chain. The structures of rice bran hemicellulose B and MGN-3 are shown in Fig. 4. MGN-3 is considered to be a biological response modifier for enhancing the immune system during and after conventional cancer treatment. Its bioactive research and clinical effect has been summarized by Ooi et al. (2017). The anticancer mechanism of MGN-3 is considered to be activation of the natural killer cells, T cells, B cells, and macrophages, either directly or indirectly (Ooi et al., 2017). All the studies including scientific research, clinical reports, and small clinical trials suggest that MGN-3 has an application value to treat cancers through upregulating the patient's immune system. So far, there is no adverse event report in the clinical applications about MGN-3.

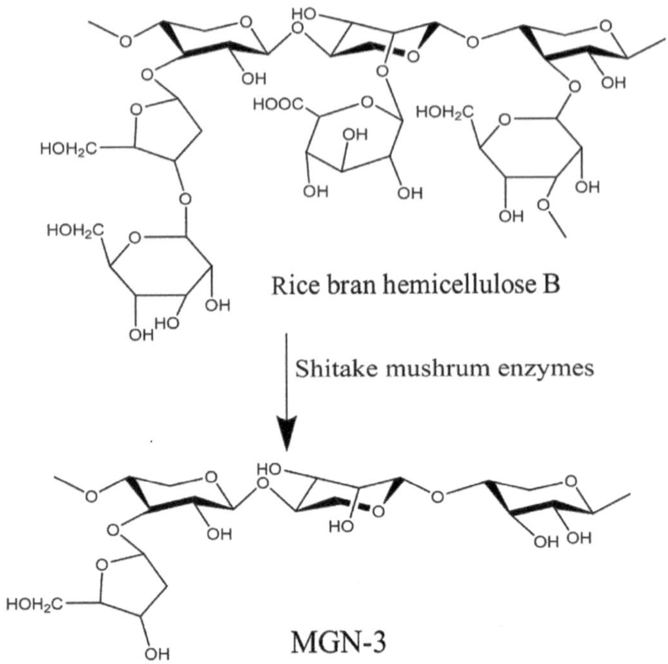

Fig. 4 The chemical structure of MGN-3. *(Based on the website: https://biobran.org/overview.)*

4.2 Preparation of Nonstarch Polysaccharides

There are two approaches to prepare NSP, namely wet extraction and dry fractionation. The wet method involves extraction of soluble polysaccharides by water, followed by the alcohol precipitation process. To improve the yield of the polysaccharides, some assistance tools could be applied, such as heating, ultrasonic, microwave irradiation, or other physical methods (Bonrath, 2004; Lianfu and Zelong, 2008; Hromádková et al., 2008; Ebringerová and Hromádková, 2002). The wet extraction has many advantages, such as efficiency and purification of polysaccharides. However, it is based on expensive chemical processes and unable to retain the major components in intact form after preparation. In addition, it may generate highly toxic effluents and waste water, which have a negative influence on the environment (Huber et al., 2006). Therefore a new environmental-friendly fractionation technology without water, namely dry fractionation, has become an important alternative. Dry fractionation does not require chemicals and does not generate waste water. Traditionally, dry fractionation includes pearling, roller milling, milling followed by air classification, and milling followed by sieving (Zhuang et al., 2018). Pearling, also called dehulling, is a process to remove outer layers of cereal grains. Milling and roller milling is used to disintegrate grains into fine particles and subsequently through shifters for fraction separation. Sieving is a separation method that separates the milled flours into different classes based on particle size (Zhuang et al., 2018). In recent years, the combined approach of ultra-milling and electrostatic fractionation has attracted intensive interest. The dry fractionation platform has been used to separate the cellulose from recalcitrant tissues enriched in lignin-hemicelluloses from wheat straw (Barakat and Rouau, 2014). This method also has been applied in the nonstarch polysaccharide preparation in DRB. Wang et al. (2016a) prepared the dietary fiber fraction using three separation routes: two-step electrostatic separation, sieving, and a combination of electrostatic separation and sieving. The dietary fiber-enriched fractions obtained from dry fractionation have similar water retention capacity and oil binding capacity compared to those extracted by enzymatic-gravimetric method (Wang et al., 2016a).

Both wet extraction and dry fractionation methods have advantages and disadvantages. Wet extraction can produce pure NSPs. However, the water and chemicals applied in a large scale generates waste water and pollutes the environment. Furthermore, the costs of chemical, enzyme, power, and dealing with the waste water is too high. In comparison, dry fractionation has many advantages, such as energy efficiency, no solvent involvement, requires low capital investments, and is environmentally friendly, but the fractions exhibit low purity and yield. The two approaches also can be used in combination with each other. The DRB can be fractionated using dry method, followed by wet extraction. This would be a good method to ensure the purity of the products without a large scale of solvent.

5. RICE BRAN PROTEINS

Rice bran protein has been recognized as nutritionally superior to other proteins especially on its reported hypoallergenicity and anticancer activity (Helm and Burks, 1996; Kawamura and Ishikawa, 2005; Shoji et al., 2001). Its protein efficiency ratio is 1.6–1.9, compared to the value of 2.5 for casein (Saunders, 1990). Furthermore, the biofunction properties of rice protein are superior to other plant-based proteins such as peanut, potato, sorghum, kidney bean, etc. (Lawal et al., 2007; Londhe et al., 2011; De Mesa-Stonestreet et al., 2012; Waglay et al., 2014; Sun and Xiong, 2014). Rice bran protein is suggested as one of the important plant-based proteins that can be applied as an ingredient in many products such as infant food, gluten-free products, and also cosmetic goods. Therefore rice protein is gaining a lot of interest in the food industry due to its unique properties, including emulsion, foaming, gel-forming, and hypoallergenic (Lawal et al., 2007; Yeom et al., 2010). More detailed information on the rice bran protein is described in the chapter "Rice Bran Protein".

6. OTHER ACTIVE PHYTOCHEMICALS IN DEFATTED RICE BRAN

There is a large amount of unsaturated lipids, iron, oxygen, and free radicals in the body of a plant. During plantation, there will be a large number of biochemical reactions. It is clear that the elements within the body must conspire to prevent the injury that occurred during extensive oxidative reactions, because it would hinder its germination and growth (Graf and Eaton, 1990). To protect against oxidative damage, antioxidant secondary metabolites are biosynthesized, such as phenolic compounds. Additionally, there is an ingredient named phytic acid that has the ability to extend seed viability by iron chelation (Graf and Eaton, 1990). There is a hypothesis asserting that phytic acid maintains iron in the Fe^{3+} oxidation state and obstructs generation of hydroxyl radical and other activated oxygen species by occupying all of the available iron coordination sites (Graf, 1983; Graf and Eaton, 1990).

Phytic acid (PA, myoinositol hexakisphosphate, IP6) is an organic compound containing phosphorus with the major chemical structure of inositol and six phosphoric acid substituents (Fig. 5) and typically accounts for 60%–90% of the total seed phosphorus (Lolas et al., 1976). There are 12 replaceable protons in the six phosphoryl groups that allow it to combined with multivalent cations and positively charged proteins. Therefore phytic acid can be found in free acid or salts. Phytate is the calcium salt of phytic acid, and phytin is the calcium/magnesium salt (Oatway et al., 2001). Complete hydrolysis of phytic acid can obtain myoinositol and inorganic phosphates (Fig. 5).

Phytic acid is very stable. It can be stable in the soil for several years and in neutral or alkaline environments for months (Graf and Eaton, 1990). It will be partly hydrolyzed in acid conditions. For example, it releases 50% of the phosphorus in 5 N HC1 at 100°C for 6 h (Graf and Eaton, 1990). Treatments of strong acids and phytase result in a mixture of myoinositol, inorganic phosphate and myoinositol mono-, di-, tri-, tetra-, penta-, and hexaphosphate (Graf and Eaton, 1990). During germination of seeds, phytic acid also can be dephosphorylated and yield myoinositol (Graf and Eaton, 1990).

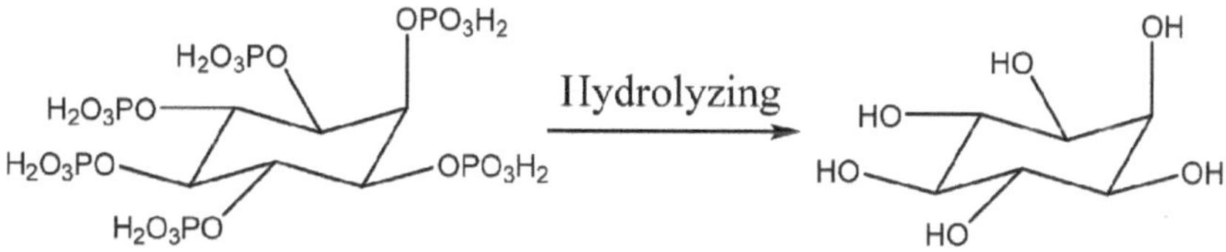

Fig. 5 Basis chemical structures of phytic acid (left) and its hydrolysis product myoinositol (right).

Phytic acid and its salts are widely spread in the cereal grains. The contents in FRB and DRB are about 6.55% and 8.70%, respectively (Lehrfeld, 1994). Most studies focused on the antinutritional effects of phytic acid. Phytic acid binds easily to minerals, proteins, and starch through formation of phosphate complexes, electrostatic charges, and hydrogen bond, respectively (Fig. 6). These bindings influence the solubility, functionality, digestion, and absorption of the nutrients.

However, many studies also pointed out that phytic acid has a broad spectrum of biological activities (Graf and Eaton, 1990). It is well studied that the iron-catalyzed reactions could protect organisms from oxidative damage (Halliwell and Gutteridge, 1985; Aust et al., 1985; Martinez-Cayuela, 1995). Ferryl and perferryl ions are important catalysts in the bioreaction procedure, generating hydroxyl radicals. As a stable chelator, the phytic acid can bind minerals such as calcium, iron, and zinc causing a decrease of their bioavailability in bodies (Ahn et al., 2003). Norazalina and coworkers reported that the phytic acid extracted from rice bran has

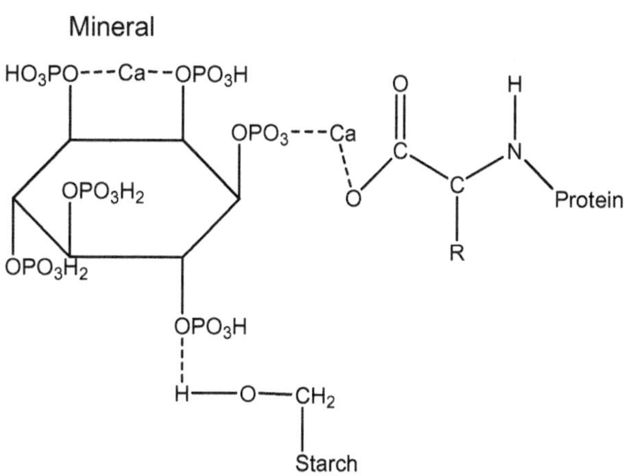

Fig. 6 Interactions of phytic acid with mineral, proteins, and starch.

the effect of reducing colon cancer risk in azoxymethane (AOM)-induced rats by reducing the incidence and multiplicity of total tumors (Norazalina et al., 2010). In addition, phytic acid also has bioactivities of neuroprotection (Xu et al., 2008), prevention of atherosclerosis (Grases et al., 2007), diabetes (Lee et al., 2006), and inhibition of the crystallization of calcium phosphate and calcium oxalate formation in urine (Conte et al., 1999; Grases et al., 2000).

DRB is an economic resource to extract and prepare the phytic acid. China is the world's largest rice producer, with an annual yield of rice bran of >16 million tons. Most of the rice bran that can be effectively used is applied to prepare the rice bran oil and phytic acid. Phytic acid is extracted using strong acid solutions, such as HCl and H_2SO_4 (Wu et al., 2009) followed by alkali precipitation. Zhang and Bai (2014) and Ren et al. (2017) extracted phytic acid from rice bran and peanut meal, respectively, using HCl solution as solvent, and further optimized its extraction process using response surface methodology. Canan et al. (2011) extracted phytic acid from rice bran using 1 M HCl solution followed by 1.5 M Na_2CO_3 precipitation. The crud extract was resuspended using the HCl solution and then diatomaceous earth was added to denature the protein and to remove other contaminants. The purity of the phytic acid is up to 64.63 ± 2.77 mg/g (Canan et al., 2011), as shown in Fig. 7.

In DRB, there are still an abundance of phytochemicals. Therefore many works focused on the extraction and analyzation of phytochemicals in DRB and tested their antioxidant activities. The DRB was extracted using

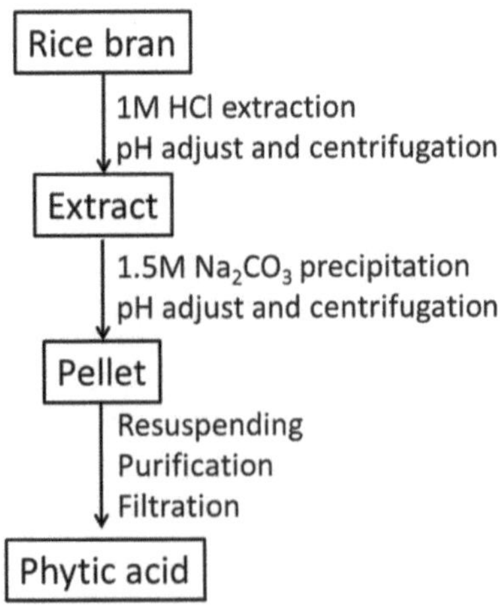

Fig. 7 The extraction and preparation of phytic acid.

methanol and yielded a crude methanolic extract (CME). CME was re-extracted with acetone to give an acetone extract (AE). The AE was further fractionated into lipophilic phase (AE-LP) with hexane extraction and a polar phase (AE-PP) with acetone (Renuka Devi et al., 2007). With the inhibiting lipid oxidation as an indicator of the antioxidant activity, the activities of the extracts followed the order AE-PP > AE-LP = AE > CME, and the activity of AE-PP is equivalent to that of butylated hydroxytoluene (BHT) at a 200 ppm (Renuka Devi et al., 2007). The scavenging effects of 1,1-diphenyl-2-picrylhydrazyl radical (DPPH·) and superoxide radical are all at the orders of AE-PP > AE > CME > AE-LP, and the scavenging activities of AE, AE-LP, and AE-PP was significantly related to the levels of total phenols (TPC) and ferulic acid in it (Renuka Devi and Arumughan, 2007a). The AE-LP was enriched in oryzanols and tocols by about 65 times compared to their contents in DRB, whereas AE-PP was enriched in ferulic acid by 70 times as compared to their contents in DRB (Renuka Devi and Arumughan, 2007b).

7. THE DEVELOPMENT PROSPECT OF DEFATTED RICE BRAN

Most reported studies on rice bran fractionations focused on the enrichment of one or two nutrients (such as starch, soluble polysaccharides, or protein), but most of other valuable components were neglected. Comprehensive development of rice bran will be important for efficient utilization. Studies on other cereals can provide important clues to achieve this purpose. The approach can recover almost all the major nutrients (β-glucan, protein, and starch) from oats simultaneously as shown in Fig. 8 (Liu, 2014; Liu and Barrows, 2017).

Traditionally, DRB have been directly developed. However, in recent years, DRB fermentation attracted intensive interest. In the fermentation process, an abundance of enzymes were generated to catalyze the nutrient. It may produce phytochemicals that did not previously exist (Ryan et al., 2011) or increase the content of such compositions (Kaur et al., 2016). Solid fermentation has advantages such as low cost, simpler process, lesser energy requirements, absence of rigorous control of fermentation parameters, less water consumed, and reduced waste water release. Therefore it will attract more attention in the production of food and functional food aspects.

With increasing health awareness, the usage of healthful rice bran oil byproducts is anticipated to become more important. Preparations of

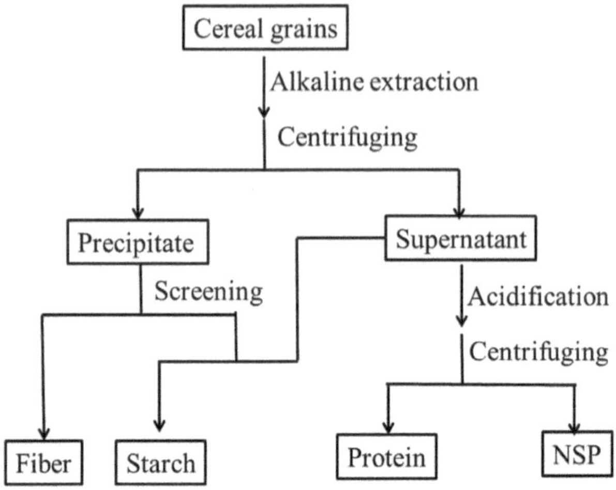

Fig. 8 A flow chart showing the comprehensive development of DRB (Liu and Barrows, 2017).

nutritional ingredients and active compositions, as well as their applications in functional food for certain segments of the populations, such as hypertensive patients, obese patients, immunodeficiency patients, and others, have considerable potential in future markets.

REFERENCES

Ahn, H.J., Kim, J.H., Yook, H.S., Byun, M.W., 2003. Irradiation effects on free radical scavenging and antioxidant activity of phytic acid. J. Food Sci. 68, 2221–2224.

Akoh, C.C., Chang, S.-W., Lee, G.-C., Shaw, J.-F., 2008. Biocatalysis for the production of industrial products and functional foods from rice and other agricultural produce. J. Agric. Food Chem. 56, 10445–10451.

Annison, G., Choct, M., Cheetham, N.W., 1992. Analysis of wheat arabinoxylans from a large-scale isolation. Carbohydr. Polym. 19, 151–159.

Antoniou, T., Marquardt, R.R., Cansfield, P.E., 1981. Isolation, partial characterization, and antinutritional activity of a factor (pentosans) in rye grain. J. Agric. Food Chem. 29, 1240–1247.

Aust, S., Morehouse, L.A., Thomas, C., 1985. Role of metals in oxygen radical reactions. J. Free Radic. Biol. Med. 1, 3–25.

Badarudin, A., Togun, H., Zubir, M.N.M., Oon, C.S., Gharehkhani, S., 2014. Sustainability and environmental impact of ethanol as a biofuel: reviews in chemical engineering. Rev. Chem. Eng. 30, 51–72.

Baldwin, M., 2001. Starch granule-associated proteins and polypeptides: a review. Starch 53, 475–503.

Bamforth, C., 1982. Barley β-glucans. Their role in malting and brewing. Brew. Dig. 57, 22–27.

Barakat, A., Rouau, X., 2014. New dry technology of environmentally friendly biomass refinery: glucose yield and energy efficiency. Biotechnol. Biofuels 7, 138.

Bechtel, D., Pomeranz, Y., 1978. Ultrastructure of the mature Ungerminated rice (Oryza sativa) caryopsis. The germ. Am. J. Bot. 65, 75–85.
Bedford, M.R., Classen, H.L., Campbell, G.L., 1991. The effect of pelleting, salt, and pentosanase on the viscosity of intestinal contents and the performance of broilers fed Rye. Poult. Sci. 70, 1571–1577.
Beloshapka, A., Buff, P., Fahey, G., Swanson, K., 2016. Compositional analysis of whole grains, processed grains, grain co-products, and other carbohydrate sources with applicability to pet animal nutrition. Foods 5, 23.
Bengtsson, S., Åman, P., Graham, H., Newman, C.W., Newman, R.K., 1990. Chemical studies on mixed-linked β-glucans in hull- less barley cultivars giving different hypocholesterolaemic responses in chickens. J. Sci. Food Agric. 52, 435–445.
Biliaderis, C.G., Tonogai, J.R., Perez, C.M., Juliano, B.O., 1993. Thermophysical properties of milled rice starch as influenced by variety and parboiling method. Cereal Chem. 70, 512–516.
Bonrath, W., 2004. Chemical reactions under "non-classical conditions", microwaves and ultrasound in the synthesis of vitamins. Ultrason. Sonochem. 11, 1–4.
Bornhorst, G.M., Paul, S.R., 2014. Gastric digestion in vivo and in vitro: how the structural aspects of food influence the digestion process. Annu. Rev. Food Sci. Technol. 5, 111–132.
Canan, C., Cruz, F.T.L., Delaroza, F., Casagrande, R., Sarmento, C.P.M., Shimokomaki, M., Ida, E.I., 2011. Studies on the extraction and purification of phytic acid from rice bran. J. Food Compos. Anal. 24, 1057–1063.
Cao, X., 2015. Extracting Rice Bran Polysaccharides by Fermentation and Bioactivity. (Master degree, thesis). South China University of Technology.
Chen, X.D., Xiao-Fu, T.U., Mao, Q.Z., Zhou, L.P., 2014. Conventional japonica Rice in Fujian Monascus traditional Chinese Rice wine technology and quality control. Liquor Making, 174–175.
Cheng, H., Liu, J., Xu, Z., Yin, X., 2012. A micro-fluidic sub-microliter sample introduction system for direct analysis of Chinese rice wine by inductively coupled plasma mass spectrometry using external aqueous calibration. Spectrochim. Acta Part B At. Spectrosc. 73, 55–61.
Choct, M., 2002. Non-starch polysaccharides: effect on nutritive value. In: MCNAB, J.M., BOORMAN, K.N. (Eds.), Poultry Feedstuffs: Supply, Composition and Nutritive Value. CABI Publishing, Oxon.
Choct, M., Annison, G., 1990. Anti-nutritive activity of wheat pentosans in broiler diets. Br. Poult. Sci. 31, 811–821.
Choct, M., Annison, G., 1992. The inhibition of nutrient digestion by wheat pentosans. Br. J. Nutr. 67, 123–132.
Collins, H.M., Burton, R.A., Topping, D.L., Liao, M.-L., Bacic, A., Fincher, G.B., 2010. REVIEW: Variability in fine structures of noncellulosic Cell Wall polysaccharides from cereal grains: Potential importance in human health and nutrition. Cereal Chem. J. 87, 272–282.
Conte, A., Piz, P., Garcia-Raja, A., Grases, F., Costa-Bauz, A., prieto, R., 1999. Urinary lithogen risk test: usefulness in the evaluation of renal lithiasis treatment using crystallization inhibitors (citrate and phytate). Arch. Esp. Urol. 52, 94–99.
Dang, J., Copeland, L., 2004. Studies of the fracture surface of rice grains using environmental scanning electron microscopy. J. Sci. Food Agri. 84, 707–713.
De Mesa-Stonestreet, N.J., Alavi, S., Gwirtz, J., 2012. Extrusion-enzyme liquefaction as a method for producing sorghum protein concentrates. J. Food Eng. 108, 365–375.
Delcour, J.A., Bruneel, C., Derde, L.J., Gomand, S.V., Pareyt, B., Putseys, J.A., Wilderjans, E., Lamberts, L., 2010. Fate of starch in food processing: from raw materials to final food products. Annu. Rev. Food Sci. Technol. 1, 87–111.

Derycke, V., Veraverbeke, W.S., Vandeputte, G.E., De Man, W., Hoseney, R., Delcour, J., 2005a. Impact of proteins on pasting and cooking properties of nonparboiled and parboiled rice. Cereal Chem. 82, 468–474.

Derycke, V., Vandeputte, G.E., Vermeylen, R., Man, W., Goderis, B., Mhj, K., Delcour, J.A., 2005b. Starch gelatinization and amylose-lipid interactions during rice parboiling investigated by temperature resolved wide angle X-ray scattering and differential scanning calorimetry. J. Cereal Sci. 42, 334–343.

Earle, K.E., Kienzle, E., Opitz, B., Smith, P.M., maskell, I.E., 1998. Fiber affects digestibility of organic matter and energy in pet foods. J. Nutr. 128, 2798S–2800S.

Ebringerov, A., Hrom Dkov, Z., 2002. Effect of ultrasound on the extractibility of corn bran hemicelluloses. Ultrason. Sonochem. 9, 225–229.

Englyst, H., 1989. Classification and measurement of plant polysaccharides. Anim. Feed Sci. Technol. 23, 27–42.

Fei, Q., Pan, M.X., 2006. Antioxidant activities of five Chinese rice wines and the involvement of phenolic compounds. Food Res. Int. 39, 581–587.

Fox, G., Manley, M., 2014. Applications of single kernel conventional and hyperspectral imaging near infrared spectroscopy in cereals. J. Sci. Food Agric. 94, 174–179.

Gopala Krishna, A.G., Raja, R.G., Bhatnagar, A., 2012. Rice bran: chemistry, production and applications—a review. Beverage & Food World. 31–36.

Goufo, P., Trindade, H., 2014. Rice antioxidants: Phenolic acids, flavonoids, anthocyanins, proanthocyanidins, tocopherols, tocotrienols, γ-oryzanol, and phytic acid. Food Sci. Nutr. 2, 75–104.

Graf, E., 1983. Applications of phytic acid. J. Am. Oil Chem. Soc. 60, 1861–1867.

Graf, E., Eaton, J.W., 1990. Antioxidant functions of phytic acid. Free Radic. Biol. Med. 8, 61–69.

Grases, F., Simonet, B.M., March, J.G., Prieto, R.M., 2000. Inositol hexakisphosphate in urine: the relationship between oral intake and urinary excretion. BJU Int. 85, 138–142.

Grases, F., Sanchis, P., Perello, J., Isern, B., Prieto, R., Fernandez Palomeque, C., Torres, J., 2007. Effect of crystallization inhibitors on vascular calcifications induced by vitamin D: a pilot study in Sprague-Dawley rats. Circ. J. 71, 1152–1156.

Guo, H., Wang, P., You, B., Xing, Y., Lee, J.D., 2007. Chinese yellow wine inhibits production of homocysteine-induced extracellular matrix metalloproteinase-2 in cultured rat vascular smooth muscle cells. Clin. Nutr. 26, 348–354.

Halliwell, B., Gutteridge, J.M.C., 1985. The importance of free radicals and catalytic metal ions in human diseases. Mol. Asp. Med. 8, 89–193.

Hamaker, B.R., Griffin, V.K., 1990. Changing the viscoelastic properties of cooked rice through protein disruption. Cereal Chem. 67, 261–264.

Hashimoto, S., Shogren, M., Bolte, L., Pomeranz, Y., 1987. Cereal pentosans: their estimation and significance. 3. Pentosans in abraded grains and milling by-products. Cereal Chem. 64, 39–41.

Helm, R.M., Burks, A.W., 1996. Hypoallergenicity of rice protein. Cereal Foods World. 41, 839–843.

Henderson, A.J., Ollila, C.A., Kumar, A., Borresen, E.C., Raina, K., Agarwal, R., Ryan, E.P., 2012. Chemopreventive properties of dietary Rice bran: Current status and future prospects. Adv. Nutr. 3, 643–653.

Hromádková, Z., Kost'álová, Z., Ebringerová, A., 2008. Comparison of conventional and ultrasound-assisted extraction of phenolics-rich heteroxylans from wheat bran. Ultrason. Sonochem. 15, 1062–1068.

Huber, G.W., Iborra, S., Corma, A., 2006. Synthesis of transportation fuels from biomass: chemistry, catalysts, and engineering. Chem. Rev. 106, 4044–4098.

Juliano, B.O., Perez, C.M., Blakeney, A.B., Castillo, T., Kongseree, N., Laignelet, B., Lapis, E.T., Murty, V.V.S., Paule, C.M., Webb, B.D., 1981. International cooperative testing on the amylose content of milled rice. Starch 33, 157–162.

Kaur, A., Jassal, V., Thind, S.S., Aggarwal, P., 2012. Rice bran oil an alternate bakery shortening. J. Food Sci. Technol. 49, 110–114.

Kaur, H., Arora, M., Bhatia, S., Alam, M.S., 2016. Soild state fermentation of deoiled rice bran for production of fungal enzymes-a review. Int. J. Appl. Pure Sci. Agri. 2, 143–157.

Kawamura, Y., Ishikawa, M., 2005. Anti-tumorigenic and immunoactive protein and peptide factors in foodstuffs. 1. Antitumorigenic protein from tricholoma matsutake. Food Cancer Prev., 327–330.

King, R.H., Taverner, M.R., 1975. Prediction of the digestible energy in pig diets from analyses of fibre content. Anim. Sci. 21, 275–284.

Klopfenstein, C.F., 1988. The role of cereal beta-glucans in nutrition and health. Cereal Foods World. 33, 865–869.

Klopfenstein, C.F., Hoseney, R., 1987. Cholesterol-lowering effect of beta-glucan enriched bread. Nutr. Rep. Int. 36, 1091–1098.

Kumar, V., Sinha, A.K., Makkar, H.P.S., De Boeck, G., Becker, K., 2012. Dietary roles of non-starch polysachharides in human nutrition: a review. Crit. Rev. Food Sci. Nutr. 52, 899–935.

Lakkakula, N.R., Lima, M., Walker, T., 2004. Rice bran stabilization and rice bran oil extraction using ohmic heating. Bioresour. Technol. 92, 157–161.

Lawal, O.S., Adebowale, K.O., Adebowale, Y.A., 2007. Functional properties of native and chemically modified protein concentrates from bambarra groundnut. Food Res. Int. 40, 1003–1011.

Lee, S.-H., Park, H.-J., Chun, H.-K., Cho, S.-Y., Cho, S.-M., Lillehoj, H.S., 2006. Dietary phytic acid lowers the blood glucose level in diabetic KK mice. Nutr. Res. 26, 474–479.

Lehrfeld, J., 1994. HPLC separation and quantitation of phytic acid and some inositol phosphates in foods: problems and solutions. J. Agric. Food Chem. 42, 2726–2731.

Lianfu, Z., Zelong, L., 2008. Optimization and comparison of ultrasound/microwave assisted extraction (UMAE) and ultrasonic assisted extraction (UAE) of lycopene from tomatoes. Ultrason. Sonochem. 15, 731–737.

Lii, C.Y., Tsai, M.L., Tseng, K.H., 1996. Effect of amylose content on the rheological property of rice starch. Cereal Chem. 73, 415–420.

Limayem, A., Ricke, S.C., 2012. Lignocellulosic biomass for bioethanol production: current perspectives, potential issues and future prospects. Prog. Energy Combust. Sci. 38, 449–467.

Liu, K., 2014. Fractionation of oats into products enriched with protein, beta-glucan, starch, or other carbohydrates. J. Cereal Sci. 60, 317–322.

Liu, K., Barrows, F.T., 2017. Wet processing of barley grains into concentrates of proteins, β-glucan, and starch. Cereal Chem. 94, 161–169.

Liu, X.-H., Ye, C.-X., Ye, J.-D., Shen, B.-D., Wang, C.-Y., Wang, A.-L., 2014. Effects of dietary amylose/amylopectin ratio on growth performance, feed utilization, digestive enzymes, and postprandial metabolic responses in juvenile obscure puffer Takifugu obscurus. Fish Physiol. Biochem. 40, 1423–1436.

Liu, Q., Cao, X., Zhuang, X., Han, W., Guo, W., Xiong, J., Zhang, X., 2017. Rice bran polysaccharides and oligosaccharides modified by Grifola frondosa fermentation: antioxidant activities and effects on the production of NO. Food Chem. 223, 49–53.

Lolas, G.M., Palamidis, N., Markakis, P., 1976. The phytic acid-total phosphorus relationship in barley, oats, soybeans, and wheat. Cereal Chem. 53, 867–870.

Londhe, S., Joshi, M., Bhosale, A., Kale, S., 2011. Isolation of quality soy protein from soya flakes. Int. J. Res. Pharm. Biomed. Sci. 2, 1175–1177.

Lu, S., Chen, L., Lii, C., 1997. Correlations between the fine structure, physicochemical properties, and retrogradation of amylopectins from Taiwan rice varieties. Cereal Chem. 74, 34–39.

Martinez-Cayuela, M., 1995. Oxygen free radicals and human disease. Biochimie 77, 147–161.

Mathlouthi, N., Saulnier, L., Quemener, B., Larbier, M., 2002. Xylanase, β-glucanase, and other side enzymatic activities have greater effects on the viscosity of several feedstuffs than xylanase and β-glucanase used alone or in combination. J. Agric. Food Chem. 50, 5121–5127.

Morrison, W.R., Azudin, M.N., 1987. Variation in the amylose and lipid contents and some physical properties of rice starches. J. Cereal Sci. 5, 35–44.

Noosuk, P., Hill, S.E., Pradipasena, P., Mitchell, J.R., 2003. Structure-viscosity relationships for Thai rice starches. Starch 55, 337–344.

Norazalina, S., Norhaizan, M.E., Hairuszah, I., Norashareena, M.S., 2010. Anticarcinogenic efficacy of phytic acid extracted from rice bran on azoxymethane-induced colon carcinogenesis in rats. Exp. Toxicol. Pathol. 62, 259–268.

Oatway, L., Vasanthan, T., Helm, J.H., 2001. Phytic acid. Food Rev. Int. 17, 419–431.

Ooi, S.L., Mcmullen, D., Golombick, T., Nut, D., Pak, S.C., 2017. Evidence-based review of BioBran/MGN-3 arabinoxylan compound as a complementary therapy for conventional cancer treatment. Integr. Cancer Ther. https://doi.org/10.1177/1534735417735379 (14 pages).

Perdon, A.A., Siebenmorgen, T.J., Buescher, R.W., Gbur, E.E., 1999. Starch retrogradation and texture of cooked milled rice during storage. J. Food Sci. 64, 828–832.

Philpot, K., Martin, M., Butardo, V., Willoughby, D., Fitzgerald, M., 2006. Environmental factors that affect the ability of amylose to contribute to retrogradation in gels made from rice flour. J. Agric. Food Chem. 54, 5182–5190.

Pluske, J.R., Montagne, L., Cavaney, F.S., Mullan, B.P., Pethick, D.W., Hampson, D.J., 2007. Feeding different types of cooked white rice to piglets after weaning influences starch digestion, digesta and fermentation characteristics and the faecal shedding of beta-haemolytic Escherichia coli. Br. J. Nutr. 97, 298–306.

Priestley, R.J., 1976. Studies on parboiled rice: part I. comparison of the characteristics of raw and parboiled rice. Food Chem. 1, 5–14.

Rafe, A., Sadeghian, A., Hoseini-Yazdi, S.Z., 2017. Physicochemical, functional, and nutritional characteristics of stabilized rice bran form tarom cultivar. Food Sci. Nutr. 5, 407–414.

Ramesh, M., Ali, S.Z., Bhattacharya, K.R., 1999. Structure of rice starch and its relation to cooked-rice texture. Carbohydr. Polym. 38, 337–347.

Rani, M.R.S., Bhattacharya, K.R., 1989. Rheology of rice-flour pastes: effect of variety, concentration, and temperature and time of cooking. J. Texture Stud. 20, 127–137.

Reddy, K.R., Ali, S.Z., Bhattacharya, K.R., 1993. The fine structure of rice starch amylopectin and its relation to cooked rice texture. Carbohydr. Polym. 22, 267–275.

Reddy, K.R., Subramanian, R., Ali, S.Z., Bhattacharya, K.R., 1994. Viscoelastic properties of rice-flour pastes and their relationship to amylose content and rice quality. Cereal Chem. 71, 548–552.

Ren, H., Li, T., Wan, H., 2017. Optimization of extraction condition for phytic acid from peanut meal by response surface methodology. Resource-Efficient Technol. 3, 226–231.

Renuka Devi, R., Arumughan, C., 2007a. Antiradical efficacy of phytochemical extracts from defatted rice bran. Food Chem. Toxicol. 45, 2014–2021.

Renuka Devi, R., Arumughan, C., 2007b. Phytochemical characterization of defatted rice bran and optimization of a process for their extraction and enrichment. Bioresour. Technol. 98, 3037–3043.

Renuka Devi, R., Jayalekshmy, A., Arumughan, C., 2007. Antioxidant efficacy of phytochemical extracts from defatted rice bran in the bulk oil system. Food Chem. 104, 658–664.

Ryan, E.P., Heuberger, A.L., Weir, T.L., Barnett, B., Broeckling, C.D., Prenni, J.E., 2011. Rice bran fermented with Saccharomyces boulardii generates novel metabolite profiles with bioactivity. J. Agric. Food Chem. 59, 1862–1870.

Saunders, R.M., 1990. The properties of rice bran as a foodstuff. Cereal Foods World. 35 (632), 634–636.
Seo, M.Y., Chung, S.Y., Choi, W.K., Seo, Y.K., Jung, S.H., Park, J.M., Seo, M.J., Park, J.K., Kim, J.W., Park, C.S., 2009. Anti-aging effect of rice wine in cultured human fibroblasts and keratinocytes. J. Biosci. Bioeng. 107, 266–271.
Shoji, Y., Mita, T., Isemura, M., Mega, T., Hase, S., Isemura, S., Aoyagi, Y., 2001. A fibronectin-binding protein from rice bran with cell adhesion activity for animal tumor cells. Biosci. Biotechnol. Biochem. 65, 1181–1186.
Sun, Q., Xiong, C.S., 2014. Functional and pasting properties of pea starch and peanut protein isolate blends. Carbohydr. Polym. 101, 1134–1139.
Takeda, Y., Hizukuri, S., Juliano, B.O., 1986. Purification and structure of amylose from rice starch. Carbohydr. Res. 148, 299–308.
Takeda, Y., Shibahara, S., Hanashiro, I., 2003. Examination of the structure of amylopectin molecules by fluorescent labeling. Carbohydr. Res. 338, 471–475.
Takeo, S., Kado, H., Yamamoto, H., Kamimura, M., Watanabe, N., Uchida, K., Mori, Y., 1988. Studies on an antitumor polysaccharide RBS derived from rice bran. II. Preparation and general properties of RON, an active fraction of RBS. Chem. Pharm. Bull. 36, 3609–3613.
Tan, Y., Corke, H., 2002. Factor analysis of physicochemical properties of 63 rice varieties. J. Sci. Food Agri. 82, 745–752.
Tanigami, Y., Kusumoto, S., Nagao, S., Kokeguchi, S., Kato, K., Kotani, S., Shiba, T., 1991. Partial degradation and biological activities of an antitumor polysaccharide from rice bran. Chem. Pharm. Bull. 39, 1782–1787.
Tester, R., Morrison, W.R., 1990. Swelling and gelatinization of cereal starches. II. Waxy rice starches. Cereal Chem. 67, 558–563.
Tiwari, U., Cummins, E., 2011. Meta-analysis of the effect of β-glucan intake on blood cholesterol and glucose levels. Nutrition 27, 1008–1016.
Topping, D.L., Fukushima, M., Bird, A.R., 2003. Resistant starch as a prebiotic and synbiotic: state of the art. Proc. Nutr. Soc. 62, 171–176.
Vandeputte, G.E., Derycke, V., Geeroms, J., Delcour, J.A., 2003a. Rice starches. II. Structural aspects provide insight into swelling and pasting properties. J. Cereal Sci. 38, 53–59.
Vandeputte, G.E., Vermeylen, R., Geeroms, J., Delcour, J.A., 2003b. Rice starches. I. Structural aspects provide insight into crystallinity characteristics and gelatinisation behaviour of granular starch. J. Cereal Sci. 38, 43–52.
Vandeputte, G.E., Vermeylen, R., Geeroms, J., Delcour, J.A., 2003c. Rice starches. III. Structural aspects provide insight in amylopectin retrogradation properties and gel texture. J. Cereal Sci. 38, 61–68.
Villareal, C.P., Hizukuri, S., Juliano, B.O., 1997. Amylopectin staling of cooked milled rices and properties of amylopectin and amylose. Cereal Chem. 74, 163–167.
Waglay, A., Karboune, S., Alli, I., 2014. Potato protein isolates: recovery and characterization of their properties. Food Chem. 142, 373–382.
Wang, J., Suo, G., De Wit, M., Boom, R.M., Schutyser, M.A.I., 2016a. Dietary fibre enrichment from defatted rice bran by dry fractionation. J. Food Eng. 186, 50–57.
Wang, L., Li, Y., Zhu, L., Yin, R., Wang, R., Luo, X., Li, Y., Li, Y., Chen, Z., 2016b. Antitumor activities and immunomodulatory of rice bran polysaccharides and its sulfates in vitro. Int. J. Biol. Macromol. 88, 424–432.
Webber, A., Hettiarachchy, N.S., Webber, D.M., Sivarooban, T., Horax, R., 2014. Heat-stabilized defatted Rice bran (HDRB) as an alternative growth medium for Saccharomyces cerevisiae. J. Food Nutr. 1, 1–6.
Wei, Z., Xiao, X., Wang, J., Wang, H., 2017. Identification of the rice wines with different marked ages by electronic nose coupled with smartphone and cloud storage platform. Sensors (Basel) 17, 2500.

Woodward, J.R., Phillips, D.R., Fincher, G.B., 1983. Water-soluble (1→3), (1→4)-β-d-glucans from barley (Hordeum vulgare) endosperm. I. Physicochemical properties. Carbohydr. Polym. 3, 143–156.

Wu, P., Tian, J.-C., Walker, C.E., Wang, F.-C., 2009. Determination of phytic acid in cereals—a brief review. Int. J. Food Sci. Technol. 44, 1671–1676.

Xu, Q., Kanthasamy, A.G., Reddy, M.B., 2008. Neuroprotective effect of the natural iron chelator, phytic acid in a cell culture model of Parkinson's disease. Toxicology 245, 101–108.

Yang, Y., Xia, Y., Wang, G., Yu, J., Ai, L., 2017. Effect of mixed yeast starter on volatile flavor compounds in Chinese rice wine during different brewing stages. LWT—Food Sci. Technol. 78, 373–381.

Yeom, H.J., Lee, E.H., Ha, M.S., Ha, S.D., Bae, D.H., 2010. Production and physicochemical properties of rice bran protein isolates prepared with autoclaving and enzymatic hydrolysis. J. Korean Soc. Appl. Bio. Chem. 53, 62–70.

Zhang, H.W., Bai, X.L., 2014. Optimization of extraction conditions for phytic acid from rice bran using response surface methodology and its antioxidant effects. J. Food Sci. Technol. 51, 371–376.

Zhang, Z., Smith, C., Li, W., 2014. Extraction and modification technology of arabinoxylans from cereal by-products: a critical review. Food Res. Int. 65, 423–436.

Zhong, J., Ye, X., Fang, Z., Xie, G., Liao, N., Shu, J., Liu, D., 2012. Determination of biogenic amines in semi-dry and semi-sweet Chinese rice wines from the Shaoxing region. Food Control 28, 151–156.

Zhuang, X., Zhao, C., Liu, K., Rubinelli, P.M., Ricke, S.C., Atungulu, G.G., 2018. Cereal grain fractions as potential sources of prebiotics: current status, opportunities, and potential applications. In: Ricke, S.C., Atungulu, G.G., Rainwater, C.E., Park, S.H. (Eds.), Food and Feed Safety Systems and Analysis. Academic Press, Cambridge.

CHAPTER 11

Rice Bran Protein: Extraction, Nutraceutical Properties, and Potential Applications

Yan Zheng, Nisi Gao, Juan Wu, Baoru Yin
Wilmar Global Research and Development Center, Shanghai, China

1. INTRODUCTION

Rice (*Oryza sativa*) is one of the world's most important cereals, acting as a staple food for a large section of the world's population, especially Asian populations (Bird et al., 2000). World rice production in 2008 was ~661 million metric tons (USDA, 2009). Milling of paddy rice yielded nearly 70% of rice (endosperm) as its major product, and some unconsumed byproducts, such as rice husk (20%), rice bran (8%), and rice germ (2%) (Kim et al., 2011). Rice bran is by far the most nutritious part of the rice kernel. The composition of rice bran is 15–22% lipids, 34.1–52.3% carbohydrates, 7–11.4% fiber, 6.6–9.9% ash, 8–12% moisture, and 10–16% highly nutritional protein (Faria et al., 2012). Based on rice bran's current production, there is a staggering 5 million metric tons of rice bran protein isolate potentially "harvested" from its base bran materials.

Recently, rice bran has attracted considerable attention owing to its high nutritional properties, such as well-balanced amino acids, hypoallergenicity, and hypocholesterolemic and hypolipidemic effects, as well as high anticancer activity, etc. (Ni et al., 2003; Yeom, 2009; Zhou et al., 2012). Besides this, rice bran protein is free from lactose, gluten, and soybeans, which renders it a good marketing opportunity in the nonallergic and healthier food sector. Rice protein is suggested as one of the most important plant-based proteins, which can be applied or used as an ingredient in many products such as infant food and gluten-free products.

There are several reports on extraction techniques of rice bran protein, including homogenization, colloid milling, microwave, ultrasonic, and enzymatic methods (Silpradit et al., 2010; Tang et al., 2002, 2003a). However, as of now, commercial rice bran protein is still rare in the market. This chapter is aimed at providing valuable discussions on rice bran protein,

Rice Bran and Rice Bran Oil
https://doi.org/10.1016/B978-0-12-812828-2.00011-1

including its various nutritional and functional properties, its application, and extraction methods for development of commoditized rice bran protein product.

2. NUTRACEUTICAL PROPERTIES AND HEALTH BENEFITS OF RICE BRAN PROTEIN

With the increasing awareness of health benefits of plant proteins in recent years, rice bran protein has gained significant attention for its nutraceutical properties including superior protein quality, hypoallergenicity, and various biological activities.

2.1 Protein Quality

Protein quality is determined by its essential amino acid composition, digestibility, and bioavailability of its amino acids. The Food and Agricultural Organization (FAO) and World Health Organization (WHO) have recommended the use of the Protein Digestibility-Corrected Amino Acid Score (PDCAAS) for evaluation of protein quality (WHO Technical Report Series 935, 2007). Lysine is the first limiting amino acid in cereal grain proteins (Young and Pellett, 1994; Chuanlai, 2010). Rice bran protein has been reported to have a higher lysine content as compared to oat, maize, and wheat protein (Juliano, 1985; Wang et al., 2015; Amagliani et al., 2017; Mota et al., 2016). Besides amino acid composition, protein quality can also be reflected in terms of efficiency of protein utilization based on true digestibility (TD) and biological value (BV). Rice bran protein and casein had similar TDs of 94.8, which is higher in comparison to whey protein isolate (92.8), soy protein isolate (91.7), and rice endosperm protein (90.8). Although rice bran protein has a lower biological value (BV) as compared to whey protein isolate, its BV is higher than casein and soy protein isolate. Due to general lysine deficiency in plant protein, PDCAAS value was lower for rice protein as compared to dairy protein ingredients. However, PDCAAS value of rice bran protein was comparable to that of plant-based soy protein (Han et al., 2015).

2.2 Hypoallergenicity

Rice is generally perceived as hypoallergenic (Gastanaduy et al., 1990; Caffarelli et al., 2010; Reche et al., 2010). In the case of a suspected food allergy testing, rice is the only grain allowed to be consumed in the elimination diet (Van Hooser and Crawford, 1989). Two proteins (14–16 and 33 kDa) that showed positive reaction with IgE antibodies were related to the rare case of cooked rice allergy (Usui et al., 2001; Matsuda et al., 2006). However, food allergy caused by rice bran protein isolates or concentrate has never been reported. This is mainly due to remarkable reduction of major allergic proteins through alkaline or enzymatic extraction steps (Ikezawa et al., 1999; Watanabe et al., 1990). In comparison to the most common allergenic source in infant formulations and weaning food, such as milk, wheat, and soy, rice bran protein is free from lactose, gluten, and soy proteins. Therefore extracted rice bran protein is a safe ingredient and can be used for the development of hypoallergenic formulas.

2.3 Health Benefits of Rice Bran Protein

Health benefits of rice bran protein are mainly connected to the isolated protein and peptides that are an important group of bioactive molecules involved in physiological functions. The most reported biological activities include antioxidant capacity, antihypertensive effect, antidiabetic activity, anticancer activity, and cholesterol-lowering effect.

2.3.1 Antioxidant Capacity

Antioxidant peptides derived from rice bran protein have strong free radical scavenging activity, which have been reported to be effective in preventing a number of age-related diseases ranging from cancer to Alzheimer's disease (Mills et al., 2004; Liu et al., 2016). It has been demonstrated that oxidative stress can be suppressed by rice protein in a series of in vivo studies (Fan et al., 2008; Hopps et al., 2010; Leopold and Loscalzo, 2009; Liu et al., 2016). Two major biomarkers, namely malondialdehyde (MDA) and protein carbonyl (PCO), were significantly reduced indicating lipid peroxidation and protein oxidation status were suppressed (Yang et al., 2012). Because hydrogen donors are abundant in rice bran protein hydrolysates, they can scavenge free radicals and terminate free radical reaction chains, thus achieving antiaging functions (Fan et al., 2008).

2.3.2 Antihypertensive Effect

Angiotensin-converting enzyme (ACE) is an enzyme that helps regulate blood pressure and acts on the rennin-angiotensin system (Iroyukifujita et al., 2000). The inhibition of ACE is usually associated with antihypertensive effect as the inhibitors can relax constricted blood vessels and lower blood pressure (Liu et al., 2017b). Rice bran protein hydrolysate with molecular weight (Mw) <4 kDa exhibited the strongest ACE inhibitory activity (Wang et al., 2017). Among four rice bran protein fractions, the hydrolysates of albumin and glutelin had higher inhibitory activities than that of globulin and prolamin (Uraipong and Zhao, 2016). A tripeptide with sequence of Try-Ser-Lys isolated from rice bran protein hydrolysate was found to be related to formation of very strong hydrogen bonds within the active pockets of human ACE. Thus rice bran protein and its hydrolysates are promising ingredients for use as therapeutics in prevention and treatment of hypertension (Liu et al., 2017b).

2.3.3 Antidiabetic Effect

The hydrolysates of defatted rice bran protein isolates obtained by two commercial proteases (Alcalase and Protamax) showed strong inhibition effect on α-amylase and α-glucosidase (Uraipong and Zhao, 2016). The inhibitory activity was found to be comparable in magnitude to the standard antidiabetic drug acarbose indicating rice bran protein hydrolysates had the potential to be developed into a dietary or nutraceutical supplement for management of diabetes (Uraipong and Zhao, 2016). An in vitro study showed, under human digestive conditions, the rice bran protein hydrolysates produced by digestive enzymes also exerted strong α-glucosidase inhibitory activities, suggesting consumption of rice bran proteins can potentially lead to generation of antidiabetic peptides (Uraipong and Zhao, 2018).

2.3.4 Anticancer Activity

Rice bran peptide fractions demonstrated in vitro potential activities against viability and proliferation of colon, breast, and liver cancer cells. Heat-stabilized defatted rice bran was treated with Alcalase to prepare hydrolysates and peptides. The gastrointestinal-resistant fractions were isolated after simulated gastric and intestinal juices treatment. When treated with the fraction (Mw <5 kDa) at 500 μg/mL, it showed significant growth inhibitory and cytotoxic effects on human colon and breast cancer cell lines (Kannan et al., 2009). After purification and characterization of the peptides from rice bran protein hydrolysates, pure peptide with amino acid sequence of Glu-Gln-Arg-Pro-Arg showed 84% inhibition on colon cancer cell growth, 80% inhibition on breast cancer cell growth, and 84% inhibition on liver cancer cells growth (Kannan et al., 2010). These results suggested that the functional pentapeptide derived from rice bran protein can be purified and developed as an antitumor agent for treating colon, breast, and liver cancer.

2.3.5 Cholesterol-Lowering Effect

The hypocholesterolemic action of rice bran protein could be caused by bile-acid binding proteins (Wang et al., 2015). The mechanism is based on sequestrant factors binding to bile acids resulting in formation of insoluble complexes, which are excreted in feces. To compensate for the excreted bile acids, more cholesterol is converted to bile acids in the liver. It promotes the expression of the low-density lipoprotein (LDL) receptor, which subsequently lowers the serum LDL-cholesterol level. Wang et al. (2015) reported OsJ_13801 protein in rice bran exerted strong ability to bind taurocholate and inhibit the micellar solubility of cholesterol, compared with casein.

3. EXTRACTION OF RICE BRAN PROTEIN

Various approaches for rice bran protein extraction have been documented since the 1970s (Chen and Houston, 1970). Many researchers reported that proteins in rice bran have low solubility profile, induced by the presence of extensive disulfide bonding, glycosylation, aggregation, and entanglements (Hamada, 1997; Gupta et al., 2008; Fabian and Ju, 2011; Phongthai et al., 2017b). Thus it is required to disrupt these crosslinking and aggregations before they can be totally solubilized and extracted. Here in this section, the main procedures reported for extraction of rice bran protein including alkali extraction, addition of dissociating agents, enzymatic hydrolysis, physical treatment, and subcritical water treatment will be introduced.

3.1 Alkali Extraction

Alkali extraction is the most common method used to produce rice bran protein, and Fig. 1 shows the typical extraction process. Nevertheless, high concentration of sodium hydroxide (NaOH) can cause hydrolysis of hydrogen, and amide and disulfide bonds in protein aggregations (Phongthai et al., 2017b).

Table 1 summarizes the major results reported in literature for alkali extraction of rice bran protein. It was reported that the yield of rice bran protein increased with extraction temperature and pH (Prakash and Ramanatham, 1994; Gupta et al., 2008; Sari et al., 2015). In terms of pH, protein purity increased linearly from pH 7.5 to 11, but decreased as pH reached 12, which was due to some nonprotein nitrogen solubilization at higher pH (Jiamyangyuen et al., 2005). The emulsion capacity also decreased with extraction temperature (Gupta et al., 2008).

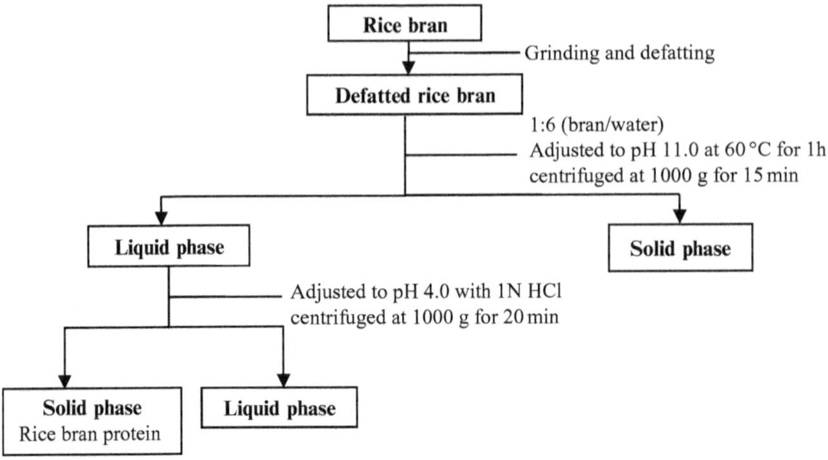

Fig. 1 Flow chart for alkali extraction of rice bran protein.

Table 1 Alkali methods for extracting rice bran protein

Condition	Source	Yield (%)	Purity (%)	Reference
pH 11	Defatted rice bran	37	85	Chen and Houston (1970)
pH 9 at 24°C	Full fat rice bran	14–20	23–31	Connor et al. (1976)
pH 6.0 for 30 min then pH 4.0	Defatted rice bran	25	–	Prakash and Ramanatham (1994)
pH 11 for 30 min then pH 4.0	Defatted rice bran	70	–	Prakash and Ramanatham (1994)
pH 9.5 for 58 min	Defatted rice bran	9.6	72.6	Jiamyangyuen et al. (2005)
pH 12 for 45 min	Defatted rice bran	12.2	70.43	Jiamyangyuen et al. (2005)
pH 10 for 1 h at 50–55°C, then pH 4.0	Defatted rice bran	–	86.2	Bandyopadhyay et al. (2008)
pH 9.5 at 30–75°C for 1 h, 1:10	Defatted rice bran	21.2–48.1	71.0–79.9	Gupta et al. (2008)
pH 11 for 60 min at 60° C, then pH 4.5	Defatted rice bran	13.2	37.6	Yadav et al. (2011)
pH 9.5 for 2 h at 50°C, then pH 3.8	Heat-stabilized defatted rice bran	32.9		Zhang et al. (2012)

Exposure of protein to extreme alkali and heat could also bring several other disadvantages: (1) severe degradation and denaturation of protein, (2) undesirable flavor and color, (3) loss of nutritional and functional value, and (4) toxic compounds such as lysinoalanine (Kelly et al., 1982; Ansharullah et al., 1997; Hamada, 1997; Sari et al., 2015). Although commercially available soybean protein isolation (SPI) is normally produced by alkali extraction, the drawback of rice bran protein preparation from alkali extraction is obvious. Therefore further research is required to explore effective methods for improving nutritional and functional properties of rice bran protein.

3.2 Enzymatic Extraction

Compared with the alkali method, enzymatic treatment is an alternative way to extract rice bran protein at mild conditions, which not only retain the nutritional characteristics but also improves the functional and sensory properties of rice bran protein. There are several kinds of enzymes to assist isolation, including carbohydrase, protease, and phytase, which can break down the interaction or bonding between proteins and other components such as starch, fiber, and phytic acid. The results of enzymatic extraction are summarized in Table 2.

Table 2 Enzymatic extraction for rice bran protein

Enzyme	Condition	Yield (%)	Purity (%)	Reference
Protease				
OpticleanL-100	4%, pH 10 and 40°C	63–93 with 0%–10% DH		Hamada (1999)
Alcalase 2.4L	0.5%, pH 8.0 and 50°C	81.4 with 7.5% DH	27.6	Hamada (2000)
Flavourzyme	0.5%, pH 8.0 and 50°C	87.6 with 8.8% DH	29.9	Hamada (2000)
Alcalase 0.6L	1%, pH 9.0 and 50°C for 2 h	68		Hanmoungjai et al. (2002)
Bromelain	6% (*w*/w), pH 9.0 and 50°C for 30 min	73.3		Apinunjarupong et al. (2009)
Alcalase 2.4L	250 μ/g, pH 10 and 50°C for 2 h	44.8		Zhang et al. (2012)
Papain	0.1%, pH 7 and 50°C for 1 h	87.0		Bandyopadhyay et al. (2008)

Continued

Table 2 Enzymatic extraction for rice bran protein—cont'd

Enzyme	Condition	Yield (%)	Purity (%)	Reference
Carbohydrases				
Viscozyme L	pH 3.8, 50°C for 5 h	57.9		Ansharullah et al. (1997)
Celluclast 1.5L	pH 3.8, 50°C for 5 h	25.9		Ansharullah et al. (1997)
Phytase	pH 5.0, 55°C for 2 h	57.3	80.2	Wang et al. (1999)
Xylanase	pH 5.0, 55°C for 2 h	54.5	81.7	Wang et al. (1999)
α-Amylase	0.15%, pH 6.5, 90°C for 45 min	48.9	24.8	Shih et al. (1999)
Hemicellulase C	0.5%, pH 4, 55°C for 2 h	53.3	14.3	Shih et al. (1999)
Cellulase AC	0.2%, pH 4, 55°C for 2 h	48.0	18.9	Shih et al. (1999)
α-Amylase	1%, pH 5.0, 30°C	45.4		Tang et al. (2003a)
Celluclast	1%, pH 5.0, 50°C	12.1		Tang et al. (2003a)
Complex enzyme				
Phytase and xylanase	1:7.5, pH 5.0, 55°C for 2 h	74.6	92.0	Wang et al. (1999)
Amylase and protease	1:10, amylase at pH 6.5, 45°C for 3.5 h, then protease at 50°C for 2 h	60.9		Tang et al. (2002)
Celluclast and alcalase 0.6L	1% cellulase for 1 h, then 1% protease at pH 9.0 for 1 h	50		Hanmoungjai et al. (2002)

Proteases have been used effectively in hydrolyzing rice bran protein for better solubility and greater extractability (Hamada, 1999, 2000; Tang et al., 2003a). Hamada (1999) reported that extractability of rice bran protein increased with the use of protease from about 60% to 93% with increasing degree of hydrolysis. Proteases can be also used in coextraction of oil and protein simultaneously. Hanmoungjai et al. (2001) used a commercial protease to get maximal extraction yields of oil and protein with 79% and 68%, respectively.

Carbohydrases are cell wall-breaking enzymes, including hemicelluloses, celluloses, α-amylase, and pectinase, or its combination, which can release proteins from the polysaccharide matrix (Ansharullah et al., 1997; Wang et al., 1999; Shih et al., 1999; Tang et al., 2003a). Tang et al. (2003a) reported that amylase, viscozyme, and celluclast can extract a maximum of 45.4%, 28.5%, and 12.1% rice bran protein, respectively. Compared with protease, carbohydrases had less efficient protein extractability, which suggest degradation of cellulosic material in bran was not enough to liberate the protein (Hanmoungjai et al., 2002).

Several studies reported a combination of proteases and carbohydrolases to maximize protein extraction (Ansharullah et al., 1997; Wang et al., 1999; Hanmoungjai et al., 2002). A study by Wang et al. (1999) shows highest protein content (92.0%) and yield (74.6%) can be obtained from rice bran by a combination of xylanase and phytase, and the extracts had similar foaming properties to egg white. The addition of protease in rice bran suspension contained cellulase, hemicellulase, pectinase, or viscozyme, which also helped to increase protein yield by 30%–40% (Hanmoungjai et al., 2002).

3.3 Sequential Solvent Extraction and Dissociating Agent

Proteins in rice bran can be classified into albumin, globulins, prolamins, and glutelins based on their solubility in water, salt, alcohol solution, and acids, respectively. Research by Hamada (1997) evaluated extraction efficiency of these mild solvents for rice bran protein. Rice bran proteins were sequentially fractionated into 34% albumins, 15% globulins, 6% prolamins, and 11% glutelins by using water, 2% NaCl, 70% alcohol, and 0.1 M acetic acid, respectively. The remaining insoluble fractions were composed of high molecular weight polypeptides with strong aggregations and crosslinking. Wang et al. (2014) also obtained rice bran protein with 13.8% yield by using 0.4 M NaCl, 80% ethanol, and 0.01 M NaOH as extraction solvents. It was revealed that sequential fractionation by mild solvents had low efficiency for rice bran protein extraction, and multiple steps for sequential extraction are also not feasible on an industrial scale.

Addition of dissociating and disulfide breaking agents, such as mercaptoethanol, dithiothreitol, sodium sulfite, sodium dodecyl sulfate (SDS), urea (<8 M), sodium stearate, and cetyltrimethylammonium bromide, have been also been reported to enhance protein extractability (Prakash and Ramanatham, 1994; Fabian and Ju, 2011). For example, 0.01 M SDS increased the yield of rice bran protein to 64% (pH 7.0) and 90% (pH 12.0), but it was hard to remove SDS from final extracts (Prakash and Ramanatham, 1994). Hamada (1997) reported that about 79% proteins can be recovered from defatted rice bran using strong dissociation solvent, AUC (0.1 M acetic acid, 3 M urea, and 0.01 M cetyltrimethylammonium bromide), which was capable of extracting all soluble portions as well as partly aggregated proteins. However, most of the dissociating agents used were not approved for food use, thus there is a need to explore food grade agents for extraction of rice bran protein used in food applications.

3.4 Physical Methods

Common physical treatments applied for protein extraction include microwave, colloid milling, homogenizing, high speed blending, freeze-thaw, high pressure, and sonication (Anderson and Guraya, 2001; Tang et al., 2002; Phongthai et al., 2017b).

Colloid milling and homogenizing reduced particle size of rice bran and influenced the distribution of protein in the recovered products (Anderson and Guraya, 2001). Prakash and Ramanatham (1994) reported that milling and differential sieving of rice bran increased the protein extractability by 15%. Sonication or ultrasonic can disrupt cell walls and molecular bonds in that high temperature and shock waves result in cavitation collapse of bubbles generated from ultrasound (Tang et al., 2002). A study by Tang et al. (2002) evaluated extraction efficiency of several physical methods, such as freeze-thaw, sonication, high-speed blending, and high pressure. It was found that yield of protein from heat-stabilized defatted rice bran was about 11%–16%.

Although physical techniques are easier and more economical to be industrialized, the protein recovery was obviously lower as compared to previous chemical and enzymatic extractions. This indicated that physical processes are not enough to disrupt the extensive aggregations in the proteins. For this reason, it is suggested that physical processing in combination with other methods, such as enzyme and alkali treatments, can be more effective to extract protein from rice bran (Tang et al., 2002). The relevant results have been listed in Table 3.

Different pretreatments of rice bran, such as defatted, acid stabilized, and heat stabilized and parboiled, also had a marked effect on protein extractability and precipitation. Research by Prakash and Ramanatham (1994) showed

Table 3 Physical treatment for rice bran protein extraction

Physical treatment	Condition	Yield (%)	Purity (%)	Reference
Milling	Milled defatted rice bran (150 μm) at pH 11 for 30 min	69.1	11.8	Prakash and Ramanatham (1994)
Acid-stabilization	Acid-stabilized defatted rice bran at pH 12 for 30 min	76	71.4	Prakash and Ramanatham (1994)
Parboiling	Parboiled defatted rice bran at pH 12 for 30 min	25	54.7	Prakash and Ramanatham (1994)
Heat-stabilization	Heat-stabilized defatted rice bran at pH 12 for 30 min	20	38.9	Prakash and Ramanatham (1994)
Colloidal milling and homogenization	7500 rpm for milling and homogenization pressure of 1.7×10^4 kPa for full-fat unstabilized rice bran	38.2	26.7	Anderson and Guraya (2001)
Freeze-thaw	Freeze for 16 h	12.0		Tang et al. (2002)
Sonication	0%–100% output of 750 W	9.5–13.5		Tang et al. (2002)
High pressure	0–800 MPa	10.5–11.1		Tang et al. (2002)
Hydrothermal cooking	150°C for 60 s at pH 9.0	50.0	52.6	Xia et al. (2012)
Freeze-thaw, sonication and amylase		44.5		Tang et al. (2002)
High-speed blending, amylase and protease		65.9		Tang et al. (2002)
Hydrothermal cooking and amylase	150°C for 60 s at pH 9.0	50	74.1	Xia et al. (2012)

protein solubility and precipitation ratio in various treated rice bran increased in the following order: Untreated > Acid stabilized > Parboiled > Heat-stabilized. And full-fat unstabilized bran had the highest soluble protein content and extractability as compared with full-fat stabilized or defatted rice bran (Anderson and Guraya, 2001). It was indicated that pretreatment involving heat may lead to denaturation and poor extractability of rice bran protein, which should be avoided during isolation process.

3.5 Subcritical Water

Subcritical water is hot water capable of maintaining its liquid state at a temperature ranging from 100°C to 374°C under high pressure. Due to the low dielectric constant and high polarity at subcritical points, it can be utilized as a suitable solvent for extraction and hydrolysis of biomass (Pourali et al., 2009; Fabian and Ju, 2011). A summary of the reported literature on rice bran protein extraction by subcritical water treatment is presented in Table 4.

Protein and amino acid yields generally increased with temperature (Sereewatthanawut et al., 2008; Hata et al., 2008). A study by Sereewatthanawut et al. (2008) shows the highest protein yield (219 ± 26 mg/g rice bran) and amino acid yield (8.0 ± 1.6 mg/g rice bran) can be obtained at 200°C for 30 min. Pourali et al. (2009) found that protein in rice bran was converted to water-soluble amino acids and further hydrolyzed to organic acid above 190°C.

Table 4 Subcritical water treatment for rice bran protein

Condition	Yield (%)	Purity (%)	Reference
200°C for 5 min	0.2 g protein/g bran		Wiboonsirikul et al. (2007)
200°C for 30 min	0.219 g protein/g bran and 8 mg amino acid/g bran		Sereewatthanawut et al. (2008)
240°C for 5 min	0.222 g protein/g bran	0.607 g/g extract	Chiou et al. (2010)
230°C for 5 min		0.33 mg protein/g extract	Chiou et al. (2013)
220°C	130 mg protein/g bran		Watchararuji et al. (2008)

The extracts by subcritical water treatment had been reported to exhibit high radical scavenging, antioxidative, emulsifying, and emulsion-stabilizing activities (Hata et al., 2008; Chiou et al., 2010, 2013; Wiboonsirikul et al., 2007), which indicated that subcritical water extraction was a promising way to improve functional properties of rice bran.

In summary, researchers are striving to seek environmentally friendly and feasible procedures for extraction and commercialization of rice bran protein using a combination of different traditional methods or novel techniques. Proteins as key ingredients in many food systems provide nutrition and functionalities to final products. Many nutritional functional properties rely on solubility of protein, such as emulsification, foaming properties, and bioactivity. The procedures for extracting protein from rice bran must be carefully selected to produce protein concentrates and isolates with desirable functional properties.

4. RICE BRAN PROTEIN APPLICATION

Apart from nutritional properties, rice bran protein as ingredient should also present suitable functionality such as solubility, water/oil holding ability, and emulsifying ability to better meet application requirements and further extend the application area (Esmaeili et al., 2016; Tang et al., 2003b; Fabian and Ju, 2011). Functional properties of rice bran protein are reported to be varied by different extraction methods, thus the influence of extraction conditions on the functionalities of rice bran proteins are reviewed and listed in Table 5 to better guide the protein application.

The application of rice bran protein has been systematically reviewed by Prakash (1996) and Ali et al. (2010), respectively. Hence, the latest development regarding rice bran protein application and commercialized products will be discussed here.

4.1 Protein Supplement

Due to great hypoallergenicity, rice bran proteins have huge potential for use as an ideal protein ingredient for infant formulations, weaning foods, and for the restricted formulations of children and adults with dysfunctions, such as lactose intolerance or celiac disease (Fabian and Ju, 2011; Khan et al., 2011). However, a concern should be noted that rice, especially rice bran, is reported to have a large amount of arsenic, and infants and children may be more vulnerable to arsenic exposure (Sun et al., 2008). The major species of total arsenic (tAs) in rice and rice bran is inorganic arsenic (iAs), which has

Table 5 Functional properties of rice bran protein as a function of extraction and modification methods

Products	Extraction, modification	Functionalities	References
Rice bran protein concentrate	Alkaline extraction (pH 9.5) and isoelectric precipitation (pH 4.5)	Maximum nitrogen solubility at pH 8.0 (URBPC 71.5%, SRBPC 50.9%)	Gnanasambandam and Heltiarachchy (1995)
Rice bran protein concentrate	Alkaline extraction (pH 9) and isoelectric precipitation (pH 4.5)	NSI: 47%–73%, water absorption: 3.87–5.6 g/g, oil absorption: 3.74–9.18 g/g, bulk density: 0.12–0.21 g/mL; Foaming stability: half-life of 42.6 h at 15% sugar concentration, emulsion stable at different pH, salt and sugar concentration	Chandi and Sogi (2007)
Rice bran protein hydrolysate	Flavourzyme (F) or alcalase (A) hydrolysis	Solubility, emulsifying activity (EA) and emulsifying ability (ES) of F hydrolysate was greater than A hydrolysate; EA of F hydrolysate was similar to casein and BSA at pH 7; ES of F hydrolysate was similar to casein and better than BSA at pH 5	Hamada (2000)
Rice bran protein concentrate	Alkaline extraction (pH 9) at 30–75°C and isoelectric precipitation (pH 4.5)	Water/oil adsorption capacity: 1.1–2.20 g/g, 1.64–6.89 g/g. Maximum foaming capacity at 60° C, emulsifying capacity 33.71%–70.85%	Gupta et al. (2008)

Rice bran protein	Precipitation of protein using carrageenan (2:1) or alginate (1:1) at pH 3.5	Foaming capacity for alginate-protein precipitate was similar to rice bran protein extract from water or subcritical extraction	Fabian et al. (2010)
Rice bran protein isolate	Autoclaving at 120°C, Alkaline extraction (pH 9) and isoelectric precipitation (pH 4), enzyme hydrolysis	Solubility: 97.4% at pH 10, elevated emulsion activity above pH 6, increased foaming capacity but reduced foaming stability	Yeom (2009)
Rice bran protein concentrate	Alkaline extraction (RBP1) or alcalase 2.4L hydrolysis (RBP2)	Solubility: RBP1 72.5%, RBP2 84.6% at pH 11; emulsifying/foaming capacity: RBP1 0.149/98%, RBP2 0.634/115%; water/oil absorption: RBP1 3.71/4.24 g/g, RBP2 4.4/5.13 g/g	Zhang et al. (2012)
Rice bran protein concentrate	Alkaline extraction from rice bran with different cultivars	Lower emulsifying ability compared to BSA, similar foaming ability to egg white. Tarom isolates had higher solubility, emulsifying and foaming stability to Shiroodi isolates	Esmaeili et al. (2016)
Rice bran protein concentrate	Microwave-assisted extraction, alcalase hydrolysis	Improved functionality including foaming ability and emulsion activity index at DH 5.04%	Phongthai et al. (2016)
Rice bran protein	Defatted, alkaline extraction and high pressure treatment	Elevated solubility and oil adsorption at 200 MPa, maximum water absorption and foaming capacity at 500 MPa, maximum emulsion stability and surface hydrophobicity at 400 MPa	Zhu et al. (2016)

been classified as a group 1 carcinogenic to humans (Meharg and Zhao, 2012; Munera-Picazo et al., 2014). The tAs and iAs of rice-based infant/baby foods are substantially elevated. Munera-Picazo et al. (2014) reported the tAs and iAs content of rice bran and rice bran-soluble products in the United States and Japan in the range of 0.71–1.98 and 0.61–1.88 mg/kg, respectively. Rice bran solubles are marketed as a supplement to malnourished children in international aid programs (Sun et al., 2008). Therefore a variety of toxicological analyses should be conducted to prove the safety of the products before their incorporation into a child's diet.

Furthermore, peptides derived from rice bran proteins have been reported to possess improved solubility and emulsifying capability, as well as the satisfying physiological function including antioxidant (Taniguchi et al., 2017; Wang et al., 2017), anticancer (Chalamaiah et al., 2018), antidiabetic capacity (Boonloh et al., 2015), and hypolipidemic functions (Zhang et al., 2016), thus they can be potentially used as other protein peptides in healthy supplements (Ali et al., 2010) and in special dietary foods.

4.2 Rice Bran Proteins as Bioactive Compounds Carrier

Food proteins, including rice bran proteins, can be used as carriers for bioactive compounds via hydrogen bonding and hydrophobic interaction (Santiago and Castro, 2016). Shi et al. (2017) reported that incorporation of rice bran protein, such as globulin and albumin, to tea catechins can enhance stability of tea catechin during digestion. Globulin and albumin are two essential components for the incorporation. Liu et al. (2017a) isolated albumin from rice bran and further fabricated curcumin-loaded albumin nanoparticles, which demonstrated enhanced in vitro stability and in vivo bioactivity. Besides that, rice bran protein could serve as a natural colorant carrier for homogeneous distribution of hydrophobic colorant in food systems (Umer Abdullah et al., 2008; Ali et al., 2010).

4.3 Flavor Enhancer

The high content of asparagine and glutamine in rice bran protein (deamidated protein hydrolysates) could also serve as flavor-enhancing ingredients in various food systems such as soup, sauce, and poultry. Hamada et al. (1998) developed a preparative separation scheme for isolation of glutamine-enriched peptides as flavor-enhancing agents from rice bran protein using protease. Kaewka et al. (2009) reported a rice bran protein hydrolysate with an intensified salty aroma. Arsa and Theerakulkait (2015) proved that the intensity of sweet, cocoa, or milk-like aroma of alcalase-hydrolyzed rice bran protein could be further improved by sugar addition and spray-drying.

4.4 Antiretrogradation

Apart from the aforementioned application, Niu et al. (2017) discovered that rice bran protein hydrolysate (protamex hydrolysis for 1 h) at a hydrolysis degree of 15.1% could effectively suppress retrogradation of gelatinized rice starch and thus has the potential to serve as a natural alternative for maintaining quality of starch-based foods (Arsa and Theerakulkait, 2015).

4.5 Rice Bran Proteins in Cereal Products

Application of rice bran proteins in bakery products has been reviewed by Ali et al. (2010). In addition to that, Yadav et al. (2011) prepared rice bran protein concentrate (RBPC) with satisfying water and oil adsorption capacity of 2.9 and 2.3 mL/g using alkaline extraction, and further developed a two-fold protein-increased biscuit with desirable overall acceptability using RBPC. Recently, gluten-free products have met increasing demand by consumers due to a deeper understanding of gluten-related celiac disease (Naqash et al., 2017). Plant protein, including rice bran protein, received increased attraction as a food ingredient in the production of gluten-free bakery products, pasta, and rice noodles. Phongthai et al. (2016) reported that gluten-free bread had improved specific volume, pore size, gas retention, and shelf life by adding 2% rice bran protein concentrate. The same research group also prepared protein-enriched gluten-free pasta with whey protein concentrate, egg albumen, rice bran protein concentrate (RBPC), and soy protein concentrate. RBPC-based gluten-free pasta presented the highest cooking loss but a cracked and noncontinuous surface (Phongthai et al., 2017a).

4.6 Rice Bran Protein-Based Film

In addition, the renewable rice bran proteins have been investigated for the fabrication of biodegradable film (Wang et al., 2012). Adebiyi et al. (2008) prepared a rice bran protein-based film with comparable functional properties to the soybean protein-based film. Shin et al. (2011) reported a 4% rice bran protein/4% gelatin composite film with extensively enhanced mechanical properties. Schmidt et al. (2014) reported an optimized rice bran protein film with the addition of phenolic extract.

4.7 Rice Bran Protein-Related Commercial Products

An increasing number of rice bran protein-based products have been launched on the market within the past few years. Some companies such as Ricebran Technologies (RBT) and Tsuno conducted clinical trials to verify the physiological function of rice bran protein-related products. RBT has developed a series of stabilized rice bran products and rice bran-derivative products including Risolubles, Ribalance, and Rifiber based on their proprietary stabilizing process and patent processing (Rukmini et al., 2003; McPeak et al., 2001). Among these, RiceBran Technologies' Proryza Gold is a hypoallergenic rice bran fiber and protein-based product that is suitable as a nutritive ingredient in healthy and nutritious food (RiceBran Technologies, n.d.).

5. CLOSING REMARKS

This chapter has presented a review of rice bran protein including their preparation methods, and nutritional and functional properties. Applications of rice bran proteins in food industry, pharmaceutical, and cosmetic areas are also presented. Even though rice bran proteins have been known for many years, a lot of work has been done on its nutritional and functional properties. Many works are still needed to improve what has been accomplished and in finding a more efficient and economically viable method for rice bran protein extraction, to realize its utilization.

REFERENCES

Adebiyi, A.P., Adebiyi, A.O., Jin, D.-H., Ogawa, T., Muramoto, K., 2008. Rice bran protein-based edible films. Int. J. Food Sci. Technol. 43 (3), 476–483.

Ali, R., Shih, F.F., Riaz, M.N., 2010. Processing and functionality of rice bran proteins and peptides. In: Mine, Y., Li-Chan, E., Jiang, B. (Eds.), Bioactive Proteins and Peptides as Functional Foods and Nutraceuticals. Wiley-Blackwell, Ames, IA, pp. 233–246.

Amagliani, L., O'regan, J., Kelly, A.L., O'mahony, J.A., 2017. Composition and protein profile analysis of rice protein ingredients. J. Food Compos. Anal. 59, 18–26.

Anderson, A.K., Guraya, H.S., 2001. Extractability of protein in physically processed rice bran. J. Am. Oil Chem. Soc. 78 (9), 969–972.

Ansharullah, A., Hourigan, J., F Chesterman, C., 1997. Application of carbohydrases in extracting protein from rice bran. J. Sci. Food Agric. 74, 141–146.

Apinunjarupong, S., Lapnirun, S., Theerakulkait, C., 2009. Preparation and some functional properties of rice bran protein concentrate at different degree of hydrolysis using bromelain and alkaline extraction. Prep. Biochem. Biotechnol. 39 (2), 183–193.

Arsa, S., Theerakulkait, C., 2015. Sensory aroma characteristics of alcalase hydrolyzed rice bran protein concentrate as affected by spray drying and sugar addition. J. Food Sci. Technol. 52 (8), 5285–5291.

Bandyopadhyay, K., Misra, G., Ghosh, S., 2008. Preparation and characterisation of protein hydrolysates from Indian defatted rice bran meal. J. Oleo Sci. 57 (1), 47–52.
Bird, A., Hayakawa, T., Marsono, Y., Gooden, J.M., Record, I.R., Correll, R.L., Topping, D.L., 2000. Coarse brown rice increases fecal and large bowel short chain fatty acids and raises large bowel starch in pigs. J. Nutr. 130 (7), 1780–1787.
Boonloh, K., Kukongviriyapan, U., Pannangpetch, P., Kongyingyoes, B., Senggunprai, L., Prawan, A., Thawornchinsombut, S., Kukongviriyapan, V., 2015. Rice bran protein hydrolysates prevented interleukin-6- and high glucose-induced insulin resistance in HepG2 cells. Food Funct. 6 (2), 566–573.
Caffarelli, C., Baldi, F., Bendandi, B., Calzone, L., Marani, M., Pasquinelli, P., 2010. Cow's milk protein allergy in children: a practical guide. Ital. J. Pediatr. 36 (1), 5.
Chalamaiah, M., Yu, W., Wu, J., 2018. Immunomodulatory and anticancer protein hydrolysates (peptides) from food proteins: a review. Food Chem. 245, 205–222.
Chandi, G.K., Sogi, D.S., 2007. Functional properties of rice bran protein concentrates. J. Food Eng. 79 (2), 592–597.
Chen, L., Houston, D.F., 1970. Solubilization and recovery of protein from defatted rice bran. Cereal Chem. 47 (1), 72.
Chiou, T.Y., Neoh, T.L., Kobayashi, T., Adachi, S., 2010. Antioxidative ability of defatted rice bran extract obtained by subcritical water extraction in bulk oil and aqueous dispersion systems. Jpn. J. Food Eng. 12 (4), 147–154.
Chiou, T.Y., Ogino, A., Kobayashi, T., Adachi, S., 2013. Characteristics and antioxidative ability of defatted rice bran extracts obtained using several extractants under subcritical conditions. J. Oleo Sci. 62, 1–8.
Chuanlai, X.U., 2010. Book review: Principles of cereal science and technology 3rd edition. Int. J. Food Sci. Technol. 45 (9), 1963.
Connor, M.A., Saunders, R.M., Kohler, G.O., 1976. Rice bran protein concentrates obtained by wet alkaline extraction. Cereal Chem. 53 (4), 488–496.
Esmaeili, M., Rafe, A., Shahidi, S.-A., Ghorbani Hasan-Saraei, A., 2016. Functional properties of rice bran protein isolate at different pH levels. Cereal Chem. 93 (1), 58–63.
Fabian, C., Ju, Y.H., 2011. A review on rice bran protein: its properties and extraction methods. Crit. Rev. Food Sci. Nutr. 51 (9), 816–827.
Fabian, C.B., Huynh, L.H., Ju, Y.H., 2010. Precipitation of rice bran protein using carrageenan and alginate. LWT Food Sci. Technol. 43 (2), 375–379.
Fan, J.J., Luo, X., Dong, Z., 2008. Extraction isolation and purification of rice bran peptides. Food Sci. Technol. 33 (12), 169–172.
Faria, S.A.S.C., Bassinello, P., Camargo, P., 2012. Nutritional composition of rice bran submitted to different stabilization procedures. Braz. J. Pharm. Sci. 48 (4), 651–657.
Gastanaduy, A., Cordano, A., Graham, G.G., 1990. Acceptability, tolerance, and nutritional value of a rice-based infant formula. J. Pediatr. Gastroenterol. Nutr. 11 (2), 240–246.
Gnanasambandam, R., Heltiarachchy, N.S., 1995. Protein concentrates from unstabilized and stabilized rice bran: preparation and properties. J. Food Sci. 60 (5), 1066–1069.
Gupta, S., Chandi Gurpreet, K., Sogi Dalbir, S., 2008. Effect of extraction temperature on functional properties of rice bran protein concentrates. Int. J. Food Eng. 4 (2), 99–107.
Hamada, J.S., 1997. Characterization of protein fractions of rice bran to devise effective methods of protein solubilization. Cereal Chem. 74 (5), 662–668.
Hamada, J.S., 1999. Use of protease to enhance solubilization of rice bran proteins. J. Food Biochem. 23 (3), 307–321.
Hamada, J.S., 2000. Characterization and functional properties of rice bran proteins modified by commercial exoproteases and endoproteases. J. Food Sci. 65 (2), 305–310.
Hamada, J.S., Spanier, A.M., Bland, J.M., Diack, M., 1998. Preparative separation of value-added peptides from rice bran proteins by high-performance liquid chromatography. J. Chromatogr. A 827 (2), 319–327.

Han, S.W., Chee, K.M., Cho, S.J., 2015. Nutritional quality of rice bran protein in comparison to animal and vegetable protein. Food Chem. 172 (3), 766–769.
Hanmoungjai, P., Pyle, D.L., Niranjan, K., 2001. Enzymatic process for extracting oil and protein from rice bran. J. Am. Oil Chem. Soc. 78 (8), 817–821.
Hanmoungjai, P., Pyle, D.L., Niranjan, K., 2002. Enzyme-assisted water-extraction of oil and protein from rice bran. J. Chem. Technol. Biotechnol. 77 (7), 771–776.
Hata, S., Wiboonsirikul, J., Maeda, A., Kimura, Y., Adachi, S., 2008. Extraction of defatted rice bran by subcritical water treatment. Biochem. Eng. J. 40 (1), 44–53.
Hopps, E., Noto, D., Caimi, G., Averna, M.R., 2010. A novel component of the metabolic syndrome: the oxidative stress. Nutr. Metab. Cardiovasc. Dis. 20 (1), 72–77.
Ikezawa, Z., Tsubaki, K., Osuna, H., Shimada, T., Moteki, K., Sugiyama, H., Katumata, K., Anzai, H., Amano, S., 1999. Usefulness of hypoallergenic rice (AFT-R 1) and analysis of the salt insoluble rice allergen molecule. Arerugi 48 (1), 40–49.
Iroyukifujita, H., Eiichiyokoyama, K., Yoshikawa, M., 2000. Classification and antihypertensive activity of angiotensin I-converting enzyme inhibitory peptides derived from food proteins. J. Food Sci. 65 (4), 564–569.
Jiamyangyuen, S., Srijesdaruk, V., James Harper, W., 2005. Extraction of rice bran protein concentrate and its application in bread. Songklanakarin J. Sci. Technol. 11 (6), 403–416.
Juliano, B.O., 1985. Rice bran. In: Juliano, B.O. (Ed.), Rice Chemistry and Technology. American Association of Cereal Chemists, St. Paul, MN, pp. 647–680.
Kaewka, K., Therakulkait, C., Cadwallader, K.R., 2009. Effect of preparation conditions on composition and sensory aroma characteristics of acid hydrolyzed rice bran protein concentrate. J. Cereal Sci. 50 (1), 56–60.
Kannan, A., Hettiarachchy, N., Narayan, S., 2009. Colon and breast anti-cancer effects of peptide hydrolysates derived from rice bran. Open Bioactive Comp. J. 2 (1), 17–20.
Kannan, A., Hettiarachchy, N.S., Lay, J.O., Liyanage, R., 2010. Human cancer cell proliferation inhibition by a pentapeptide isolated and characterized from rice bran. Peptides 31 (9), 1629–1634.
Kelly, R., Robbins, Ballew, J.E., 1982. Effect of alkaline treatment of soy protein on sulfur amino acid bioavailability. J. Food Sci. 47 (6), 2070–2071.
Khan, S.H., Butt, M.S., Anjum, F.M., Sameen, A., 2011. Quality evaluation of rice bran protein isolate-based weaning food for preschoolers. Int. J. Food Sci. Nutr. 62 (3), 280–288.
Kim, S.P., Yang, J.Y., Kang, M.Y., Park, J.C., Nam, S.H., Friedman, M., 2011. Composition of liquid rice hull smoke and anti-inflammatory effects in mice. J. Agric. Food Chem. 59 (9), 4570–4581.
Leopold, J.A., Loscalzo, J., 2009. Oxidative risk for atherothrombotic cardiovascular disease. Free Radic. Biol. Med. 47 (12), 1673–1706.
Liu, Y., Wang, Z., Li, H., Liang, M., Yang, L., 2016. In vitro antioxidant activity of rice protein affected by alkaline degree and gastrointestinal protease digestion. J. Sci. Food Agric. 96 (15), 4940–4950.
Liu, C., Yang, X., Wu, W., Long, Z., Xiao, H., Luo, F., Shen, Y., Lin, Q., 2017a. Elaboration of curcumin-loaded rice bran albumin nanoparticles formulation with increased in vitro bioactivity and in vivo bioavailability. Food Hydrocoll. 77, 834–842.
Liu, Y.Q., Strappe, P., Shang, W.T., Zhou, Z.K., 2017b. Functional peptides derived from rice bran proteins. Crit. Rev. Food Sci. Nutr. 1–8.
Matsuda, T., Nakasa, M., Alvarez, A.M., Izumi, H., Kato, T., Tada, Y., 2006. Rice-seed allergenic protein and hypoallergenic rice. In: Mine, Y., Shahidi, F. (Eds.), Nutraceutical Proteins and Peptides in Health and Disease. Taylor and Francis Group, London, pp. 493–511.
McPeak, P., Rukmini, C., Cherukuri, R.S.V., 2001. Supportive Therapy for Diabetes, Hyperglycemia and Hypoglycemia. US6303586B1.

Meharg, A.A., Zhao, F.-J. (Eds.), 2012. Arsenic and Rice. Springer, Dordrecht.

Mills, E.N., Jenkins, J.A., Alcocer, M.J., Shewry, P.R., 2004. Structural, biological, and evolutionary relationships of plant food allergens sensitizing via the gastrointestinal tract. Crit. Rev. Food Sci. Nutr. 44 (5), 379–407.

Mota, C., Santos, M., Mauro, R., Samman, N., Matos, A.S., Torres, D., Castanheira, I., 2016. Protein content and amino acids profile of pseudocereals. Food Chem. 193, 55–61.

Munera-Picazo, S., Ramírez-Gandolfo, A., Cascio, C., Castaño-Iglesias, C., Signes-Pastor, A.J., Burló, F., Haris, P.I., Carbonell-Barrachina, A.A., 2014. Arsenic in rice-based infant foods. In: Watson, R.R., Preedy, V.R., Zibadi, S. (Eds.), Wheat and Rice in Disease Prevention and Health. Benefits, Risks and Mechanisms of Whole Grains in Health Promotion. Elsevier Science and Technology, Oxford, pp. 377–391.

Naqash, F., Gani, A., Gani, A., Masoodi, F.A., 2017. Gluten-free baking: combating the challenges—a review. Trends Food Sci. Technol. 66, 98–107.

Ni, W., Tsuda, Y., Takashima, S., Sato, H., Sato, M., Imaizumi, K., 2003. Anti-atherogenic effect of soya and rice-protein isolate, compared with casein, in apolipoprotein E-deficient mice. Br. J. Nutr. 90 (1), 13–20.

Niu, L., Wu, L., Xiao, J., 2017. Inhibition of gelatinized rice starch retrogradation by rice bran protein hydrolysates. Carbohydr. Polym. 175, 311–319.

Phongthai, S., D'amico, S., Schoenlechner, R., Rawdkuen, S., 2016. Comparative study of rice bran protein concentrate and egg albumin on gluten-free bread properties. J. Cereal Sci. 72, 38–45.

Phongthai, S., D'amico, S., Schoenlechner, R., Homthawornchoo, W., Rawdkuen, S., 2017a. Effects of protein enrichment on the properties of rice flour based gluten-free pasta. LWT Food Sci. Technol. 80, 378–385.

Phongthai, S., Homthawornchoo, W., Rawdkuen, S., 2017b. Preparation, properties and application of rice bran protein: a review. Int. Food Res. J. 24 (1), 25–34.

Pourali, O., Asghari, F.S., Yoshida, H., 2009. Sub-critical water treatment of rice bran to produce valuable materials. Food Chem. 115 (1), 1–7.

Prakash, J., 1996. Rice bran proteins: properties and food uses. Crit. Rev. Food Sci. Nutr. 36 (6), 537–552.

Prakash, J., Ramanatham, G., 1994. Effect of stabilization of rice bran on extractability and recovery of protein. Mol. Nutr. Food Res. 38 (1), 87–95.

Reche, M., Pascual, C., Fiandor, A., Polanco, I., Rivero-Urgell, M., Chifre, R., Johnston, S., Martín-Esteban, M., 2010. The effect of a partially hydrolysed formula based on rice protein in the treatment of infants with cow's milk protein allergy. Pediatr. Allergy Immunol. 21, 577–585.

RiceBran Technologies. (n.d.). Rice Bran Product Information. Available from: https://www.ricebrantech.com/our-products/product-information. (Accessed 19 March 2018).

Rukmini, C., Patricia, M., Cherukuri, R.S.V., Lynch, I., Qureshi, A.A., 2003. Method for Treating Hypercholestrolemia, Hyperlipidemia, and Atherosclerosis. US6558714B2.

Santiago, L.G., Castro, G.R., 2016. Novel technologies for the encapsulation of bioactive food compounds. Curr. Opin. Food Sci. 7, 78–85.

Sari, Y.W., Mulder, W.J., Sanders, J.P., Bruins, M.E., 2015. Towards plant protein refinery: review on protein extraction using alkali and potential enzymatic assistance. Biotechnol. J. 10 (8), 1138–1157.

Schmidt, C.G., Cerqueira, M.A., Vicente, A.A., Teixeira, J.A., Furlong, E.B., 2014. Rice bran protein-based films enriched by phenolic extract of fermented rice bran and montmorillonite clay. CyTA—J. Food 13 (2), 204–212.

Sereewatthanawut, I., Prapintip, S., Watchiraruji, K., Goto, M., Sasaki, M., Shotipruk, A., 2008. Extraction of protein and amino acids from deoiled rice bran by subcritical water hydrolysis. Bioresour. Technol. 99 (3), 555–561.

Shi, M., Huang, L.Y., Nie, N., Ye, J.H., Zheng, X.Q., Lu, J.L., Liang, Y.R., 2017. Binding of tea catechins to rice bran protein isolate: interaction and protective effect during in vitro digestion. Food Res. Int. 93, 1–7.

Shih, F.F., Champagne, E.T., Daigle, K., Zarins, Z., 1999. Use of enzymes in the processing of protein products from rice bran and rice flour. Mol. Nutr. Food Res. 43 (1), 14–18.

Shin, Y.J., Jang, S.-A., Song, K.B., 2011. Preparation and mechanical properties of rice bran protein composite films containing gelatin or red algae. Food Sci. Biotechnol. 20 (3), 703–707.

Silpradit, K., Tadakittasarn, S., Rimkeeree, H., Winitchai, S., Haruthaithanasan, V., 2010. Optimization of rice bran protein hydrolysate production using alcalase. Asian J. Food Agro-Ind. 3 (2), 221–231.

Sun, G.X., Williams, P.N., Carey, A.M., Zhu, Y.G., Deacon, C., Raab, A., Feldmann, J., Islam, R.M., Meharg, A.A., 2008. Inorganic arsenic in rice bran and its products are an order of magnitude higher than in bulk grain. Environ. Sci. Technol. 42 (19), 7542–7546.

Tang, S., Hettiarachchy, N.S., Shellhammer, T.H., 2002. Protein extraction from heat-stabilized defatted rice bran. 1. Physical processing and enzyme treatments. J. Agric. Food Chem. 50 (25), 7444–7448.

Tang, S., Hettiarachchy, N.S., Eswaranandam, S., Crandall, P., 2003a. Protein extraction from heat-stabilized defatted Rice bran: II. The role of amylase, celluclast, and viscozyme. J. Food Sci. 68 (2), 471–475.

Tang, S., Hettiarachchy, N.S., Horax, R., Eswaranandam, S., 2003b. Physicochemical properties and functionality of rice bran protein hydrolyzate prepared from heat-stabilized defatted rice bran with the aid of enzymes. J. Food Sci. 68 (1), 152–157.

Taniguchi, M., Kameda, M., Namae, T., Ochiai, A., Saitoh, E., Tanaka, T., 2017. Identification and characterization of multifunctional cationic peptides derived from peptic hydrolysates of rice bran protein. J. Funct. Foods 34, 287–296.

Umer Abdullah, S., Badaruddin, M., Asad Sayeed, S., Ali, R., Riaz, M.N., 2008. Binding ability of allura red with food proteins and its impact on protein digestibility. Food Chem. 110 (3), 605–610.

Uraipong, C., Zhao, J., 2016. Rice bran protein hydrolysates exhibit strong in vitro alpha-amylase, beta-glucosidase and ACE-inhibition activities. J. Sci. Food Agric. 96 (4), 1101–1110.

Uraipong, C., Zhao, J., 2018. In vitro digestion of rice bran proteins produces peptides with potent inhibitory effects on alpha-glucosidase and angiotensin I converting enzyme. J. Sci. Food Agric. 98 (2), 758–766.

Usui, Y., Nakase, M., Hotta, H., Urisu, A., Aoki, N., Kitajima, K., Matsuda, T., 2001. A 33-kDa allergen from rice (Oryza sativa L. Japonica). J. Biol. Chem. 276 (14), 11376–11381.

Van Hooser, B., Crawford, L.V., 1989. Allergy diets for infants and children. Compr. Ther. 15 (10), 38–47.

Wang, M., Hettiarachchy, N.S., Qi, M., Burks, W., Siebenmorgen, T., 1999. Preparation and functional properties of rice bran protein isolate. J. Agric. Food Chem. 47 (2), 411–416.

Wang, S., Marcone, M.F., Barbut, S., Lim, L.-T., 2012. Fortification of dietary biopolymers-based packaging material with bioactive plant extracts. Food Res. Int. 49 (1), 80–91.

Wang, C., Li, D., Xu, F., Hao, T., Zhang, M., 2014. Comparison of two methods for the extraction of fractionated rice bran protein. J. Chem. 2014 (4), 1–10.

Wang, J., Shimada, M., Kato, Y., Kusada, M., Nagaoka, S., 2015. Cholesterol-lowering effect of rice bran protein containing bile acid-binding proteins. Biosci. Biotechnol. Biochem. 79 (3), 456–461.

Wang, X., Chen, H., Fu, X., Li, S., Wei, J., 2017. A novel antioxidant and ACE inhibitory peptide from rice bran protein: biochemical characterization and molecular docking study. LWT Food Sci. Technol. 75, 93–99.

Watanabe, M., Miyakawa, J., Ikezawa, Z., Suzuki, Y., Hirao, T., Yoshizawa, T., Arai, S., 1990. Production of hypoallergenic rice by enzymatic decomposition of constituent proteins. J. Food Sci. 55 (3), 781–783.

Watchararuji, K., Goto, M., Sasaki, M., Shotipruk, A., 2008. Value-added subcritical water hydrolysate from rice bran and soybean meal. Bioresour. Technol. 99 (14), 6207–6213.

WHO/FAO/UNU, 2007. Protein and amino acid requirement in human nutrition. Report of Joint WHO/FAO/UNU Expert Consultation. Albany, NY. WHO Technical Report Series, United Nations University.

Wiboonsirikul, J., Kimura, Y., Kadota, M., Morita, H., Tsuno, T., Adachi, S., 2007. Properties of extracts from defatted rice bran by its subcritical water treatment. J. Agric. Food Chem. 55 (21), 8759–8765.

Xia, N., Wang, J., Yang, X., Yin, S., Qi, J., Hu, L., Zhou, X., 2012. Preparation and characterization of protein from heat-stabilized rice bran using hydrothermal cooking combined with amylase pretreatment. J. Food Eng. 110 (1), 95–101.

Yadav, R.B., Yadav, B.S., Chaudhary, D., 2011. Extraction, characterization and utilization of rice bran protein concentrate for biscuit making. Br. Food J. 113 (9), 1173–1182.

Yang, L., Chen, J.H., Xu, T., Zhou, A.S., Yang, H.K., 2012. Rice protein improves oxidative stress by regulating glutathione metabolism and attenuating oxidative damage to lipids and proteins in rats. Life Sci. 91 (1), 389–394.

Yeom, H.-J., 2009. Production and physicochemical properties of rice bran protein isolates prepared with autoclaving and enzymatic hydrolysis. J. Korean Soc. Appl. Biol. Chem. 53 (1), 62–70.

Young, V.R., Pellett, P.L., 1994. Plant proteins in relation to human protein and amino acid nutrition. Am. J. Clin. Nutr. 59 (5), 1203s–1212s.

Zhang, H.J., Zhang, H., Wang, L., Guo, X.N., 2012. Preparation and functional properties of rice bran proteins from heat-stabilized defatted rice bran. Food Res. Int. 47 (2), 359–363.

Zhang, H., Wang, J., Liu, Y., Gong, L., Sun, B., 2016. Rice bran proteins and their hydrolysates modulate cholesterol metabolism in mice on hypercholesterolemic diets. Food Funct. 7 (6), 2747–2753.

Zhou, K., Sun, S., Canning, C., 2012. Production and functional characterisation of antioxidative hydrolysates from corn protein via enzymatic hydrolysis and ultrafiltration. Food Chem. 135 (3), 1192–1197.

Zhu, S.M., Lin, S.L., Ramaswamy, H.S., Yu, Y., Zhang, Q.T., 2016. Enhancement of functional properties of rice bran proteins by high pressure treatment and their correlation with surface hydrophobicity. Food Bioprocess Technol. 10 (2), 317–327.

INDEX

Note: Page numbers followed by *f* indicate figures and *t* indicate tables.

M

N

O

P

Q

R

S

Kolibri 360 GmbH
Ferdinand-Jühlke-Straße 7
99095 Erfurt, Deutschland
produktsicherheit@kolibri360.de